# THE FIVE HORSEMEN OF THE MODERN WORLD

—

DANIEL CALLAHAN

# THE FIVE HORSEMEN OF THE MODERN WORLD

Climate, Food, Water, Disease, and Obesity

COLUMBIA UNIVERSITY PRESS
NEW YORK

Columbia University Press
*Publishers Since 1893*
New York   Chichester, West Sussex
Copyright © 2016 Daniel Callahan
All rights reserved

Library of Congress Cataloging-in-Publication Data
Callahan, Daniel, 1930– , author.
The five horsemen of the modern world : climate, food, water, disease,
and obesity / Daniel Callahan.
p. ; cm.
Includes bibliographical references and index.
ISBN 978-0-231-17002-4 (cloth : alk. paper)
ISBN 978-0-231-54152-7 (e-book)
I. Title.
[DNLM: 1.  Public Health. 2.  Ecosystem. 3.  Environmental Health.
4.  Global Health.  WA 100]

RA418
362.1—dc23

2015034495

Columbia University Press books are printed on permanent
and durable acid-free paper.
This book is printed on paper with recycled content.
Printed in the United States of America

c 10 9 8 7 6 5 4 3 2 1

Cover design: Milenda Nan Ok Lee

*For*
*Mary Crowley, Susan Gilbert, Ellen Theg, Lynn Traverse,*
*Friends and colleagues*

# CONTENTS

## III  TOWARD THE FUTURE:
## PROGRESS, HOPE, AND FEAR

—

# PREFACE

I need first to identify myself to provide a rationale for taking on a book of this kind, a comparative study of some global crises that I call the five horsemen: global warming, food shortages, water shortages and quality, chronic illness, and obesity. I call them the five horsemen because they can remind us of the biblical four horsemen of the Apocalypse, traditionally described as conquest, war, famine, and death, all ancient evils. My five are not the only threats to our present lives but are high and notable on that list. While each drew some attention in earlier times, they have jumped to public and policy attention in a striking way over the past forty to fifty years, and they have some uniquely contemporary features, only faint or absent earlier.

My professional home for most of my career has been medical ethics and health policy, focused on the latter good. As a philosopher by discipline I early moved away from problems traditional for that field, bearing on theories of ethics and ideas about the nature of the human good. Those are surely important, but I found myself drawn to understanding the impact of culture and politics on how we find that good, working from the ground up, so to speak. I also became something of a wanderer, moving from topic to topic over the years but always looking for themes and threads they shared, often unnoticed, not so common among those who become more specialized in their work.

That's the way it was with this book, leading me from the good of the body to the good of the of planet, noticing how often they overlap. For much of the 1980s and through the early 2000s I was engrossed in the debates on health care that had broken out most contentiously in the United States, but in many other countries as well. The rising cost of that care was a leading issue, in great part brought on by aging populations but also by how much great medical and technological progress had introduced terrible dilemmas in the care of the dying; progress itself became both a social and individual problem. The debate over the provision of care to the uninsured in the United States was a focal point, bringing out great divisions among the public and the political parties. Eventually the Affordable Care (ACA) legislation was passed, but only narrowly, by Congress; it is now threatened by a Republican majority determined to kill or severely cripple it.

While watching those struggles I was also drawn to the rising anxiety about global warming, well covered by the media. Along the way, I began noting news stories about global water and food challenges and chronic illness and obesity. They shared many traits: similar kinds of political and ideological arguments, like those I had found in health care: the play of personalities and public opinion; scientific disagreements; and the powerful role of vested interests, particularly on the part of industries threatened by the necessary losses that change could bring.

What stuck most tenaciously in my mind was an unusual feature: all five of them are getting worse not better as global problems, despite decades of expensive and concerted efforts to deal with them. Is it really true, and if so, how could that be? I have not been able to find any global crises of similar magnitude in terms of death, morbidity, or projected destruction that have proved so recalcitrant to change, and already with us. As Gregg Easterbrook showed with solid evidence over a decade ago in his book *The Progress Paradox: How Life Gets Better While People Feel Worse*, human life in general has been improving, even if to many of us it seems to be deteriorating.[1] An important baseline for that judgment can be found in the eight global development goals established by the UN Millennium Project in 2005, which aim by 2015 to

- Eradicate extreme poverty and hunger
- Achieve universal primary education
- Reduce childhood mortality
- Promote gender equality and empowerment of women

- Improve maternal health
- Combat HIV/AIDS, malaria, and other diseases
- Ensure environmental sustainability (including global warming)
- Develop a global partnership for development. [2]

Remarkably, while few of the goals will be fully met by 2015, the United Nations could point to progress in most of them over the ten-year program, some striking—such as a 50% reduction of extreme poverty. There is one exception: "global emissions of carbon dioxide ($CO_2$) continued their upward trend and those in 2011 were almost 50% above their 1990 level."[3] A dour judgment of the other four horsemen would fit as well, even if less dramatically. While progress has been made in reducing food shortages and malnutrition, I list food shortages among my five horsemen because the future projections for food availability are mixed and borderline.

What is it about the horsemen that makes them so hard to move forward? That is the question this book tries to answer, appraising along the ways the various strategies employed to do so. By the end of this book I propose what I think is the answer to that question, peeling off the many layers that obscure it from sight. That means examining all the strategies used over the years to grapple with the horsemen; and they turn out to some extent, but not all, to be similar. As stressed again and again in much of the global-warming literature, but noticeable in all of the books and articles written about my horsemen, it can be extremely challenging to sharply distinguish between facts and values, particularly in making the move from scientific knowledge to policy and action.

We all bring to that effort values of one kind or another, shaped by our upbringing, our politics, our experience, our social classes or circles, and many subtle factors as well. We are urged to be aware of those values and to take account of them in our judgments. Here are some of mine, some I was aware of well before working on this book, and others I discovered while writing it. I have been a lifelong Democrat, open to a strong government role but more in the middle of the road than far to the left. I have never been a booster of the magic of the market or drawn to getting rid of it either, but I am favorable to strongly controlling and regulating it. It should be our servant, not our master. What I discovered about myself as the book moved along is that I am not someone much moved by the danger of some tactics valuable for taking up arms against the horsemen, especially when they are statistically rare or speculative. What counts is the greater danger of not acting. I have not been impressed, that is, with the supposed great dangers posed by nuclear power plants, genetically modified food, or fracking, for instance. They can be risky, but not enough to undo the good they can bring. I am thus a soft opponent of fracking, opposed yet aware of the loss of potential jobs in poor areas.

A pervasive problem with each of the horsemen is that they all lack a single, simple, clean, and acceptable path to successfully manage them; no magic bullet, that is, to use a constantly encountered phrase. They all require balancing gains and losses, costs and benefits, one kind of value over another, clashing emotions and temperament, ideological fights, and fickle, fluctuating public opinion (or sometimes persistent public opinion of an unhelpful kind).

It was hard to devise a method for dealing with that combination, requiring that I explore a wide, intimidating range of puzzles that can defy anything that looks like a plausible methodology; and I could not find a suitable one. I had to invent my own, drawing in part what I already knew something about and educating myself about that which was new to me. I had two general aims. I needed to lay out and summarize the nature of the problems posed by each of the horsemen in some reasonably nonpartisan way, assuming many readers would be familiar with (and even expert in) some but probably not all of the horsemen. I also needed to compare them, by taking each apart so I could stand them side by side, bringing out both the similar and different strategies employed to deal with them. That approach required that I avoid undue repetition between points made in part I and taken up again in different ways in part II; and also that the reader would recall enough of what I said in part I to minimize the need to refresh the reader's memory to make sense of my analysis in part II.

The plan for this book reveals what I devised. It has three parts: (1) chapters 1–5, a summary of the state of the question for each horsemen; (2) chapters 6–9, analysis of strategies to manage them; and (3) chapters 10–11, seeking solutions.

## PART I: MAPPING AN IRREGULAR TERRAIN

Each of my five horsemen has a somewhat different history, a different set of actors, and require a different way of framing their challenges. I use the word "somewhat," because they all reveal similarities as well; for instance, each displays an accelerating severity beginning in the 1970s, each is bedeviled by the science and policy relationship, each reveals ideological splits, and each reflects different academic and research disciplines, with specialized university programs, meetings and organizations, and journals. Inevitably, they also have infighting among specialists, requiring the outsider to detect and understand those factions. My aim in these opening chapters is to capture in a fair way the state of the question with each of the horsemen in a way that those specialists find tolerable. Whenever possible I found experts in those disciplines to vet my efforts.

## PART II: EXAMINING THE PATHWAYS THROUGH THE THICKETS

I think of each of those topics as cross-cutting. They cut through and bear on each of the horsemen. World population is now over 7 billion and projected to rise to 11 billion. More people means more demand on everything: energy, food, water, natural resources and space, most notably. Aging populations are now a feature of all nations, with those over sixty years of age globally only recently coming to exceed those under five. The fastest aging increase is now in the developing countries, putting a whole set of new burdens on every such society: health care, the ratio of retirees to workers, and the economic needs of the retired. The growth of gross domestic product, GDP, has long been sought in developed countries and is a leading reason for their prosperity. There has also been a counter-current, seeking to reduce GDP growth, and coming from advocates in affluent countries, a quest unattractive to poor countries. The developing countries have worked hard to keep their GDPs growing and many have, especially India and China, reducing poverty as a result. A major tension in global warming has been the reluctance of some countries to reduce their carbon emissions, a threat to the industrial progress they have made but a threat to global welfare also.

Chapter 7: The Technology Fix: A Way Out?

Technology might well be called almost everyone's dream solution. It can be used to create energy but also to reduce the global harm done by its generation. It can relieve water and food problems, just as it has been the bedrock of modern medical progress and the main weapon used to combat chronic disease. Only obesity has failed to find significant technologies that answer its needs and can be effectively applied. Most important, technological research and innovation have a history of being used for most of the great discoveries of modern life, improving health, economic growth, and the quality of life. Technology is familiar, cherished, and often idolized, giving it a leg up on other strategies. When well managed, technological innovation can created jobs and wealth, a notable attraction. There is at present with the five horsemen, however, the desperate need to find money for research on and implementation of technology, now in short supply.

Chapter 8: A Volatile Mix: Public Policy, the Media, and Public Opinion

At the core of this chapter is a question: Once a problem has been recognized, how best to get meaningful action? My answer is that it takes (1) a combination of public education and persuasion to get a problem on the public agenda, (2) the collection of public opinion to determine public understanding of an issue and how much people are willing to give up to help out, and then (3) for the media to take an interest and publicize it as a problem—and finally to use all of these together as an incentive for legislators and administrators to take action. I spend a fair amount of space in this chapter trying to understand as sympathetically as possible both outright denial or minimizing of a problem and the equally great need to understand the ways in which, even with that recognition, there is too weak a response to make progress in bringing about change. I am also interested in how to gain the necessary emotional intensity, and with what language, needed to engender change.

Chapter 9: Law and Governance: Managing Our Public Planet and Our Private Bodies

At some point or other, every one of the horsemen will need the help of a governance mechanism and matching laws and regulations to bring about change. There are three broad levels for doing so—the global, regional or national, and local levels—and often enough at all levels simultaneously. Effective governance at the global level has been hard to come by, stymied by the tenacious grip of national sovereignty. There are lesser strains at the national and regional levels, but also an inviting territory for industry opposition, for effective lobbying to influence legislators, and for the force of public opinion to be felt. Social movements come into play as well, a necessary ingredient.

## PART III: TOWARD THE FUTURE: PROGRESS, HOPE, AND FEAR

Chapter 10: Progress and Its Errant Children: More Is Never Enough

While all of the horsemen have some historical roots, they all became most prominent in planting their contemporary roots in the post–World War II era, most intensively during the 1970s. It was then that they rapidly became what I call "life as usual," simply the way we unwittingly came to live, first in the developed countries, then aspired to by the poorer countries. Economic growth and spreading affluence created and sustained the momentum, easily becoming entwined with the intensification of market values and capitalism. Underneath all of them is the most tenacious root of all, the potent and enduring idea and value of progress, that human life ought always to get better, that it is has no natural stopping point and should never cease aspiring to move ahead. Globalization plays a large part in this push for more, and more, and more. Can the culture of unfettered and unlimited progress be eliminated or at least pacified? Possibly.

Chapter 11: The Necessary Coalition: Social Movements, Legislatures, and
      Business

As happened so often along the way as this book progressed, global warming stole the show, with more colorful actors, media and public opinion attention, and initiatives for change than any of the other horsemen—and indeed more than the other four together. That benefit has not, unfortunately, led to much greater bottom-line success with global warming than the others, the horsemen's quieter siblings. In this chapter I provide a set of criteria to measure success and then apply it to each of the horsemen. Those looking for hope, which is just about everybody, will not find too much of it in that assessment. But I do have my own offering in that direction.

If industry has so often been obstructive in helping things more forward, as is well emphasized in this book, I belatedly discovered a major development in business. It was not coming from the usual suspects, who give no sign of changing, but from a wide range of corporations working under the umbrella term "sustainability," forging new relationships with environmental efforts. I have come to believe that a coalition is necessary and attainable. It is that of social movements pushing up from the bottom, global governance and pressure moving from the top down, with decisiveness at the middle level, that of the national and the regional levels. All of that needs, however, some powerful shot in the arm, and my candidate is business, which has the money and clout to cut through and invigorate all the levels. It can be done, and I lay out a plausible pathway to do it.

# ACKNOWLEDGMENTS

I n the nature of the case, I came into this wide-territory book project with a mixed repertoire of knowledge and background. It ranged from knowing a lot about some of the issues, having a smattering of knowledge about others, and a vast desert of knowledge about a few. It was obvious I would need help, especially when I did not personally know the pertinent experts, or what to read, or how to make my way through a thicket of debates that seemed to make little room for amateurs. I solved those problems by writing to total strangers who know what I did not and then asked them to be my Good Samaritans, picking me up on the side of the road, and pointing me in the right direction. I was pleased and relieved that so many responded and gave me help. With a few, I think, I was seen as a kind of menace, but they still helped me. In the end, there were three groups who assisted me: journalists and bloggers (with whom I had no personal contact), experts in various fields with whom I did interact, and some people who are not experts on my topics but astute lay readers who know what makes sense and what does not.

My research associate Ellen Theg, MBA, deserves special praise. She worked with me nearly two years on this book in various ways, but her key contribution—making use of her business degree and experience—was to carefully look at the role of business and industry, the good and the bad. Her work shows especially in the last chapter, where I make a case for the necessity of a partnership with business, using a mass of information she culled indicating that such a partnership is beginning to emerge, even if it has not attracted the public attention it deserves. I could not have written that without her.

I gained an immense amount of insight and timely reports from journalists, bloggers, and some fine websites, most notably on global warming, which saw a steady, often overwhelming, stream of new reports and studies. I made good use of the massive number of illuminating books and reports available, but there is usually a time lag between their publication and what has happened in the meantime.

I needed both. Among the journalists, Justin Gillis and Cora Diamond of the *New York Times* were almost daily fare and as good with the back story as with the front story. Andrew Revkin's website, Dot Earth, is outstanding in its coverage and nuance, and Yale's Environment 360 website was invaluable as well, drawing nicely on scientists and policy experts as well as gifted journalists such as Fred Pearce. Bill McKibben is triple winner, taken seriously for his research, his journalism, and as an organizer and activist. His website, 350.org, with its global reach, displays all those traits. Mark Bittman and Michael Pollan have written incisively on food and obesity, as have Steven Solomon and Charles Fishman on water, and Julian Cribb on food.

I have been most fortunate in finding the Good Samaritan experts, whose writings, conversation, and reading of draft chapters both enlightened me and kept me out of trouble. It is not usual to cite the writings of those whose help one received, but in this case I profited both from their books and their comments on my drafts:

Lawrence O. Gostin, director of the O'Neill Institute for Global and National Health Law, Georgetown University, and the author of *Global Health Care*; Spencer Weart, former director for history of physics of the American Institute of Physics, and the author of *The Discovery of Global Warming*; Tony Allan, professor of geography, Kings College, University of London, and the author of *Virtual Water*; Jennifer Clapp, professor of global health governance, University of Waterloo, Canada, and author of *Food*; John Bongaarts, vice president and distinguished scholar, the Population Council, and editorial committee member, *Population and Development Review*; Richard Jackson, senior associate, Center for Strategic and International Studies, and author of *Global Aging and The Future of Emerging Markets*; Barron Lerner, M.D., professor of medicine, New York University Langone School of Medicine, and author of *The Good Doctor*; Rebecca Puhl, deputy director, Ruud Center for Food and Obesity, University of Connecticut; Robert Keohane, professor of international affairs, Woodrow Wilson School, Princeton University, and author of *Power and Governance in a Partially Globalized World*; Abigail Saguy, associate professor of sociology and gender, University of California–Los Angeles, and author of *What's Wrong with Fat?*; Daniel Sarewitz, professor, School of Life Sciences and School of Sustainability, Arizona State University, and author of *The Techno-Human Condition*; Timothy Hoffman, Green Owl and Watts Capital and New York City Metro director for the Clean Tech Open.

I conclude finally with my friends and colleagues, who work professionally with none of the five horsemen but were thoughtful and sensitive readers of my drafts:

Mary McDonough, Harvard Medical School Program in Bioethics; Laura Haupt, associate editor, the *Hastings Center Report*; Frank Trainer, board of directors, the Hastings Center, and former director for fixed income, Sanford C. Bernstein; David Roscoe, chairman, Hastings Center board of directors, and retired executive for RiskMetric Group.

Finally, there is my wife Sidney. She astutely drew on the experience of our long marriage and her profession as a psychologist to offer comfort, therapy, and sympathy. While writing this book I ranged from ranting and raving to quiet desperation in trying to make sense of the five horsemen—whose horses have defied bridles or saddles and at times seem blind. Sometimes the question was whether they were blind or I was, and Sidney helped me to know the difference, I think.

# I

# MAPPING AN IRREGULAR TERRAIN

# 1

# OUR OVERHEATING, FRAYING PLANET

The unique and dangerously threatening problem of global warming, often called climate change, is that it carries with it all the hazards of the biblical four horsemen: pestilence, war, famine, and death. Two of those horsemen embody seemingly permanent features of life itself. Death always has been our fate as human beings, but is now more likely because of struggles triggered by potential famines and dwindling water resources. Pestilences, never far away, will, with temperature changes in different parts of the world, have new breeding grounds.

Global warming will likely affect all of us one way or another, but some more than others, some from one part of the threat, some from other parts. But no country can avoid it, whatever the differences. It is all too easy to think of global warming as someone else's problem, occurring elsewhere on the globe or affecting only future generations. The fact that most of us in our daily lives do not experience firsthand the impacts of climate change can make it seem unreal and easy to put out of our minds. This is not the only reason it has been hard to deal with global warming effectively. It requires a a profound alteration in the way we live our modern lives, changing much we have cherished and worked hard to achieve. We thought the earth was ours for the taking, not noticing that it requires care and nurturing, much as our own bodies do. The British social scientist Anthony Giddens, surveying global policy efforts, concluded "that we have no politics of climate change," that is, no reasonably clear road map agreed upon by scientists, national legislators, and global multilateral agencies. One cannot have a road map without determining what kind of terrain one must traverse.[1]

As a way making of sense of that last demand—and in line with the overall theme of this book, that of comparing and contrasting the *five* horsemen—I will focus on the main schools of thought and emotion in the global-warming debate. My aim is to sort out the competing proposals, arguments, and factions who work on or have an interest in global warming, withholding (for the most part) my own judgment for now on the controversial and multifaceted issues (saving them for part II of the book). There are, broadly, at least three streams of opinion and judgment on global warming—that of the minimizers or rejecters, that of those who see us on the way to disaster, and that of those who believe we have time to make changes to adapt to what we cannot wholly avoid—and many gradations in between.

## EARLY HISTORY

Climate change was established as an area of inquiry by scientists and has been in their hands ever since, even though they have been joined by a host of others over the years. It is important to understand that history, which goes back a long way. Like most other important developments in science, it has moved along step by step, layer by layer, to where it is today. Progress has been made by outstanding individuals, by teams of scientists working together, and by many kibitzers on the sidelines. Although climate change research began with largely theoretical speculations, of interest mainly to scientists, it soon became a matter of public interest. Those early speculations began to extend to the environmental and social realms: global warming could be a threat to our common welfare. That gradual shift also brought climate change into the arena of laws, policy, partisan politics, economics, and media attention. Finding a good fit between science and all those external factors has never been easy, and that struggle continues unabated.

It began in the early decades of the nineteenth century with a small number of scientists.[2] One of them, the Frenchman Jean-Baptiste Fourier, wondered in 1827 what determined the temperature of celestial bodies such as the earth. He noted the similarity between the atmosphere on earth and that within a cold frame, the latter trapping warm air.

That observation came to be called the "greenhouse effect." About 1860 an English scientist, John Tyndall, speculated that small changes in the earth's atmosphere could lead to a loss of heat. A Swedish chemist, Svante Arrhenius, estimated in 1896 that a doubling of carbon dioxide in the atmosphere could increase the average global temperature by 5°C–6°C (9°F–10.8°F), which is close to recent projected figures. Around 1940 G. S. Callendar in England noted the impact of fossil fuels on the increase of $CO_2$ in the atmosphere.

Thereafter, the pace of science increased, as did industrial change and global-warming. In the 1920s oil fields blossomed in Texas and the Middle East, bringing large supplies of cheap energy to supplement inexpensive coal. By the 1930s a global-warming trend for some decades was recorded. Earth Day, initiated in 1970, aimed to arouse interest in the environment, including the atmosphere. By 1977 scientific agreement determined that global warming was a serious danger for the coming century. In 1979 the U.S. National Academy of Sciences found it likely that a doubling of $CO_2$ was well on the way and would lead to a 1.5°C–4°C (2.7°F–7.2°F) rise in global warming. By the late 1970s into the 1980s, various energy and conservative political interests became alarmed and began to put together an organized and well-financed opposition to the rising public and legislative efforts.

By this time global warming had become an international concern, leading to a burst of global agencies, particularly at the United Nations, under whose auspices the Intergovernmental Panel on Climate Change (IPCC) was established. The IPCC became the leading collaborative organization of scientists dedicated to climate change study, issuing periodic reports updating scientific research and data. Those reports, well covered by the media, gradually came to be both central to moving global-warming response efforts along and a main target of climate deniers and minimizers. In 1992 a conference in Rio de Janeiro produced the UN Framework Convention on Climate Change (UNFCCC), stimulating U.S. efforts to block its recommendations. Another conference in 1997 led to the creation of the Kyoto Protocol, which set targets for reducing greenhouse gas emissions and establishing international treaties. The protocol was signed by many nations, but not the United States, and carried no obligations. Despite subsequent meetings to advance those goals, no treaties were signed, and the principal opposition came from the United States, China, and India, which together account for close to 50% of global warming. An important conference to develop a successor to the Kyoto Protocol is planned for Paris in December 2015. IPCC reports  2013, 2014, and a 2015 summary report, have provided the scientific foundation for the 2015 meeting. The expectations for that meeting are very high, especially because of encouraging shifts in the Chinese effort, notably a plan to put in place a cap and trade policy to lower carbon dioxide emissions in 2017.

## SCIENTIFIC CERTAINTIES

Let me note right off that careful scientists avoid speaking the language of "certainty" in describing the findings of their research. They much prefer to speak in terms of probabilities, of "likely," "highly likely," or "unlikely." Moreover, the term "global warming" has itself generated contention, with "climate change" as an alternative. Technically, "global warming" refers to the long-term trend of rising global temperature, and "climate change" to the climate changes produced by the rising temperature. As a matter of rhetoric, however, global warming has often been understood as a more forceful, evocative phrase. And that is why I have chosen to use it, even if the milder term "climate change" has often been favored by the UN. I find the word "certainty" similarly conveys a heavier emotive weight, but I think it should be acceptable for general public use.

One of the most cited passages in the recent literature is from a 2007 scientific report of the IPCC: "Warming of the climate system is unequivocal, as is now evident from observations of increases in global average air and ocean temperatures, widespread melting of snow and ice, and rising global average sea level."[3] Some skeptics challenged the certainty of that widely held scientific judgment, arguing that the rise in $CO_2$ was not the cause of global warming, which they saw as a random occurrence.[4] However, for hundreds of thousands of years the $CO_2$ content of the atmosphere was below 290 parts per million (ppm), but by 2013 it had reached 400 ppm and is increasing at an average rate of 2% a year. The rate of increase in the twentieth century was ten times faster than in all the centuries since the last ice age. The year 2010 showed the biggest increase in $CO_2$ emissions ever recorded, a 5.9% increase. The last twenty years have been the warmest for thousands of years, and 2014 was the warmest year since records began being kept in 1880 (figure 1.1). At the present rate of warming, the average surface temperature of the globe could rise by 2°C (3.6°F)—or more—above the preindustrial level, beyond which is truly dangerous territory. The IPCC's fourth assessment report, released in 2007, contended that a 450 ppm level of $CO_2$ could allow a 2°C limit, but that "achieving that target could mean reducing global emissions by up to 85% over 1990 levels by 2050."[5] Everything, that is, was moving in the wrong direction.

But $CO_2$ is not the only greenhouse gas, though it is by far the most important. The others of consequence are methane, 1.8 ppm ($CH_4$), nitrous oxide, 0.3 ppm ($N_2O$), and chlorofluorocarbons, 0.0001 ppm.[6] Another category of air pollutants is black carbon (commonly called soot) coming from mineral dust and land degradation; household fires and other biomass burning; and other sources, such as particulate matter, including volatile organic chemicals and airplane contrails. Urbanization, overgrazing, desertification, irrigation, deforestation, and dairy farming also play significant roles.

Global warming has already slightly raised sea levels, caused a melting of Arctic sea ice sufficient to open seasonal navigation across the top of the globe, and has resulted in a striking decline of mountain glaciers. The climate changes have been variable and are expected to remain so from region to region, with some feeling their effects more than others. The number and range of the likely climate impacts are themselves overwhelming: fewer cold days and warmer and more frequent hot days and nights; heavy and more frequent precipitation almost everywhere; an increase in areas subject to drought; a reduced risk of mortality due to cold and an increased risk of heat-related mortality; increased incidences of extreme high sea levels and fiercer storms.[7] Moreover, often overlooked, ocean acidification is another consequence of carbon emissions, posing a threat to fisheries.

There are likely to be serious impacts on health, on agriculture and food supply, on the use and availability of water and on water quality, on ecosystems, and on coastal areas from storms and rising sea levels.

(a)

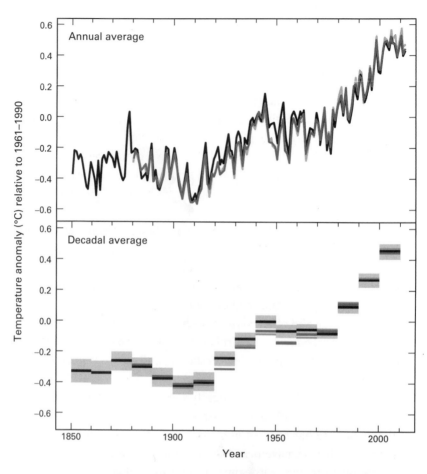

(b)

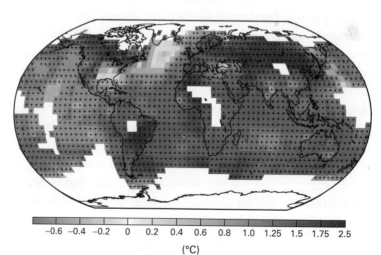

Quite apart from the direct impact on life, the potential economic impact under a business-as-usual (BAU) model—that is, do nothing—is staggering. The cost projections range from a minimum increase of 5% of gross domestic product (GDP) annually now and in the future and, depending on different types of calculations, an increase of between 11% and 14%.[8]

## SCIENTIFIC UNCERTAINTIES

The astute climate observer may well observe that under the previous subsection labeled "Scientific Certainties," I managed to include many descriptions of the wide range of environmental damage global warming will bring. But as some commentators have wryly noted, beyond the consensus (by most scientists) that global warming "is unequivocal," much else is uncertain. The territory of probabilities ranges from very likely to likely, unlikely, or possible. Taken literally, my phrase "will bring" should have read "could bring." But in most of our lives, few important things are unequivocally certain. Aspirin will probably help your aching shoulder, but may not. Your new car will almost certainly make it through a short trip, but something could go wrong. Your house will probably not burn down, but it is wise to have fire insurance. In many aspects of our lives, sometimes the most important, we must act in the face of uncertainty. We may harm ourselves in some ways if we do act, but then we may harm ourselves in other ways if we do not act.

Climate change puts all such everyday decision making to a severe test. There are reasons to be wary of apocalyptic predictions. If we do not act now to cope with it, nothing too serious may happen tomorrow or even for some decades. The severe, widespread, and record-breaking U.S.

**FIGURE 1.1**

(*a*) Observed globally averaged combined land and ocean surface temperature anomaly, 1850–2012; (*b*) observed change in surface temperature, 1901–2012. (Intergovernmental Panel on Climate Change [IPCC] Working Group I and Thomas F. Stocker, *Fifth Assessment Report: Climate Change 2013: The Physical Science Basis—Summary for Policymakers*: IPCC; Intergovernmental Panel on Climate Change, 2013) droughts of 2011 and 2012, and again in 2014–2015, though surely alarming, may be nothing more than regional variations (as many think is the case in California). The fact that the earth's temperature has been steadily and of late rapidly rising is no guarantee it will continue to do so. No way may be found to significantly lower $CO_2$ emissions. And with regional climate, variations in some areas will be a benefit and in others a loss—a global wash, so to speak.

But uncertainty about the future is a key feature of good science. There is uncertainty about uncertainty, but over the years the climate agencies and commissions, as well as the scientists, have taken it seriously. It is a well-established part of the field's professional culture of research and policy. And with good reason. There are well-organized and well-financed groups of skeptics ready to pounce on their findings. There have been some publicized scientific errors of projections and predictions, and often a failure to make clear the difference between them. Raising the level of danger is sometimes a deliberate practice on the part of global-warming activists, and yet endless hand-wringing is hardly helpful in the face of the need to mobilize the public and governments to act strongly and speedily. It is not an easy balance to strike.

There are many interrelated areas of uncertainty about global warming: the difficulty of developing robust projections of future climate trends; sensitivity (how much warming and how fast it will occur because of carbon building); the actual impact of present or proposed policies; the relationship of scientific evidence and the fashioning of effective policy; and the communication to the public and legislators of plausible and meaningful ways to understand the problem and respond to it. As a nonscientist I am in no position to pass judgment on the quality of the science in general or particular research findings. My impression, having looked at research in various fields over the years, is that climate science is a well-established field with rigorous standards—and particularly successful in bringing groups of scientists together from many nations to see where a consensus can be discovered and where further research is needed. Unlike, say, theoretical physics or astronomy, climate scientists know their work has a direct bearing on human welfare; it has consequences. More reliable projections and improved modeling efforts are under way—and they always will be. That is the essence of good science.

## SCIENTIFIC UNCERTAINTY AND POLICY

A more vexing problem is that of assessing scientific uncertainty as a part of policy analysis. Mike Hulme has nicely described three models of the decision-making process.[9] One of them was called the "decisionist model" by the German sociologist Max Weber. Facts alone can never determine what policy should be. It is for politicians to determine the goals of policy and for scientists to establish the best means of achieving those goals. Another is the "technocratic model." The politician depends on the scientist to do value-free science to establish the facts and from there to provide impartial guidance to politicians. Still another is the "co-production model": "both the goals of policy and the means of securing emerge out of joint scientific and non-scientific considerations. . . . This co-production model is sympathetic to framing knowledge in terms of risk, in which uncertainties are inherent and visible."[10]

Hulme does not choose among the three models. They can, he says, "coexist within the same nation at the same time" and among countries with different cultures. And that is why we disagree: "because we have different understandings of the relationship of science to other things." They include truth, risk, and "the legitimate role of knowledge in policy making." Daniel Sarewitz, in an important article with the provocative title "How Science Makes Environmental Controversies Worse," carries Hulme's argument a step further. He contends that the sheer quantity of research findings allows support for almost any ideological or political position. "Moreover," he writes, "when political controversy exists, the whole idea of 'reducing uncertainty' through more research is incoherent because there never will be a single problem for which a single, optimizable research strategy or solution path can be identified."[11] A theme that arises constantly now in climate policy research and in assessing lay knowledge is the need of all parties for self-awareness, an understanding of their own values and those of others, and how those values relate to policy.

While I believe Hulme and Sarewitz perceptively describe why there is disagreement on science and policy, holding in the end that the search for definitive predictive knowledge and final firm policies is illusory, I cannot fail to note that Sarewitz at least comes, in the end, back to what

Hulme describes as the "decisionist model." He concludes his article by saying "politics helps us describe the direction to step, science helps the eyes to focus."[12] But it was not politicians or policy experts who discovered global warming. Scientists called their attention to it and then began offering policy ideas for managing it. Moreover, when I address the role of technological innovation later in this chapter, Hulme and Sarewitz will reappear as participants in developing the 2010 "Hartwell Paper."[13] In the name of incremental and oblique approaches, that report rejects mainline top-down governmental approaches in favor of vigorous technological ones. It is a case of scientists moving from their scientific knowledge to advocate for some major policy shifts. If the policy they call for is incremental and oblique, the language they use in calling for it is firm and decisive.

In a telling response to an article by a group of skeptical scientists in the *Wall Street Journal*, the Yale economist William Nordhaus cited a number of scientific certainties: the planet is warming, human influences are an important contributor, and $CO_2$ is an atmospheric pollutant. Yet he also agrees with the emphasis placed on scientific uncertainties by the skeptics, saying they have grown over the years precisely as the result of research. Nordhaus then turns that argument on its head: "if anything, the uncertainties would point to a more rather than less forceful policy—and one starting sooner rather than later—to slow climate change."[14]

## INDUSTRY

A large number of industries have a stake in global warming. Some will profit from efforts to control damaging emissions (wind energy); others will suffer threats to profits and jobs. It is on the latter that I will concentrate at this point. Their profits come from the enormous range of industries that produce energy—and its potential threats from all of the actual and proposed regulatory ways to reduce its emissions. Assorted technological innovations could reduce the emissions and threats, but as we will see later, they will require government subsidies and large industry investments that are so far not forthcoming. As John Houghton has succinctly put it: "Industry currently accounts for nearly one-third of

worldwide primary energy use and about 1/4 of carbon dioxide emissions, of which 30% comes from the iron and steel industry, 27% from
non-metallic minerals (mainly cement) and 16% from chemicals and pet-
rochemical production."[15]

The U.S. Environmental Protection Agency (EPA) has detailed the
main sources of American emissions: electricity production (32%, more
than 70% from burning fossil fuels); transportation (cars, trucks, 18%);
industry (fossil fuels, 20%); commercial and residential buildings (10%);
and agriculture (10%). Electricity production is 32% from residential and
commercial buildings, 29% from industry, and 10% from agriculture.[16]
Globally, the Climate Disclosure Project undertook to persuade the
world's 500 largest corporations to disclose their emissions, and some
403 agreed to do so.[17] The fifty largest emitters of those 500 produced 73%
of total 2013 emissions. The top five emitters showed an annual increase
of 2.3% between 2009 and 2013.[18]

In sum, corporate global-warming emitters bear a heavy responsibility for their emissions and, accordingly, fear efforts to reduce them:
they are in the crosshairs. That pressure has hardly gone unnoticed.
Going back to the late 1970s these industries have fought back, supported by wealthy philanthropists of a conservative Republican bent, the
affected industries themselves, a variety of trade and other associations
and think tanks, and a sympathetic conservative portion of the media.
The U.S. Chamber of Commerce, the largest general trade association,
was an early leader of that effort, helped along by massive donations
from the philanthropists (notably the Koch brothers) and the industries. The weapon of choice was that of casting doubt on the quality of
the scientific research and the motives of the global-warming scientists,
particularly the hundreds who worked on the IPCC reports. Some of
the opponents were outright rejecters of global-warming, while others
were more subtle minimizers ("It is a concern but not a serious one").
The Union of Concerned Scientists found some twenty-eight large corporations and lobbying groups in a leadership position in those efforts.[19]
Another study found 140 foundations in that camp, together with those
already mentioned, and estimated there was an annual expenditure of
$1 billion on such efforts.[20] One early important organization in those
efforts was the Global Climate Coalition.

The historians Naomi Oreskes and Erik M. Conway have written two insightful books pertinent to the role of industry. One of them, *Merchants of Doubt*, details the efforts of a small number of scientists to systematically cast doubt on the harm of tobacco smoking and aspects of global warming, The other book, *The Collapse of Western Civilization: A View from the Future*, tries to imagine how generations far in the future will try to make sense of the failure of the present generation to take global warming seriously, with a central destructive role assigned to industry and free market advocacy.[21]

## PUBLIC OPINION

What exact impact those negative efforts have made is not clear, but the persistent attacks on global-warming science and policy must have influenced public opinion. A late 2011 survey by the Pew Research Center showed (1) that some 77% of those surveyed believed there is "solid evidence the earth is warming . . . because of human activity" but that number had declined to 57% by 2009 (although it had risen to 63% by 2011); and (2) that global warming was taken to be a "very serious" problem by 43% in 2006, declining to 35% in 2009, but going up to 38% in 2011 (with only 27% in 2011 taking it to be "somewhat serious in that year"). There is also a partisan political divide, with 77% of Democrats, 63% of independents, and 43% of Republicans believing that the earth has been getting warmer in recent decades and only 19% of Republicans attribute that rise to human activity.[22] Whether the well-publicized droughts in the Midwest and Southwest will make a difference in public opinion remains to be seen (and many experts are doubtful that global warming is causing those droughts, though it may be intensifying them). Other public opinion surveys have shown that Americans tend to think global warming is not a local problem, perceiving it as a danger to be faced by those in distant places and mainly nonhuman in its impact.[23] Other polls have found it to be of low priority as an important national interest. A 2014 Gallup Survey showed some increased concern, but not significantly changed.[24]

Meanwhile, a 2010 Pew survey found that 70% of people in China, India, and South Korea were willing to pay more for energy in response to global warming. A telling 2008 survey of Portland, Oregon, and Houston, Texas, found that 98% and 92% of those surveyed, respectively, were aware of climate change, but that less than 50% had changed their behavior as a result (e.g., reducing gasoline consumption and home energy usage, recycling).[25] Age and level of education were important variables, with the younger and more highly educated being among those willing to act. A useful 2009 survey of public understanding of "climate change" and "global warming" among residents in the south of England, which cited evidence from other countries as well, found a different response to those two phrases. "Global warming" is more effective in creating concern than the phrase "climate change," and far more likely to refer to human causes.[26]

## GOVERNANCE AND INTERNATIONAL EFFORTS

Although knowledge about global warming goes back many decades, it was not until the 1980s that political interest in the problem began to emerge in a serious way. An important point of departure was the formation of the IPCC, the International Energy Agency, the UN World Meteorological Organization, the UN Environmental Programme, and the UN Framework Convention on Climate Change (FCCC). The first meeting of the IPCC was in 1988, and it issued a number of important reports in the following years. The degree of scientific consensus achieved by that international panel has been vital in influencing governments. Its work was awarded a Nobel Prize in 2007. Another important event, drawing on the work of the IPCC, was the 1992 UN Conference on Environment and Development, which formulated the FCCC. That convention was signed in 1992 by more than 160 nations (and was ratified by the U.S. Senate).

In 1997 what came to be called the Kyoto Protocol, a complex agreement, was reached. It became the main foundation for ensuing international policy. It dealt with the greenhouse gases to be covered and rules for monitoring, reporting, and compliance. At a meeting in Marrakesh in 2001, it was confirmed that 120 countries had ratified the protocol, with the final details worked out at the meeting—but not the United States, which withdrew. A key feature established national commitments to reduce greenhouse gas emissions between 2008 and 2012 by 5% below 1990 levels. It was also agreed that global temperature should rise to no more than 2°C above its preindustrial level. Various mechanisms were established to help the signatory countries meet these goals. The following were the most important areas of action:

- Rapid reduction in tropical deforestation and increase in afforestation
- Aggressive increase in energy-saving and conservation measures
- Rapid movement to sources of energy free of carbon emissions of greenhouse gases, for example, through carbon capture and storage and renewable energy sources
- Some relatively easily achieved reductions in emissions of greenhouse gases other than $CO_2$, especially methane.[27]

The Kyoto Protocol had three special means of achieving emissions reductions:

- Joint country implementation: industrialized countries can reduce emissions in the territories of other countries
- The Clean Development Mechanism permits industrialized countries to implement projects that reduce emissions in developing countries
- Emissions trading ("cap and trade") allows industrialized countries to purchase units of emissions from other industrialized countries that find it easier to curb emissions.[28]

## KYOTO AND POLITICS

A great shortcoming of the protocol was that the signatories called on only industrialized countries to reduce emissions, leading the United States to reject it. The U.S. Senate in 2001 voted down participation in the protocol by a vote of 95 to 0, and George W. Bush said, "For America, complying with those mandates would have a negative economic impact, with lay-off of workers and price increases for consumers." Canada, Australia, and India eventually dropped out as well. The emissions limitation for developed countries only remained contentious until the December 2014 Lima Accord, in which all nations agreed to make reductions, although of unspecified and voluntary amounts. While some 65% of the world's emissions were part of the original protocol, that number had declined to 32% in 2002, influenced by the strong economic growth of the countries not covered by the protocol (particularly China, which went on to become the world emissions leader).

What have the results been? From 1990 to 2009 emissions in Russia and western Europe declined between 36% and 15%, while they increased by 206% in China, 244% in India, 172% in the Middle East, and 7% in the United States. In 2010 the global increase was, as noted above, 5.7%, the largest single-year increase recorded. There was widespread agreement that in light of such figures, the Kyoto Protocol has been a failure, and a strikingly dismal one at that. By contrast, the 1987 Montreal Treaty on Substances That Deplete the Ozone Layer has been considered a great success.[29] It brought about a 90% decline in the use of ozone-depleting substances over two decades beginning in the 1980s and particularly stabilized the stratospheric ozone over Antarctica (known as the ozone hole), attracting wide public attention. At the same time it is not held up as a good model for the larger emissions problem: it was not "wicked," as the larger global-warming challenge is; it involved only a few industries; and it met little industry or political opposition. As for the Kyoto Protocol, in the words of American economist Scott Barrett, "To be successful, a treaty must fulfill three conditions . . . it must attract broad participation . . . it must insure compliance . . . and it must do both of these things even if it asks its parties to change their behavior substantially. It is easy for a treaty to meet one or two of these conditions. It is very hard for a treaty to do all of them. Kyoto fails to do any of them."[30]

In light of the Kyoto failure there have been proposals over the years for a much stronger international organization, along the lines of the World Trade Association, with binding regulatory power. At the same time, from other quarters (and rejecting top-down bureaucracies) have come calls for a more deliberative democracy, accountable to the public and transparent in its activities and regulation setting. Those latter conditions have rarely been met by top-down, bureaucratic agencies. Mike Zajko, a Canadian sociologist, has said that "climate science can be more robust and credible if it is removed from the center of what are largely political disagreements."[31] But it is implausible to think that policy efforts that require the cooperation of all the nations of the world, with their different cultures, interests, and politics, can bypass political disagreements—unless of course they all agree that failure is not an option.

A 2009 FCCC conference in Copenhagen began to show the strains of global governance, failing to achieve any significant advances on the Kyoto Protocol—with global warming unabated—and placing in jeopardy the very idea of such an international mechanism. After the Copenhagen failure, the 17th FCCC conference in Durban in 2011 was then seen by many frustrated climate scientists and environmentalists as a critical test for the future of the protocol. In many respects it failed that test, achieving little agreement on many important issues, but it did have two noteworthy outcomes: an agreement to work toward a new global treaty in the future and the creation of a new climate fund. The new treaty would move beyond the older agreements, which committed only industrialized countries to cutting emissions, and include all countries. Beginning in 2015 a program of emissions reductions would be developed and put in place in 2020, setting a sliding scale for rich and poor nations. As with most agreements, the emissions reductions to be required and mechanisms for enforcement were not worked out, repeating a common hazard of earlier failures—agreeing to agree in the future. The Green Climate Fund, whose target will be to amass a fund of $100 billion a year, will aim to help developing countries in adapting to climate change and developing clean energy sources. Just how that kind of money can be found was left unclear.

The role of the United States in the future of the global effort is no less clear. President Obama gave less and less attention to global warming as his administration progressed, and conservative opposition remains at a high and toxic level. But Obama increased his efforts in 2014. As the world's second-largest climate polluter, the United States is poorly situated politically to take a global leadership role.

## ENERGY EFFICIENCY AND ENERGY DECARBONIZATION

Energy makes the modern world function, whether through the consumption of coal and oil or the production of electricity from those and other sources. And it is energy production that accounts for the bulk of the human contribution to global warming. The industrialized countries, using by far the most energy, are the main source of pollution, but the developing countries, dependent upon wood, dung, rice husks, and what are called "biomass" sources, make a difference also. Some 10% of the world's energy comes from those sources, but they are being employed by one-third of the world's population. The burning of that biomass through domestic cooking with open fires (with only 5% of the heat reaching the insides of pots) causes many serious, often fatal, health problems, particularly for children.[32] Simple and inexpensive stoves would greatly alleviate that problem, but of course that will not reduce the deforestation that results from burning wood, thus contributing its own share of $CO_2$.

The developing countries need new and modern sources of energy, with an estimated 1.5 to 2 billion people the most needy. As we will see, while many environmentalists have called for a slowing or halting of economic GDP growth, developing countries have responded by saying they need to see that growth continue for reasons both of fairness—why should they be denied the kinds of benefits now enjoyed by the affluent?—and for health and other quality of life reasons as well. For their part, the developed countries have some nonenvironmental reasons to want to reduce dependence on oil, notably a heavy need for imported oil, which is not only economically volatile but subject to the vagaries of the politics of oil-producing countries. That has been hard to accomplish but no less so than dealing with their own dependence upon coal, the least expensive and most easily available climate-polluting fossil fuel. But by early 2015 oil prices had begun to drop; a global glut had appeared.

As the environmental scientist Roger Pielke Jr. has stressed, there are only two ways to decarbonize the economic activity that drives atmospheric pollution: to improve energy efficiency or to decarbonize the energy supply. The means of improving energy efficiency are well known, such as improving automobile engine efficiency to reduce mileage per gallon and to do the same for coal-fired power stations. Many countries, including the United States, are already making fair progress with auto efficiency but have far to go with power stations. The decarbonization of the energy supply, as with much energy efficiency, also requires technology but of a more extensive and innovative kind. The demand for energy is always growing, even with moderate economic advances, and the requirement for cleaner and more efficient provision of energy in the future is all the more urgent. Unfortunately, greater energy efficiency has the potential of generating more energy demand and thus neutralizing or overcoming any efficiency benefits—the smaller, more efficient automobiles sold in China have made more people able to buy them, and they have done so in a dramatic, traffic-jamming, and polluting way. As air conditioners have become cheaper and more efficient, the demand for them has dramatically risen in India, requiring more electricity.

In sum, while there need not be an either-or choice between energy efficiency and energy decarbonization, it is the latter that many advocates now emphasize. That emphasis means a great investment in technological innovation (which will bring about mitigation of harms) and, for some, a greater emphasis on adaptation, that is, finding ways of better living with and surviving the wide range of damages likely to result from continued global warming (the dikes of Holland serving as the iconic model).

## TECHNOLOGICAL FIXES

The technological possibilities are many, ranging from the already available to those that are highly speculative, even scientifically utopian, and they range from those aiming to improve efficiency and those focused on decarbonization:

*Efficiency improvement*: building efficiency and auto fuel economy; greater efficiency of electrical appliances in homes and offices; home insulation; hybrid automobiles combining internal-combustion engines with electric drive trains and improved battery technology; and hydropower from dams and tides

*Decarbonization*: replacing coal with natural gas; capturing $CO_2$ at coal power plants; capturing $CO_2$ during hydrogen production; creating devices to capture and store $CO_2$; nuclear fission; wind or tidal power; fuel cells for gasoline in hybrid cars; biomass fuel; geoengineering (altering the earth's climate, solar-radiation management, stratospheric particle injection of sulfur dioxide particles into the stratosphere); removing $CO_2$ from the atmosphere and burying it in the ocean or deep in the earth; and soot reduction[33]

## Criteria for Assessing Technological Fixes

There are multiple obstacles standing in the way of these ideas: the cost of developing and deploying the technologies; coordinating the role of governments and the private sector in doing so; the potential hazards of some of the more speculative innovations, particularly geoengineering; and the difficulty of calculating their actual impact if developed. Daniel Sarewitz and Richard Nelson have directly taken on that problem and have proposed three relevant criteria:[34]

The technology must largely embody the cause-effect relationship connecting problem to solution. The use of vaccines for disease control offers an example of a technology to control a variety of diseases wherein the cause-effect relationship is clear; various methods of teaching reading have not yielded similarly clear results.

The effects of the technological fix must be assessable using relatively unambiguous or uncontroversial criteria. Despite opposition to the use of vaccines over their history, their medical effectiveness has met the test of efficacy and low risks.

Research and development is most likely to contribute decisively to solving a social problem when it focuses on improving a standardized technology that already exists. Thus solar power and wind power, with a track record and steady, incremental, improvements, are more promising for use and deployment than speculative geoengineering ideas (such as the direct removal of $CO_2$ from the atmosphere).

If these three criteria can be met, Sarewitz and Nelson conclude, they help "to solve the problem while allowing people to maintain the diversity of values and interests that impede other paths to effective action."[35] Given the ethical and economic struggles over HIV, it is easy to understand the quest for a vaccine, and the career of Steve Jobs and Apple illustrates the power of incremental innovations based on a strong theoretical foundation developed by others.

Yet the obstacles to technological fixes to global warming are formidable. They will take years to develop and disseminate and will involve a wide range of actors (government and private) who will not always have identical aims and interests. A useful 2009 study by British climate scientists clearly details the possibilities and obstacles. The authors reject at the outset the invoking of the Manhattan and Apollo projects as useful models: "they were designed, funded, and managed by federal agencies to achieve a specific technological solution for which the government was effectively the sole customer." In contrast, "both the industries developing and producing these solutions and sectors in which the technologies will be deployed comprise a very heterogeneous group . . . and they must demonstrate their cost-effectiveness, ease of operation, and reliability in systems that may be in operation for decades."[36] They then go on to offer a wide range of strategies that might well be effective, being careful to note the many difficulties that must be overcome. It was hard to come away from this important study with confidence that its sensible proposals would be easily achieved, much less meet the apocalyptic deadlines for the near future specified by many climate analysts,

## AFTER KYOTO?

Despite the two agreements reached at the Durban meeting, participants and observers alike judged the meeting a failure. Since that was the judgment passed on the previous meeting in Copenhagen, it is hardly surprising that new strategies are being proposed. One of them, mentioned earlier, is a new international agency with much greater power and ability to force change. Another, and worth some attention, was a report by a group of climate scientists and science policy experts who met in the United Kingdom in 2010: "The Hartwell Paper: A New Direction for Science Policy After the Crash of 2009."[37] Its starting premise was that, because of the failure of the Kyoto Protocol strategy to produce any discernible reduction of global emissions, a new model is needed to "set climate policy free to fly at last." The Kyoto Protocol model has failed, because "it systematically misunderstood the nature of climate change" and "it had emissions reduction as the all encompassing goal." "Climate change," the paper argues, cannot be addressed by any single governing, coherent, and enforceable thing called "climate policy." Instead, policy should be "politically attractive," "politically inclusive," and "relentlessly pragmatic" and aim to achieve three goals:[38]

- To ensure that the basic needs, especially the energy demands of the world's growing population, are adequately met. . . . [E]nergy that is simultaneously accessible, secure and low cost
- To ensure that we develop in a manner that does not undermine the essential functioning of the earth system, in recent years most commonly reflected in concerns about accumulating $CO_2$ in the atmosphere, but certainly not limited to that factor alone
- To ensure that our societies are adequately equipped to withstand the risks and dangers that come from all the vagaries of climate, whatever may be their cause.

They call this a radical reframing of the climate issue, not fully repudiating the Kyoto goal of reducing $CO_2$ emissions. Instead, they shift the emphasis to accelerated decarbonization, to eradication of short-lived agents such as black soot, to intensified mitigation and adaptation efforts, to diversified energy supplies beyond fossil fuels and the technological innovations to make those alternatives cheaper, to push efforts for protection of tropical forests—and to funding these efforts with hypothecated (dedicated) carbon taxes to pay for the innovations.

As the Hartwell Report puts it: "A slowly rising but initially low carbon tax has the advantages of avoiding negative growth effects. We are aware that as a general rule politicians in general and Ministries of Finance in particular hate the principle of hypothecation, because it ties their hands."[39] But that apparent limitation, the report says, is its virtue, "removing the issue from the political arena." It cites as an example of the feasibility of what they call an "indirect" approach to global warmth the establishment of a National Clean Energy Fund to be funded by a tax on domestic and imported coal.

This proposal has much to commend it, but would it not take an international organization like the FCCC or the United Nations to give the idea a platform and an established route to influence sovereign governments? It also requires a strong act of faith to think that innovative technologies can (1) be put in place quickly, while (2) being equitable and affordable, (3) meeting the Sarewitz/Nelson criteria for technological fixes, and (4) breaking through the political and bureaucratic barriers that often greet new technologies and their dissemination. What the paper calls the indirect, obliquity of its approach may require some blunt battering rams along the way.

## ENVIRONMENTALISM, GDP, AND GLOBAL JUSTICE

In the late 1980s a Japanese environmental scientist, Yoichi Kaya, articulated what came to be known as the Kaya Identity. There are, the identity holds, only four ways to reduce the accumulation of $CO_2$ in the atmosphere: (1) reduce population, (2) reduce per capita GDP, (3) become more efficient, and (4) switch to less carbon-intensive sources of energy. As the preceding section on technology showed, there is one group of leading climate scientists and environmentalists who hold that the only feasible and pragmatic way to move forward is by technological innovation to achieve goals (3) and (4). One of its leaders, Roger Pielke Jr. has also formulated what he holds to be an "iron law": "reducing economic growth [GDP]or limiting development as a means to address increasing carbon dioxide is simply not an option. . . . Even if per capita wealth were to stay constant, a growing population alone implies a rising GDP and thus rising emissions."[40] The Hartwell Paper took this position as well. Along with the protection of a growth of GDP, the idea of population control is put to one side as simply unlikely.

Before directly discussing environmentalism, I want to note an odd display of the politics and infighting of global warming. Two members of the Hartwell group, Michael Shellenberger and Ted Nordhaus, were the earlier authors of a scathing attack on environmentalists.[41] Among other things, they said that despite environmentalist organizations investing "hundreds of millions of dollars into combating global warming, we have little to show for it"—a judgment that is exactly the same as the Hartwell group passed on the Kyoto Protocol. It ridiculed that movement's use of term "reframing" the nature of the global-warming problem, exactly the same term which the Hartwell group used to define their aim, and scoffed at the environmentalists' effort to "focus attention on technological solutions," which of course is exactly what it did.

There are many ways of describing environmentalism, and I simply stipulate my understanding: it is the conviction that human beings have done great harm to the natural world in which we live and from which we draw much of our subsistence, that humans must learn how to live sustainably in harmony with nature and cease believing we can with impunity master and dominate it, and that from an environmental perspective human life on this globe may be crippled or even destroyed unless we become nature's stewards. Its interests have included, among other things, getting rid of air and water pollution, preserving forests and nonhuman species, organic farming, protecting the oceans and wilderness areas, taking responsibility for the welfare of future generations—and stopping global warming.

## GDP Growth

I want to focus my attention on what seems to me the most important and vexing issue, that of GDP growth. Reducing or eliminating that GDP growth (albeit with some country variance) has been a central plank in much environmental work and applicable to help solving or significantly reducing global warming. That idea was initiated some years ago when an informal group called the Club of Rome in the early 1970s commissioned a group of scientists, led by Dennis Meadows at MIT, to examine the long-term consequences of population growth and the use of depletable natural resources.[42] The result was a 1972 report, *Limits to Growth*, that sold more than a million copies and attracted considerable international attention. The combination of that population growth and the finite resources of the planet will require increased capital to keep pace, in the process diverting money from other important needs. A formula often cited by environmentalists, called "IPAT," is used as a way of measuring the impact of population and resource consumption on the planet: impact = population × affluence × technology. The impact is that of depleting finite natural resources while at the same time leaving an expanding and hazardous "ecological footprint," a major part of which is global warming. Affluence and the way of life that goes with it, mostly in ways that make use of technology to exploit natural resources, is what ends up harming the environment. The attraction of the end-of-growth movement, which spawned a large number of later editions and many other books (right up to the present), was that it caught some of the rebellious spirit of the

1960s as well as persistent concerns thereafter about the state of our society. Critiques of affluence, commercialization, excessive individualism, and consumerism were gaining ground as well, and attention was being called to visible features of modern life, such as air and water pollution, that were beginning to spoil it. The Club of Rome report, along with Rachel Carson's book *Silent Spring*, did for environmentalism what Betty Friedan's *The Feminine Mystique* was doing for feminism. The "green" movement was beginning and continues to this day. Global warming was gradually given a more prominent role as part of it.

But as might have been expected, critics of the limits to growth movement came along. Bjørn Lomborg, one of the most important and tenacious of the critics, updated earlier writings in a July/August 2012 article in *Foreign Affairs*, which the magazine editors boiled down to a long headline nicely capturing the gist of the article: "Forty years ago, the Club of Rome warned humanity that by chasing ever-greater economic growth, it was sentencing itself to a catastrophe. The predictions proved to be phenomenally wrong-headed but its malign effects persist."[43] Lomborg focused much of his attention on the claim that many basic nonrenewable resources, predicted to run out or see their cost rise to great heights, have not in fact done so or become too costly: technological advances in extractions of those resources have kept their supply plentiful and affordable. He does not say they will never run out, only that they will not in the near (or in some case long) run. Lomborg ends his article by noting that it is economic growth that will massively improve the health and social and economic welfare of the Chinese and Indians and those in poor countries, just as it has the developed countries. What the world needs is economic growth, "more of it, not less."[44] He does not take up the population issue at all. Pielke, we might recall, did not make a direct case for economic growth, only that is unthinkable that developing countries or justice considerations would accept such an idea.

If the political right and left have their own versions of an acceptable environmentalism, or a pursuit of the Kyoto agenda, or rejecting both in favor of technological fixes (one of whose aims is to circumvent ideological struggle), one other approach should be mentioned. In a 2011 article, "Capitalism vs. the Climate," in the *Nation*, Naomi Klein, a skilled climate journalist, takes on right-wing climate skeptics who are given to seeing the whole global-warming fuss as the belief "that climate change is a Trojan horse designed to abolish capitalism and replace it with some kind of eco-socialism."[45] The conservative focus is not on the quality of the science behind it (which is dismissed) so much as its political meaning, which is an assault on capitalism and with it human freedom.

On the first charge Klein says they are right, and on the second she says that global warming is the far greater danger to freedom. The deniers recognize that sharp and rapid reduction of global emissions is needed and have concluded "that this can be done only by radically reordering our economic and political systems in ways antithetical to their 'free market' belief systems." They are absolutely right, she says. There is an echo of the Club of Rome when she writes that "the central fiction" on which "our economic model is based is that nature is limitless, that we will always be able to find more of what we need."[46]

## PARADIGMS AND MYTHS

I will close this chapter by making use of what Mike Hulme calls four "myths," to which I will add two of my own. Hulme call them myths "in that they embody fundamental truths underlying our assumptions about everyday or scientific reality." I would prefer the word "paradigm" myself, that of a model that reflects a reality one is trying to capture.[47] In any event, here are his four:

*Lamenting Eden.* A theme that marks a great deal of environmental analysis in general, often invoked in global-warming debates, is the loss of the natural purity, of nature untainted by humans—Eden before the advent of man. We despoil nature by using it only for our human ends, ruining it by carelessly considering its existence only as something to be mastered and exploited. At the same time we are dependent upon the goods of nature for our own survival. Its exploitation has begun to destroy its nurturing benefits. The term "stewardship" is often invoked as the alternative to self-interested exploitation. For the good of nature itself and the future good of the human race we need to take care of it. Global warming is a part, but not by any means the only part, of a larger vision of environmentalism.

*Presaging apocalypse.* This is the belief that we have moved so far along the road of the destruction of nature, that disaster is on its way, even the end of human life, if not soon, at least in a few generations. We may survive, but there is little hope that future generations will. There are already too many of us, a limit to natural resources to meet our needs, and a tipping point in the near future that will not allow us to be assured of survival. There has been long debate on the technical question of the speed of global warming, when a tipping point might be reached, and when the threat to life would be most disastrous.

*Constructing Babel.* This model, taken from the biblical story of the construction of a tower to the heavens in the aftermath of a flood, is meant to convey the idea of a permanent place in the heaven to be spared such disasters. Its modern version is that technology that will save us from global warming and now being vigorously pursued. It has the advantage of offering a solution to global warming that can avoid the large-scale social and economic costs of serious mitigation.

*Celebrating jubilee.* The attraction of this model is that the changes necessary in the way we live our lives to cope with global warming open a much wider door to a future world of greater justice and equality for the poor and downtrodden. It can provide meaning to life in a way that has been harmed by the emphasis on consumption and the benefits of the market, capitalism, and wealth accumulation. It is an opportunity to bring new meaning to our lives, and a sense of community with others and with nature now absent.

I want to add two paradigms of my own:

*The magic of the market.* There is a fine line between the denial of global warming and the magic of the market. But *if* global warming must be dealt with, *then* the full freedom of the market should be unleashed. Market freedom, it is said, will allow people to make their own choice in responding to it, not falling prey to the dangerous belief that faceless bureaucrats and regulatory impositions can make the best decisions for everyone. Solar panels are advertised to home owners like me because they are good for the environment and will save me money on my electricity bills. It is a choice I make, not government.

*The compatibility of economic growth, emissions control, and profit.* This paradigm seems a bit like jubilee turned upside down, with the market thrown in: the best of all worlds in a capitalism-dominated global economy, doing good and making money.

The solar panel example is part of this paradigm, but there is more to it than that. It has the larger goal of changing the way we think about the economy and the environment, using technological innovation to transform global warming and the way we live our lives, and putting choice in our hands not in the hands of the government. It is jubilee in a different key.

# 2

# FEEDING A GROWING POPULATION

––––

## How, and with What Kind of Food?

The inclusion of starvation among the biblical four horsemen has fearsome historical credentials. It has always shadowed human development and civilization, remaining today as part of the folklore of the ancestors of millions around the globe. My own history can be traced back to the Irish potato famine of 1845 to 1852, which my father's family fled to the United States to escape, part of the fortunate million able to migrate and leaving behind the unfortunate bodies of the millions who died. The global famines of the past few centuries saw the deaths of tens of millions of people, most notably in China, Russia, the Sahel, and India. Over the centuries, few regions of the world have been spared. The causes of the famines can be traced mainly to drought, floods, crop failures, occasional genocide, and harsh or inept government policies—singly or in some combination.

Famine on that scale has been uncommon in recent decades, but malnutrition and the death rates that go with it have been tragically present. The cost and availability of food had reached a critical stage in 2008–2012, with a reasonable expectation of a potential international crisis in 2013.[1] That did not happen but seems to be a temporary reprieve only. Why did it not come about? A thorough review of the global food situation answered that question: "in mid-2008, there was a rapid rise in food prices, the cause of which is still being debated, that subsided when the world economy went into recession."[2] Nonetheless, despite the end of that crisis, there is every likelihood of its inevitability in the future, with population growth rates rising faster than annual food production rates. That is my rationale for continuing to count it as one of the five horsemen. The earlier crisis revealed, however, many of the persistent if fluctuating features of world food availability. They include a decline in wheat and rice stocks (mainly a consequence of poor weather); sharp rises and declines in oil prices; increased demand, a function of growing populations and a rising consumption of meat in developing countries; market commodities speculation and unbalanced trade policies; a falling off of agricultural development; and a decline in food aid from rich to poor countries, in part a reflection of economic recessions.

## FOOD SECURITY: HIGHS AND LOWS

Nonetheless, a remarkable feat in the twentieth century was the enormous increase in international food availability. Stimulated by the work of Norman Borlaug and the Green Revolution from the 1940s until the 1970s. It brought scientific research to bear on agriculture and nutrition, benefitting many millions who had earlier scrabbled for food. The Green Revolution created enormous gains in crop yields, great improvement in the quality and quantity of corn, a large increase in the global wheat harvest, and faster-growing farmed fish. It brought increased food security for hundreds of millions of people in developing countries—and a base for overall economic development in China and India.[3] The Malthusian specter of population growth outrunning food production was effectively banished. Beginning in the 1970s food prices declined markedly for over thirty years. Fears of famines almost vanished.[4] Global food trade grew rapidly, bringing affordable food to previously isolated countries and regions (figure 2.1).

It was a false dawn. By 2005 food prices began a sharp rise, increasing by 75% between 2005 and 2008. Average global per capita food consumption and food production had indeed improved, but because of disparate and unequal access food insecurity was by no means abolished. Participants from 180 countries taking part in a 1996 meeting of the United Nations Food and Agricultural Organization (FAO) pledged to cut the number of people suffering from hunger (narrowly defined as shortage of calories) to half its 1990 level by 2015, from 845 million to 422 million. In 2014 it was only down to 800 million.[5] Some countries, notably China, saw a decrease of 100 million hungry people, but food insecurity increased in the Middle East by 93%, in central Asia by 45%, and in sub-Saharan Africa by 26%.

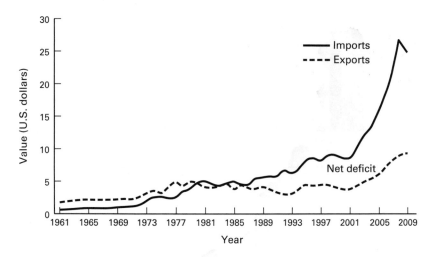

**FIGURE 2.1**

Agricultural trade balance of least-developed countries, 1961–2009. (Food and Agriculture Organization of the United Nations, 2011—TradeSTAT, cited in *Fueling the Food Crisis, the Cost to Developing Countries of U.S. Corn Ethanol Expansion, October 2012*: ActionAid USA, 2012: 15.)

Those discrepancies are a classic example of overall improvements and benefits in general, but with inequitable distribution for many (figure 2.2). While there are regional differences, average life expectancy in developed countries is seventy-nine years. The comparable figure for developing countries is sixty-seven years. While malnutrition is not the only source of trouble for those living in the poorest countries, it is a major part: they are the most directly harmed by food shortages. Their populations spend 70% to 80% of their income on food, compared with 10% for those in industrialized countries. Children are the most affected, but malnourished mothers are subject to high mortality rates, reduced physical strength, lower educational achievement, and a shorter life expectancy. Around 130 million babies are born annually, with some 8 million children stillborn or dying during the first four weeks of life in developing countries. The children among the 900 million malnourished have a high mortality rate, with some 54% of their deaths associated with malnutrition. Those children who live are likely to suffer from impaired cognitive development, stunted adolescence, and increased risk of adult chronic disease.[6]

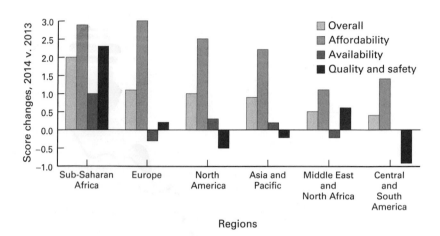

**FIGURE 2.2**

Improving global food security: score changes, 2014 versus 2013. (*The Economist* Intelligence Unit, "Global Food Security Index 2014, Regional Findings: 26," http://foodsecurityindex .eiu.com)

Moreover, if the malnutrition is not remedied during the first fourteen to eighteen months of life, those long-term impacts will be unavoidable. For the parents of those children malnutrition has a wide range of harmful social impacts on income, employment opportunities, and physical capacities. Malnutrition, in brief, represents a great hazard for all age groups.

## WHY IS THERE A PROBLEM? CAUSES AND EFFECTS

While there are many causes of food shortages, both in production and distribution, a place to start is global warming, whose tentacles reach into this domain just as they do with many others. Broadly put, global warming increases the potential for drought, a rise of sea levels with consequent flooding, water scarcity for agricultural needs, and diminishment of dietary diversity and food availability. Many regions of the world will be vulnerable, as will many different populations (notably subsistence farmers, coastal populations, fisherpeople). The 2013 Intergovernmental Panel on Climate Change report estimated that global warming could reduce food production by 2% each decade, while the demand for food will increase by 14% a decade.[7]

But global warming is only part of the problem. While there is some overlap, two useful summaries of the causes of food shortages bring out the complex breadth of the challenge. One of the most striking statistics is that of the changing relationship between world farmed area and world population growth. Between 1990 and 2005 there was a 1.8% increase in farmed area and a 23% population growth.[8] Two good lists of the various causes of food problems, one presented by Cristina Tirado and colleagues and Anuradha Mittal and the other by Julian Cribb, nicely sum up the range of the problem. I begin with Tirado and Mittal:[9]

*Decline in growth of agricultural production.* Between 1970 and 1990 agricultural production rose by an average of 2.2% a year but has declined to 1.3% since 1990. That decline reflects a reduction of government support for agricultural production in developing countries, a decline in agricultural research, and the impact of climate change and growing water scarcity.

*Increased energy costs for agricultural production.* Except in the case of agroecological farming, food production requires fuel and fertilizers. The rising cost of both over the past decade for U.S. export foods has been in the range of 15%–20%, with a doubling of energy costs for fuel and fertilizers and no foreseeable reduction of any magnitude in the future. Higher energy costs encourage the use of less fuel and fertilizer, cutting crop yields.

*Decline in global stock of grain.* Global stocks of available stored grain are an important component of food relief aid in the face of environmental threats and international food aid needs. Those stocks are adversely affected by weather events, liberalization of international food markets, and concentration of food trade in the hands of a few transnational corporations. Global stocks of grain, important in times of acute crisis, declined from 31.2% of production from 1999 to 2000 to 16.5% in 2007–2008. But by 2014 those stocks had risen to a fifteen-year high.[10]

*Biofuels.* An earlier demand for and high cost of fuel (now declining), notably oil, led to the development and use of biofuels, mainly ethanol, at present derived from corn. The expansion of the production of ethanol has directly led to a decline of corn stock levels for food and an increase in the cost of corn, as well as the reduction of foodstuff from other crops as a result of expanded acreage for ethanol production.

*Reduced state regulatory role in agricultural production.* In a policy that seemed reasonable at the time, the International Monetary Fund (IMF) decided in the 1980s and 1990s to encourage the demise of the regulatory role of governments in developing countries through what were known as marketing boards. It was the duty of such boards to prevent price volatility, to control prices to cover the cost of production, to insure a small profit, and to keep a stock of commodities that could be released when there was a bad harvest. The IMF's effort to eliminate those boards in the name of efficiency and corruption avoidance, however, did not work, and useful state mechanisms, though troubled, should have been maintained.

*Removal of agricultural tariffs and resulting import surge.* Developed countries, with their heavily subsidized agriculture, have engaged in "dumping" foods on the market below the costs of production. This is harmful to subsistence farmers in developing countries,financially unable to compete in the agricultural market, and has driven developing countries to move from being net exporters of food to large importers. That combination—in tandem with the fact that some forty or more developing countries are caught in a *shift to export crops*, with a majority dependent upon a single crop (e.g., sugar, coffee, bananas)—means that developing countries are subject to price fluctuations and a weakening of developing their own indigenous food supply.

Here is Julian Cribb's list:[11]

*Population growth and increased consumer demand for dietary improvement.* The population of the world passed the 7 billion mark in 2011. While population growth has slowed a bit in recent decades, it is still increasing by 80 million or so a year and could reach 11 billion by 2100. At the same time, save for some areas with extreme malnutrition, most other developing countries have improved economically, benefiting their growing populations. And those people have come to expect and demand dietary improvement, which means more meat, milk, fish, and eggs. That trend requires considerably more grain, increasing agricultural needs.

*Diminishment of freshwater.* Some 70% of the world's available freshwater is already consumed by farmers, heavily through irrigation, and the remaining 30% is under pressure from large urban centers.

*Land scarcity.* Arable land around the globe is diminishing, a function of population growth, urban and suburban spreading, and industrial pollution. Along with diminishing arable land comes a concomitant *loss of soil nutrients.* Soil erosion alone exceeds all the nutrients provided by fertilizer.

*Oceans.* Long-term overfishing around the world is a significant threat to the supply of fish, only some of which can be offset by aquaculture (which itself can do some environmental damage). Some parts of the world, especially Southeast Asia, are particularly dependent upon fish for nutrition, and overfishing and ocean acidification are a special threat.

*Decline of agricultural productivity research and technology development.* Developed countries have gradually reduced their research support of agricultural productivity, and that decline in turn has led to a decline in private investment. The USAID agency, the premier U.S. development agency, has cut its aid by 75% over the past two decades. The kind of research that earlier led to the Green Revolution and enormous productivity gains in poor countries has, so to speak, withered on the vine. The U.S. government program, Feed the Future, currently reaching nineteen developing countries and meant to be inclusive, uses five criteria for selection: level of need, opportunity for partnership, potential for agricultural growth, opportunity for regional synergy, and resource availability.

*Economics, politics, and trade.* This important combination will be dealt with later.

## MEAT AND BIOFUELS

Among this long list of contributors to the world food problem, two are worth further attention in the context of the aims of this book: the use of land and other resources for the production of more protein, milk, meat, and poultry; and the production of biofuels as a substitute for oil. In the case of protein, it is a desire for a different diet in developing countries that is spurring production, whereas biofuel production is an example of solving one problem and simultaneously creating another.

The cultivation of land to raise livestock, heavily driven now by demand for meat in developing countries, has been identified by the UN FAO as the source of 18% of global warming via the generation of $CO_2$.[12] At the core of that contributor to global warming is the raising, feeding, and marketing of cattle for meat. Some 8% of the world's freshwater is used for that enterprise, mostly through irrigation. Average annual per capita meat consumption in the United States, though leveling off, is 271 pounds, and 163 pounds in Europe. The U.S. per capita consumption of corn is some 432 pounds a year, much of which is used as feed for cattle and chickens. In China the average meat consumption is only 119 pounds, but rising rapidly (as is also the case in India). In the 1930s grains and vegetables made up 97% of the Chinese diet, a figure that has now declined to 67% and that continues to drop. Meat consumption is expected to double between 1995 and 2020, and some 80% of the world's agricultural land will be used to develop feed for animals and land for their pasturage. The demand for milk is an important part of the cattle story as well.

Yet if agricultural productivity research has declined, there is one set of data that tells an important story. As Cribb puts it: "If you factor in the amount of grain needed to produce meat, a single hectare of land can produce 29 times more food in the form of vegetables than in the form of chicken meat, 73 times more than pork, and 78 times more than beef."[13] This discrepancy is not a harsh deliverance of nature but a disturbing insight into the productivity consequences of the richer diet that captivates those with rising incomes and the changing food tastes that accompany them.

While much of the focus on meat and dairy products has been on the obesity their consumption can bring, there seems little doubt that up to a point meat and dairy consumption results in improved health because of an increase in intake of protein and a number of valuable minerals and vitamins (e.g., iron, zinc, vitamin A).[14] Much of the gain in height and longevity in recent years can be traced to meat and poultry consumption. It is once more the thrice-told tale that will appear again and again in this book: the too-much-of-a-good-thing phenomenon, where benefits shade off into threats.

Biofuels represent a different but related story, one in which an apparent solution to one social problem creates another.[15] In this instance, even as the global availability of good agricultural land is declining, the need to reduce dependence upon oil, particularly for gasoline and fuel for transportation, is growing. By 2006, 40% of the U.S. corn crop was devoted to ethanol production, yet it replaced only 3% of fossil fuel consumption. That production competes with food for soil nutrients, energy, pesticides, land, and water. The earlier era of cheap foods is coming to an end in other parts of the world as large sections of wetlands and forest are being clear-cut to make way for sugar cane and palm oil farms, also used as alternatives to petroleum. Moreover, all those fuels—called first-generation biofuels, by virtue of fertilizer and transportation costs—take as much energy to produce as they yield in the end.[16] A new second generation of biofuels, most still in the research and development phase (such as algae farms), offer the promise of better alternatives to oil than ethanol and biodiesel.[17]

## THE INTERNATIONAL FOOD MARKET

All the main global food roads in the world lead to currently industrialized agriculture and the international food trade. In 2008 global food sales came to an estimated $8 trillion.[18] Agriculture accounts for some 6% of the total international GDP, and the food welfare of 41% of the world's population is dependent on food trade. For the poorest countries it can run as high as 50% of GDP or higher. In industrialized countries, by contrast, agriculture accounts for 2.4% of GDP, with only 4% of their populations doing agricultural work.

While international food trade has gone on for centuries, it accelerated after World War II. There was a surplus of food production in the United States and a recognition that the international food trade—building on the political and scientific power of countries with food to sell and a market for it—beckoned business and investors. In the 1970s government support in developed countries for food trade, building on the foundation of large-scale industrial agriculture, began to move even more rapidly. International food corporations gained the benefit of agricultural subsidies, favorable trade rules, considerable control over export prices, and the increased needs of developing countries for imported food (in great part because many of them were exporting food for commercial benefits at the expense of local small farmers). Added to those powerful forces was the increased "financialization" of the international food trade, that of commodity market trading and speculation of a kind indifferent to what is being traded or its human impact: it is simply a source of money for smart traders.

The net global impact of these commercial, investment, and industrialized agricultural developments has been profound, with a wide range of global ramifications. Cheaper food imported from developed countries can undercut more expensive indigenous agriculture. The seemingly commonsense idea that water scarcity in some countries can be met by imports from water-rich countries has its own downside.[19] Transportation costs for worldwide food trade raise the cost of food and contribute to global warming. The ability of large global exporters to specify quality, appearance, and selling dates of food leads to considerable waste, as does the overstocking of supermarkets. Globalization of food favors large suppliers over small ones and subsidized over unsubsidized agriculture.

Food manufacturers and supermarket developers only infrequently invest in agricultural research or the efforts of small farmers to become more efficient and sustainable. And while the large chemical, biotech, and fertilizer companies carry out vigorous research, their aim is economic benefit, but not for rural societies and the common good. Many of the world's poorest countries have become reliant on imported food, a trend matched by the increase in the number of malnourished people. Volatility of prices and food supply is another side effect of this financialized, profit-driven international food market.[20] As Jennifer Clapp has put it, "The prioritization of agricultural production has been perceived as a national security interest in the industrialized countries, even as it has an impact on the ability of the world's poorest countries to feed themselves."[21]

## THE POWER OF THE FOOD INDUSTRY

Underlying many of those features of the food market, the influence of the IMF in the 1980s was significant, pushing hard for what was known as "structural reform"—that bland term for the aggressive introduction of market ideas and practices—and came to dominate many developing countries. It rested on two premises: the need to rapidly increase food production and the deregulation of food markets to bring that about.[22] Prior to that time, the dominant value was that of national self-sufficiency, aiming to strengthen the economic situation for small landowners and farmers, enabling them and their countries to meet their food needs on their own. The logic of structural reform and the neoliberal ideology on which it rested was to overturn that earlier way of thinking. In its place was the belief that an international free market—lowering prices for consumers, increasing incentives for producers, and seeking the comparative advantages of countries with different economic strengths—would achieve a good balance of costs and benefits. In practice this theory meant that a country that could successfully export a few crops, even at the cost of weakening agricultural self-sufficiency. It could then import from the international food market with sufficient money to pay for what they needed.

That was the theory, but it did not work out that way. Countries heavily dependent upon imported food as a result of this process became subject to international price fluctuations that commonly occur in the food market.[23] At the same time they saw their national food self-sufficiency undercut by imported foods sold at prices lower than those their farmers could charge. The net result, as Noah Zerbe notes, is "that the market, rather than the state, becomes the prime guarantor of food security, a phenomenon we might rightly conceptualize as the marketization of food security."[24] The unfortunate net result of this policy was a "combination of state *and* market failure." By 2008 the price of imported foods for developing countries had risen sharply, becoming a key ingredient in the food shortage then and remaining at a stable level subsequently—but with an expected increase in the future.

## TECHNOLOGY

The constant lament in the recent global food literature about the decline of agricultural research is surely fueled in part by a nostalgia for benefits comparable to those affected by the Green Revolution. At the Green Revolution's beginning in the 1940s about one-third of the world's population died of malnutrition or the diseases associated with it. By the beginning of the twenty-first century that figure had declined to one in eight. The Green Revolution was a critical part of that stunning improvement. If the estimated 900 million now suffering from malnutrition seems horrendous, it is not only because that figure has risen of late but also because of a belief that it should have continued downward. The Green Revolution had a dramatic impact on agricultural productivity in rich and poor countries alike.

There are many reasons why research has declined, but it is hard to fault Julian Cribb's judgment that in large part the very success of the Green Revolution led to "complacency and neglect."[25] There had long been complaints that it led to an excessive use of pesticides and helped advance the dominance of monocultures in many developing countries, but its benefits far outweighed its potential harms. I would note also, however, a similarity between the hope and enthusiasm for technology as a solution for global warming, and the potential role of technology in food production. There seems little doubt that the advent of genetically modified organisms (GMOs) has been important since its introduction and rapid embrace in the mid-1990s. But there has been in global-warming abatement efforts no technological success comparable to the Green Revolution in general or to genetically modified foods in particular.

Despite continued ethical and scientific controversy over the potential harm of GMOs, the steady rise of use of genetically modified crops in most countries, save for Africa, has managed to overcome that obstacle. Although somewhat over 50% of the global area used for genetically modified crops is in the United States, primarily large farms, over 90% of the farmers planting genetically modified crops are small landholders in developing countries. The two primary means of modifying plants are to make them herbicide tolerant (resistant to chemical weed killers) and insect resistant (reducing the need for pesticides). While a variety of crops have been investigated for GM possibilities, corn, cotton, soybeans, and canola have remained dominant.[26] That said, it is hard to find utopian technological projections comparable with the Green Revolution. The picture is more one of incremental improvements, efficiencies in agricultural practices more generally and in technological innovations.

## GOVERNANCE: WHERE IS THE POWER?

There is no one global organization to organize and implement international food policy and security. There are, to be sure, a number of important international organizations, and nongovernmental organizations (NGOs) that deal with food problems and policies: the World Health Organization (WHO), the FAO, World Food Organization, the Consultative Group on International Agricultural Research, the IMF, the World Bank (WB), the World Food Program (WFP), the UN Children's Fund, and the World Trade Organization (WTO).

On paper, the WTO would seem to have the most clout, with its power to set international trade rules, but it has done poorly in dealing with trade-distorting farm subsidies.[27] The WFP has over the years been very successful in providing famine relief but has not been helped by U.S. demands that American food producers and transporters use American assets only for relief efforts instead of cash allowances that would allow needy countries to find less expensive help. The international organizations are influential and often effective, but they all share the weakness that they usually have no decisive authority over the nations on whose behalf they work. In at least one case, that of the IMF using of its considerable power to force structural adjustments, such efforts proved harmful.

As good a symptom as any of the problems of the international agencies—in the absence of sanctions and penalties in advancing its ideas and proposal—is the shortfall in pledged money in the past few years. At a meeting in Italy in 2009 the eight richest nations pledged $20 billion over three years for food security and agricultural development. That was shortly followed by another pledge by the G20 nations (those with the largest economies) in Pittsburgh, with $900 million pledged, more than half from the United States. But so far the United States has provided only $67 million of its pledged $475 million, and Congress cut a request for the remaining $400 million to $100 million. Other countries are not doing well either, and one-third of the original $20 billion pledge turned out to be money pledged earlier than 2009. Former American secretary of agriculture Dan Glickman said, "What's needed is sustained leadership from government, NGOs, and the private sector. I see it from [Bill] Gates and some companies. But the big resources have to come from business, and that's much harder now."[28] It is necessary, of course, to point out that the pledges were made during a major recession and President Obama was not helped by a hostile Republican Party and a sharply divided Congress.

## ALL GLOBAL AGRICULTURE IS LOCAL

If national sovereignty is an obstacle for any and every kind of international governance and surety—as I noted in chapter 1 was the case for global warming—agriculture is an intensely local activity. While there is international trade and research, in the end food is produced within individual countries and is thus subject to the sway and power of their cultures, histories, and governments. Efforts were thus made over the years by the various food-oriented international agencies to influence the food policies of national governments. Per Pinstrup-Andersen and Derrill Watson have noted three phases of these efforts: (1) to let food markets alone in the 1930s, but then (2) for governments to intervene heavily in those markets from the 1940s through the 1970s, and (3) in response to the problems and disappointments of structural adjustment pushed by the World Bank and IMF in the 1980s, to find a better balance between government and the market to address the failures of both, finding a middle way between government command and control policies and unregulated market practices.[29]

"In general," Pinstrup-Andersen and Watson conclude, "the greater the risk of government failure and the greater the complexity of the subject, the more desirable it is for government to use a light hand; while the more certain it is that very specific action is required and the more pervasive market failures are, the more likely that direct regulation will be preferable."[30] That judgment, however, must be understood in the context of the low priority given to food production and malnutrition in many developing countries; the pervasiveness of corruption and poverty; the indifference of often dictatorial governments to endemic poverty; and the bias toward urban populations, which now represent about 50% of the citizenry of developing societies.

Over the years international food aid from affluent to poor countries has tried to compensate for those shortcomings and particularly to respond to food emergencies. That aid was to increase greatly after World War II, in part because of large agricultural surpluses in some countries, notably the United States—that were driven by the emergence of an international food industry that sought profit from the production of food—and in part for political and military motives. By the 1970s food aid came to be seen as an important part of an overarching need for economic development in developing countries, but by the 1990s—stimulated by food emergencies and the goal of eradicating hunger—aid for humanitarian rather than development reasons became dominant.

Various shortcomings for that policy soon became evident and are now exacerbated by diminishing food surpluses and the difficulty of gaining reliable fulfillment of promised contributions to meet fluctuating food needs. As Raymond F. Hopkins concludes after examining that history, "Addressing food and hunger crises as unique is unworkable; they must be placed within the search for international policy-makers' solutions to these larger concerns."[31]

Economic inequities and gross disparities of every kind mark the difference between rich and poor countries. But few are quite so striking and ironic as the fact that a lack of a good diet dooms hundreds of millions to malnutrition, a number that is slow to recede—while a poor diet of a different kind dooms an even larger and also expanding number to excess weight and obesity. The title of Raj Patel's book *Stuffed and Starved* perfectly catches that irony. Anthony Giddens said of global warming that "we have no politics of climate change," and it is no less fair to say that "we have no politics of food."[32]

But is it correct to use the word "doomed" for both groups of victims? I will respond to that question pertinent to obesity when I come to that chapter, but in the meantime I will note one important difference. Obesity can be traced to the way we live our modern lives and the way our culture, politics, and economics shape that kind of life. We are as a people spoiled by affluence, complicit in that life in a way that cannot be said of those suffering from malnutrition. We have choices and they do not. In a rich country, I can decide to do something about my weight, even if it might be hard and I might be faced with a variety of obstacles. No such option is open to those malnutrition victims in sub-Saharan Africa and Southeast Asia. They cannot escape food shortages, high food costs, drought, flooding, or the desertification of their land, nor can they as individuals just change their lives in a healthy direction. They are the victims of external forces. But are they doomed; that is, can something be done about those forces?

## HOPEFUL REMEDIES, HARDHEADED REALITIES, PERVASIVE DISAGREEMENTS

Or to put the question differently, who will save them? One answer might be that, if agriculture is understood to be ultimately local, it will have to be their own countries. They will have to find ways to live with their own food resources. Other than erratically and capriciously, the rest of the world cannot be counted upon to provide sustainable food security. Yet no country is fully food sufficient, even if not in dangerous ways. Even the United States has to import some food from time to time, as does every other country. At the other extreme, in a world where climate and natural resources are not equally hospitable, nothing less than international help and intervention can give everyone a more or less equal chance at that security. But as we have seen, the leading international agencies concerned with global food security and resources are not strong, lacking the power to impose solutions, to force countries to honor their pledges, or to reach agreement on important issues.

There are, however, some significant areas of agreement, the most important of which is that current governance and policy pertinent to global food security is not working. "Change is needed," is a constant refrain. There is also agreement that there are three necessary conditions for food security: "(1) Availability of sufficient supplies; (2) access to enough food, either through sufficient income or by the provision of adequate safety nets; (3) nutritional wholeness—access to a complete and healthy diet; and (4) price stability."[33] Four other areas of mainline accord are: global agricultural productivity improvement; reducing agricultural protectionism; managing national interfaces with unstable world markets; and national and international food safety nets. Alex McCalla singles out long-term productivity growth as the most critical need. "We can," he writes, "probably survive without a WTO agreement, but we cannot if we do not learn how to produce more with less."[34]

But, as with too much of life and policy, the devil is in the details. There is disagreement in particular on the place of technology and the market that introduces discord and, ultimately, different possible future directions. The differences are rarely flat either/or choices or policies but instead involve a tilt one way rather than another, with a few at the far ends of a left-right spectrum.

## TWO CONFLICTING MODELS

I have found the analysis of Jennifer Clapp of the University of Waterloo in Canada particularly insightful in illuminating two important conflicting models of global food policy, each different but also influenced by the other.[35] One is a fresh push on the present lines of more scientific and corporate involvement. The other is a more equitable approach that incurs less ecological damage.

The "dominant response" (as Clapp puts it) can be boiled down to some basic premises. They are reflected succinctly in the summary of the consensus principles and propositions sketched by Alex McCalla, commanding agreement among most of the policy actors. But by its very brevity that consensus hides as much as it illuminates. The main force behind it obscures the combined power of the various global actors who have been "captured by key agents, the mercantilist aims of states, the development goals of international organizations and private foundations, and the financial objectives of investors."[36] Lest this be seen as some kind of organized conspiracy built on self-interest alone, Clapp notes the obvious benefits that the dominant response rests on: year-round access to fresh fruits and vegetables, distribution of surpluses from rich to poor countries when food crises arise, improved food safety standards—and I would add the example of the Green Revolution, which made good use of all those agents in producing major agricultural gains and remains today a model of improved agricultural technology.

The response to the various recent food shortage crises as well as more endemic problems of food security has been to stay on the same course "but with some refinements to the rules of the game to address its most obvious weaknesses. . . . What is needed is in fact to continue along the current lines of more scientific and industrialized agriculture, more integrated global food markets, more corporate involvement in the food system, and a more active financial actor engagement in the system."[37] At the same time as these goals were being enunciated, they received the support of the G8 and G20 countries.

But that support incorporated a response to the harsh criticism of the dominant model. It introduced an alternative model. The 2009 L'Aquila Joint Statement on Food Security—accepted by those countries in backing their pledge of $20 billion over three years for agricultural investment—said: "We see a comprehensive approach as including . . . emphasis on private sector growth, small holders, women and families, preservation of the natural resource base, expansion of employment and decent work opportunities, knowledge and training."[38]

The tension reflected in the L'Aquila statement, between the dominant and competing models, was brought out most forcibly in 2008 in two reports by a pair of important international groups, the World Bank and the International Assessment of Agricultural Knowledge, Science and Technology for Development (IAASTD). The WB report emphasized both the importance of agriculture for poverty reduction and economic growth and the imperative need for increased agricultural productivity and better governance at all levels, local and global. Both reports note the relationship between environment and agriculture and want to insure that environmental and agricultural needs are not competitive.

Although sensitive to some of the harms done by the Green Revolution and the importance of biotechnology, the WB report is much more enthusiastic about the biotechnology benefits for small-scale farming in developing countries. Biotechnology should be the next stage of the new Green Revolution. Yet it is understood that this kind of technology can only be advanced by the main actors in the conventional model, government support in developed countries and corporations with the resources to engage in the necessary research. All of this will be enhanced by a reduction of trade barriers in developing countries. One can hardly ignore the fact that the WB report moves in a direction within the comfort zone of the economic and cultural interests of G8 countries.

The IAASTD report moves well beyond the conventional model, stating that the present situation, dominated by corporations and the market, promotes inequities and other harms to small-holder farms and embodies a much too narrow understanding of agriculture, far better construed to incorporate livelihood needs, environmental sensitivity, and cultural nuance. While the IAASTD report takes biotechnology seriously (but also endorses alternative methods) it does not support it with the kind of emphasis found in the WB document. That led to a withdrawal of support from some technology companies and a failure of the United States, Canada, and Australia to give full support to it because of its stance on agricultural trade and technology.[39]

## COUNTERCURRENTS: HIGH TIDES, STORM SURGES, OR TSUNAMIS?

Of scholarly critiques, measured political condemnations, ideological attacks, and outraged moral indignation directed at the conventional food model there is no end. There are also some paroxysms of hope and optimism from its critics about movements to change that model. It is hard to determine whether that assault, to use some oceanic metaphors, represents only a high tide, lasting a short time only; a temporary storm surge, powerful and dangerous but not wholly transformative; or a tsunami, capable of decisively wiping out all in its path, radically altering the terrain. It is no less difficult to determine whether the optimism of many is whistling in the dark, designed to rally the troops and spread revolutionary fervor, or firmly based in reality. Trying to get that all straight, including the many combinations, compromises, and alleged conspiracies that dot this terrain is not easy but is necessary to chart the future.

At one end of the food continuum are all the political and economic forces that support the conventional model: governments of developed countries, elite foundations and NGOs, transnational food industries, and considerable political support for markets and technologies. At the other end is a focus on small farmers, environmental values, local food, community cooperation, and a rejection of industrialized agriculture and the corporate and political interests behind it. And then there are all the mixtures in between. Industrial agriculture yes, but focused on better governance; a reduction of subsidies and tariffs; greater sensitivity to the culture and welfare of farmers; and a food industry more geared to environmental and justice values. At the other end of the spectrum are individuals and organizations, often of a grassroots kind, who share one overriding conviction: the capitalistic values behind the conventional model are the greatest threat to just and sustainable food security. For many who believe, like it or not, that global capitalism is not going to be extirpated, the challenge is to regulate it, control its hungry sweep, and counter its influence with indigenous organizations, local values, and environmental sensibilities.

One of the enduring strengths of the conventional models, ironically, is that the global food problem is in the end not fully global. Is the title of Julian Cribb's book *The Coming Famine* and his judgment that "future food shortages are a far bigger threat than global warming" excessively alarmist?[40] Not if it is understood that the harshest impacts will be on some poor countries, not every country. That imbalance is noticeable in a 2015 UN report on world hunger that shows a decline in global malnutrition overall and a 2012 UN report that details the highly uncertain future for the food security of poor countries.[41] Again and again, it can be noted, the center of the "global" troubles turns out to be not the United States or Europe or Latin America but sub-Saharan Africa and Southeast Asia, with a few other less troubled spots. Moreover, as an added twist, it should be noted that both China and India—two countries that have fully embraced market values—are not on the lists of the endangered countries, even if they feel some pressure to worry that they might be. In sum, none of the G8 or G20 countries is having a food security crisis—although all feel the impact, harmful to the less well off, of much higher food prices. Most of those countries, moreover, accept capitalism and the role of industry in national welfare. Having tasted the fruits of capitalism, even if it engenders corruption, the chance that China will repudiate it is even more unlikely than a repudiation of communism.

The moral I draw from this analysis is a simple brute fact: the capitalistic societies of the industrialized countries have a strong upper hand in resisting any radical anticapitalist movements. Increasingly, with the addition of China and India into their ranks, capitalist countries now command a majority of the world's population. I am not making a case here for the market, only pointing out where the current power lies and how difficult it will be to displace it (see chap. 10).

## ASSESSING THE "GLOBAL FOOD MOVEMENT"

That much said, I find it hard not to sympathize with those pressing for an alternative future. In 2006 I wrote a book critical of the large and dominating role of the market in American medicine, its seductions for troubled universal health-care systems in other countries, and its embrace of health care during the 1980s and 1990s as part of the failed structural reform efforts of the World Bank.[42] Moreover, in the early 1970s I came to know and admire Frances Moore Lappé, whose revolutionary book *Diet for a Small Planet* was still being circulated in its original mimeographed form and whose husband Mark then worked for me. A 2011 issue of the *Nation* magazine was heavily devoted to the food problem, with Lappé as the opening author.[43]

Listing a wide range of efforts around the world to change the conventional model, calling it the "global food movement," Lappé wrote, "It is at its heart revolutionary, with some of the world's poorest people in the lead. . . . It has the potential to transform not just the way we eat but the way we understand our world, including ourselves. And that vast power is just beginning to erupt."[44] Citing a rich array of organizations and movements, she is impressed by their love of the land, their pursuit of justice for poor farm workers, their embrace of Fair Trade, and its "unity of healthy farming ecology and social ecology" and the way in which that "social ecology transforms the market itself: from the anonymous, amoral selling and buying . . . to a market . . . structured to ensure fairness and co-responsibility."[45]

She has nothing good to say about GMOs, increased production, or industrialized agriculture. "Aligning food and farming with nature's genius," she concludes, "we realize there's more than enough for all."[46] A sympathetic commentator on Lappé's article, Raj Patel, argues that a "truly democratic food system will need to rewrite the rules of the finan-cial system. That can't happen without naming and confronting capitalism as the enemy of food sovereignty."[47]

## SOFT POWER AND STRONG POLITICS

Two other commentators, however, inject some cool-headed realism into the Lappé picture. Michael Pollan speaks of the movement's "gains in the soft power of cultural influence and its comparative weakness in conven-tional political terms,"[48] while Eric Schlosser had observed that that "without a fundamental commitment to social justice the estimated 1–2% who eat organic food will be indistinguishable from the 1–2% who control almost all of this country's wealth and power."[49] Gaining a "fundamental commitment to social justice" is a tall order in a country not notable for that virtue. Pollan, however, offers a novel twist in speculating that the health-care insurance industry, burdened with the cost of chronic illness—and exacerbated by unhealthy American diets—will see the necessity of joining forces with the global food movement: "we simply can't afford the healthcare costs incurred by the current system of cheap food."[50] Alternatively, I cannot help saying, it could also just raise insurance premiums, copayments, and deductibles, a long-established industry defense against rising health-care costs.

The global food movement draws upon some well-founded objections to the industrialized food industry. It is far more beneficial to developed countries than to poor countries still dependent on small farms and local food now threatened by monocultures and the need to import food as small farms weaken. The latter can simply be wiped out or severely weak-ened by the cheaper food the food industry can produce and market, aided, of course, by government subsidies.

## GUILT BY ASSOCIATION AND GUILT BY VIRTUE

But there is a confusion also in those criticisms and attack, both a guilt by association and virtue by association. The guilt is to lump together all industry food practices, especially scientific and technological, in the critique of capitalism: if the food industry and private institutions such as the Ford, Rockefeller, and Gates foundations support a new technology-dominated Green Revolution, then suspicions automatically arise that capitalism lies behind that advocacy. If many of the alternative food movements are opposed to GMOs, skeptical of a second Green Revolution, and favorable to local food markets and organic farming and the empowerment of poor farmers, then they are automatically virtuous. Big is bad and small is beautiful.

The political scientist Robert Paarlberg, in his book *Food Politics: What Everyone Needs to Know*, is particularly helpful in demythologizing both the endemic guilt by association and virtue by association. In the early pages of his book he makes clear that the commercial food marketplace is worrisome and that "the entrenched power of lobbyists working on behalf of traditional commercial farming and the industrial food industry" are a force to be feared.[51] At the same time, and later in the book, he does not hesitate to support genetically modified crops and food, noting, despite public anxiety, the absence of any verified harms they have caused. Yet public opinion polls, I would add, have routinely showed an overwhelming desire for GMO labeling of food, even though its presence would—absent solid evidence for or against (and mainly the former)—provide no useful information at all on the safety of GMOs. Paarlberg notes that Americans have shown no resistance whatever to genetically engineered pharmaceuticals via recombinant DNA (gene-splicing) techniques.

Organic food, a favorite of many, was soon taken over by the food industry and brought into supermarkets. Moreover, crop yields from organic farming are lower and more expensive and do not offer a way out of massive malnutrition. "It is no longer possible," Paarlberg states, "to feed the world with farming systems that exclude the use of synthetic nitrogen fertilizer."[52] And Paarlberg provides backup evidence for that claim. As for local food supporters (given the name *locavores*), he argues that local food will do little to slow climate change and will not, in any case, be anywhere near adequate in reducing global malnutrition. Local food will be necessary and desirable in many poor countries, but it will require scientific and other resources well beyond local capacities.

In sum, Paarlberg contends (successfully in my eyes) that for all of its faults, which he does not minimize, global food security will require using the science and technology developed by capitalists and the industrialized food industry to meet coming food needs; but that they can be adapted to provide greater support for small farmers in developing countries, do so in environmentally friendly ways, and meet the coming food needs. That food industry will require greater regulation and a change in its culture from a profit-driven orientation to one more sensitive to human needs, social responsibility, and affordable food.

While it is exceedingly unlikely that capitalism will be overthrown and almost certain that many practices and the scientific skills of the industrialized food industry will remain necessary, it is good that there are counterpressures from below working to humanize it. Most of the needs that Raj Patel enumerates at the conclusion of his book should be part of that effort: we should *transform our tastes* away from unhealthy foods; *eat locally and seasonably* (when we can); *eat agroecologically*, favoring organic food (when affordable); *support local business* (if it has not gone the way of the local bookstore); acknowledge that *all workers have dignity* and there should be *living wages for all;* and *end subsidies to agribusiness.* That is not all Patel proposes, but those seem to me the most feasible (with some emendations I put in parentheses).[53]

But, I might be asked, is "feasible" the right standard to be used in judging countercultural movements? More than once in that literature there are references to other radical cultural change that took decades to mature, dismissed at the start but finally accepted after a long struggle: women's rights, the abolition of slavery, gay rights. It just takes time and perseverance. That is a good response, but one can still ask: If capitalism is abandoned and the technological possibilities of the dominant model put aside, could there really be enough for all, for the 11 million people the earth may eventually have? I return to this issue in chapter 10.

# 3

# WATER

▬

## Not Everywhere and Not Always Fit to Drink

ext to air, the human need for water is undeniable and inescapable. We cannot live without it, not only for nourishing the agriculture that gives us food, but no less for the flora and fauna that is necessary for all life. Water is inside us as well. Some 83% of our blood is water, and the cells of our body are awash with it. As a glance at any globe and the immensity of the oceans make clear most of our planet is salt water (97.5%), but save for expensive desalinization technologies, not available for human use. Salt water does not taste good and plants do not like it either. The oceans often creep up on rivers that end in oceans when their freshwater flow diminishes. We are dependent upon freshwater on land from many sources, from rivers and streams, from the aquifers under the soil (called groundwater), from rain, from the melting of glaciers to feed rivers, and from rainwater in the atmosphere. A threat of shortage of water from any of those sources directly affects human welfare and economic stability. Droughts killed millions of people in the past, and dirty water kills millions even now. Nearly every major industry in every society requires water for its survival. Rapid population growth together with increased food consumption and changing diets now places enormous pressure on available freshwater, the use of which is now heavily dominated by agricultural needs.[1]

Despite its importance for just about everything, it is often hard for those of us living in industrialized countries to see water as a global problem, much less one with implications for our own welfare. Most industrialized countries are in regions where water is readily available (even if there can be droughts in Texas and California and water shortages in many southwestern parts of the United States). It is a different story in the world's two most populated countries, China and India, and in most Middle Eastern nations, and in southern and Southeast Asia (where droughts can be as deadly as floods) (figure 3.1).

Whether one wants to go as far as the skilled scientific writer Fred Pearce puts it in the subtitle of his book *When the Rivers Run Dry: Water, the Defining Crisis of the Twenty-First Century*, it is at least as good a claimant for such an alarmist description as the other of my five horsemen. As the Princeton hydrologist Justin Sheffield and his environmental

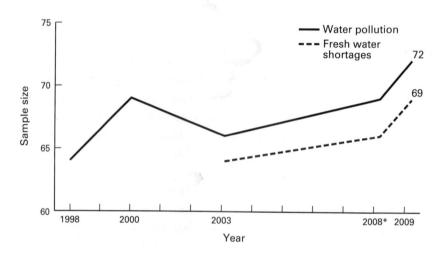

**FIGURE 3.1**

Seriousness of environmental problems: "very serious" average of eleven tracking countries*, 1998–2009.

* Asked of half of sample in 2008

Tracking countries include Brazil, Canada, China, France, Germany, Mexico, Nigeria, Turkey, United Kingdom, United States.

*Note*: Sample size *n*=1000 in all countries; Mexico 2008–2009 national sample; 1998–2006 urban sample (Globescan and Circle of Blue, *WaterViews: Water Issues Research* [Traverse City, Mich., August 17, 2009], www.circleofblue.org/waternews/wp-content/uploads/2009/08/circle_of_blue_globescan.pdf: 6.)

engineering colleague Eric F. Wood write, "Competition between meeting our basic needs, achieving a better standard of living and sustaining the environment make this [water] one of the great challenges of the 21st century . . . driven by potential increases in drought severity and frequency . . . by population increase, land use and climate change." Two charts they present show a striking trend. Global irrigated land—where some 70% of water use comes from—went from 125 million acres in 1900 to 750 million in 2000; and in many places it is running out, either due to expanded use or by soil erosion or desertification. During the same period industrial and municipal use nearly doubled.[2]

## WATER FOOTPRINT AND VIRTUAL WATER

The term "water footprint" has been used to calculate the water required to sustain a population. There are four determinants of that footprint: GDP-influenced volume of consumption (GDP growth increases food and industry consumption); the pattern of consumption (e.g., high vs. low meat consumption); influence of climate; and agricultural practices (especially the way water is harvested and used).[3] Eight countries—India, China, the United States, the Russian Federation, Indonesia, Nigeria, Brazil, and Pakistan—account for 50% of the total global footprint, with the highest three countries being India (13%), China (12%), and the United States (9%). Comparing the per capita footprints, however, it is obvious that while their per capita usage is less than that of the United States, the water demands and problems for those two countries are much greater.

In 1993 the geographer Tony Allan of Kings College, London, introduced the concept of "virtual water." By that phrase he meant the amount of water used to raise food in one country that is then imported and consumed in another country. Some countries are water rich and other countries are water poor, making importation necessary. But all of the importation need not be for food alone. The leather that is used to make shoes comes from the cattle raised to provide it, drawing on water as well. A shirt made of cotton that was raised with water from irrigation in one place can be sold in another place. Professor Allan offers a simple example of virtual water use, that of an ordinary American or British breakfast consisting of coffee, toast, bacon, eggs, and an apple. That meal uses about 800 gallons of virtual water.[4] In assessing global water availability, virtual water use has to be factored in. Water-poor countries can require goods and food made in water-rich countries; they have little other choice.

Global warming is one important driver of water problems, the source of droughts, floods, and the decline of glacial runoff. Just how much global warming can be held responsible for that is itself controversial in pinpointing the causality of individual disasters (droughts and floods have long been a part of the human story). But persistent patterns, particularly with well-established rising temperatures, give global warming a central place of importance. There is no serious disagreement, however, about the role of growing populations and increased urbanization in developing countries and the corresponding need for water, and in particular, the place of water for irrigation in agriculture, which consumes most available freshwater. But accessibility and consumption are not the only problems. The cleanliness of water is equally important, with pollution from human, agricultural, and industrial pollution the leading source of bad health and often death in developing countries (figure 3.2). The estimate is that 1 billion people lack access to clean water, with some 2.5 billion lacking water for sanitation. Some 80% of all infectious disease in poor countries can be traced to dirty water.[5] The phenomenon of easily available and clean tap water and flushable indoor toilets in industrialized countries is still a dream for millions of people in poorer parts of the world. At the same time, increased consumption of meat and milk in the developing countries means an increase in the corn necessary to feed them, and that in turn requires considerably more water, usually from irrigation, far more than for rice.

There are many ways of cataloguing and assessing the impact of water problems, some of them having an almost iconic status as readily visible and dramatic symptoms of what we are up against. At the beginning of his book Fred Pearce tells us that he was drawn to the topic of water when he heard that "the Nile in Egypt, the Yellow River in China, the Indus in Pakistan, the Colorado and Rio Grande . . . all were reported to be trickling into

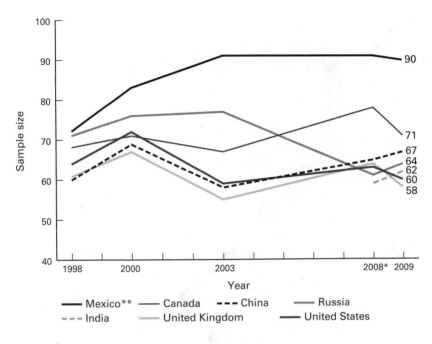

**FIGURE 3.2**

Seriousness of water pollution: "very serious problem," select countries, 1998–2009.

* Asked of half of sample in 2008

*Note*: Sample size *n* = 1000 in all countries; Mexico 2008–2009 national sample; 1998–2006 urban sample

(Globescan and Circle of Blue, *WaterViews: Water Issues Research* [Traverse City, Mich., August 17, 2009], www.circleofblue.org/waternews/wp-content/uploads/2009/08/circle_of_blue_globescan.pdf: 6.)

the sand, sometimes hundreds of miles from the sea. . . . Israel is draining the Jordan River into pipes before it reaches the country that bears its name. . . . The wells have been drying up, too. Half a century of pumping on the Great Plains of the United States has removed water that it will take 2,000 years of rain to replace. In India, farmers whose fathers lifted water from wells with a bucket now sink boreholes more than half a mile into the rocks—and still find no water."[6] If those examples caught his eye, they should catch ours as well. China, India, and the western and southwestern United States together exemplify the water problem, although the American situation is far less dire than that of the other two countries.

# THREE CASE EXAMPLES: CHINA, INDIA, AND THE WESTERN UNITED STATES

## China

The Chinese Yellow River is a good starting place. Sometimes called "China's sorrow," or its "joy and sorrow," the river—close to 4,000 miles in length and the sixth-longest river in the world—is famous for its terrible floods over the centuries, accounting for millions of deaths. The Japanese destruction of a Yellow River dike in the 1938 Sino-Japanese war killed an estimated 890,000 people, and with many fragile dams and dikes, the threat of floods from their collapse is still a serious hazard. But that threat is more than matched by droughts and the excessive removal of the water along its length, already reduced by glacial runoff from a Tibetan Plateau to the South China Sea. At one edge of that plateau there were once thousands of lakes, but half are now gone, replaced by pasturelands. If that area suffers, the situation downstream is even worse. Close to half a billion people depend upon it for drinking water and growing their food.

But as Fred Pearce puts it, recounting his visit to the river and interviews with Chinese officials, "In the river's middle reaches, irrigation canals are running dry, fields are being abandoned, and desertification is generating huge dust storms that spread east, choking lungs in Beijing, closing schools in Korea . . . state irrigation projects dotted along the river cover 30,000 square miles and soak up the majority of the river's water."[7] The river, famous for its silt (the source of the "yellow" in its name), sees a constant buildup of silt behind hundreds of dams. Then there is the drain of water from aquifers, forcing wells to be dug ever deeper (sometimes to 200 feet), not only to get to the water itself but to get past the industrial and farm waste pollution and reach drinkable water. Nearly one-half of Chinese groundwater was pumped out in northern China in the twentieth century and at present rates of withdrawal most of it will be gone by 2035.

## India

India easily manages to display as many if not even more water challenges than China. One important difference is that India has only recently begun to take its water problems seriously and, although it is a more open, democratic country than China, it has an inefficient and often corrupt government to deal with them. With its authoritarian regime, China has taken more decisive steps (even though they are still far from adequate). India's history with water goes back to the days of British colonialism, when a massive campaign of irrigation, much of it for growing cotton but in anticipation of water shortages as well, was undertaken. The net result over the decades was that India has more irrigated land than any other country in the world. That land is watered by some 60,000 miles of canals that cover the entire country. Yet after many years of rapid population growth, inefficient government, excessive dam building, and an unregulated use of groundwater to complement the canals, India has a real crisis. It has, at 1.1 billion people, some 17% of the world's population but only 4% of the available freshwater. One of the results of rapid population and economic growth has been the emergence of a sharply two-tiered society, with about 25% being members of the affluent class (with adequate access to clean water), and the rest of the population (with little access) mired in poverty.

As Steven Solomon notes, "India's poor, rural farmers are twice dispossessed: first by being located far from prime water sources occupied by the wealthy and powerful, and second by being last to be served due to the high cost of transporting water to them."[8] The situation in urban areas shows the same pattern: clean, ample water available to the affluent but not easily available water for the poor, millions of whom have to queue up daily to pay to get clean drinking water from tankers. Over 650 million people lack tap water for drinking; around 700 million lack indoor toilets and less than 10% of urban sewage is treated.

A standard source of water in India has for millennia been the seasonal monsoons, concentrating 80% of the rain into a few months only; and those same monsoons frequently create massive floods and landslides that have killed thousands of people. But if not excessive, the monsoons provide water for crops and replenish the rivers and groundwater. The failure to take efficient advantage of the monsoons has been a major reason Indian farmers have exploited the available groundwater by digging wells. India went from 800,000 wells in 1975 to some 22 million now, and they are increasing at a rate of 1 million a year. The comparatively inexpensive wells are drilled with the help of government subsidies. Half of its irrigation water comes from that source, but to get it the drills must steadily go farther down, and in parts of India the wells have already run dry and in other parts drilling over 1,000 feet down is now necessary. "As a nation," Steven Solomon judges, "it amounts to a slow-motion act of hydrological suicide."[9] It is worth noting, however, the Global Water Partnership's statement that the extraction of groundwater has increased food security, has been mainly pro-poor, and has been particularly beneficial for small farmers.[10] Once again, as can be seen in most of my five horsemen, a trade-off dilemma displays itself.[11]

A 2012 report by the Indian government on its water situation is notable for its candor and thoroughness.[12] It lists thirty failings of its water practices, not hesitating to spell out particular needs, most notably calling for a centralized government role. While "it is recognized that states have the right to frame suitable policies, laws, and regulations," the report goes on to note the need for a "national legal framework of general principles on water. . . . Water needs to be managed as a community resource."[13] Most significantly, it addresses the lack of regulation of groundwater withdrawals and cites the need to modify an 1882 act that "appears to give proprietary rights to a land owner on groundwater under his/her land."[14] How well this realism and sense of urgency will be implemented is uncertain. Community efforts to deal with the groundwater issue through voluntary self-regulation, called a "unique experiment," has been much touted, but one study found that the results have shown it has failed in a number of ways.[15] Voluntary programs for the collection of rainwater in tanks and ponds, a comparatively simple idea, appear to be more successful. The use of groundwater, however, is particularly important and far harder to deal with—and it is not being addressed. The World Bank has determined that India has only twenty years before it reaches a "critical condition," when the demand will outstrip supply.[16]

## United States

As a general proposition the United States has more than adequate water, but far more in the eastern than the western part of the country. Beyond the Mississippi River in the semi-arid high plains of Nebraska, Kansas, and Texas, the rainfall diminishes to erratic and undependable. West of the Rockies most of the valleys and lowlands were originally dry deserts, with most of the southwestern states (notably Arizona and New Mexico) getting less than 7 inches of rain a year. Apart from a few rivers, particularly the Rio Grande and the Colorado River, most of the surface water comes from mountain runoffs from melting snow in the spring and summer. Irrigation and dams played a vital part in allowing the West to turn unlikely land into a bonanza.

As Stephen Solomon sums it up: "Water scarcity, in short, was the defining geographical condition of America's far west. . . . The Far Western deserts were miraculously transformed into the richest farmland on the planet."[17] But it was also the case that "the struggle for water was inseparable from the naked contest for power and wealth," dramatized in a number of movies (of which *Chinatown*, based on a political battle for the Los Angeles water supply, is my favorite). The Hoover Dam (also known as Boulder Dam) was a turning point. Completed in 1936 on the Colorado River it was the world's largest dam and behind it was the 110-mile-long Lake Mead, able to store twice the annual flow of the river. It has generated prodigious amounts of electricity and water over 200 million acres, and the cities of Los Angeles, San Diego, Phoenix, and Las Vegas make use of it. Lake Mead is now at its lowest level ever. Meanwhile, on the high plains, water from the Ogallala aquifer produces around one-fifth of total U.S. farm irrigation.

But a price was soon to be paid for those benefits, emerging in recent years. The 2000-mile Rio Grande, running from the snowmelt in the Colorado mountains to the Gulf of Mexico, has seen four-fifths of the water taken for irrigation, with much also lost to waste and evaporation. Little water reaches the sea any more. Cities like El Paso in Texas and Juarez in Mexico take their water from that trickle, and both have chronic water shortages. The Colorado River has been suffering a comparable irrigation drain. Not only did the flow diminish the further south it went, barely reaching the sea at all, it is also beset by increased salinization, making its lower reaches often useless for irrigation. The High Plains Aquifer came to face a comparable problem. Groundwater was removed from it over the years at a rate ten times faster than it was being replenished by rainwater. In some parts of California groundwater tables have dropped by 400 feet or so and, as in China and India, wells have to be dug to constantly deeper levels. It must be noted and underlined, however, that most people in the United States have clean tap water, and that so far even in western states the only common form of rationing is to limit the watering of lawns; the same can be said of European countries and much of Latin America. The extended 2013–2015 drought in California has led to increasingly tight water restrictions.

## SOURCES OF WATER DEMAND AND POLLUTION

My brief survey of the problems of three countries provides a good glimpse of many of the causes of water shortage and pollution. Let me sum them up and add a few other considerations before moving on. At the most basic level, as an International Water Management Institute report puts it, "Fifty years ago the world had fewer than half as many people as it has today. . . . They consumed fewer calories, ate less meat, and thus required less water to produce their food. . . . They took from our rivers a third of the water than we now take."[18] Climate change as a causal factor was not a dominant note fifty years ago. But it has to be taken seriously now in its impact on water availability, on increasing desertification, and on regions particularly dependent on the melting of mountain glaciers and snow. Food demand, driven by population increase and rising appetites for meat, is surely important. The growth of cities, with some 3.5 billion living in them and the numbers constantly rising, is consuming farmland and buying water otherwise used for food production. Excessive extraction of groundwater, commonly unregulated, is of obvious importance, as is the pollution of freshwater from industrial, agricultural, and urban waste. Agricultural inefficiency, poor farming practices, and water wastage from leaks from irrigation create their share of water loss. Desertification removes water altogether, and salt and acidity can ruin it for farming use.[19]

Water pollution is a global hazard. While the U.S. Clean Water Act, enacted in 1972, is considered a model federal program with considerable success over the years, pollution remains a persistent problem. Heavy industry and its waste are a major source of river pollution, as is agricultural runoff, with fertilizer and pesticide residue among the pollutants. European and Asian nations face similar hazards. Up to a point, rivers naturally clean themselves, but that mechanism is easily overwhelmed. The greatest water damage is occurring in the developing countries, where 90% of sewage is discharged untreated, and human excreta is one of the main pollutants. The World Health Organization (WHO) has compiled some fearsome information on water pollution.[20] One billion people lack safe drinking water, and 2 million people die annually of diarrheal disease from unsafe water. Cholera is reportedly found in fifty countries, in great part transmitted by dirty water.

There are a wide array of water pollutants: organic matter, notably feces and food waste; toxic organic compounds, pesticides, and pharmaceuticals; heavy metals, especially cadmium, zinc, and copper; pathogens and microbes such as *Salmonella* and *Shigella*; and nutrients such as nitrogen and phosphorous.[21] Most of the industrial waste, some 80%, comes from developed countries, with over 50% in developing countries from food-industry pollution. Rapid urbanization and accelerating industrialism are relatively new sources of pollution. The latter are a major cause of damaged rivers. In China only about 10% of environmental laws are enforced. Its Haung River is filled with pollutants of every kind, industrial waste, untreated sewage, pesticides, and fertilizers.[22] One-third of the river's fish species have become extinct in recent years. Industrial spills happen all over the world.

## GOVERNANCE: INTERNATIONAL AND REGIONAL

There are a large number of international organizations interested in water, some that bring together representatives of different countries and agencies, and some NGOs. The World Water Council is an NGO drawing on many organizations and individuals and is the sponsor of the World Water Forum, holding a meeting every three years and drawing thousands of participants. While officially nonpartisan and open to multiple values, it has stimulated the Alternative Water Forum, with representatives from 150 organizations from thirty countries. The Alternative Water Forum advertises itself as an opponent of the World Water Forum, which it believes is a captive of market forces with the goal of the financialization and privatization of water resources. Other NGOs are the World Water

Assessment Program, the International Water Resources Association, and H2O (requiring participant organizations to work in at least two countries). The UN-Water Program, coordinates all freshwater issues for thirty UN organizations. The UN Environment Programme has a unit focused on water policy and strategy. The Global Analysis and Assessment of Sanitation and Drinking-Water is produced every two years by the WHO on behalf of UN-Water.

While there are, then, a number of global organizations, governmentally oriented as well as NGOs, there are no organizations that have binding legal authority or global governance.[23] Moreover, in a familiar refrain, one study of drought and desertification in southern Africa found that top-down policy formation working with UN policy recommendations is inadequately adapted to local social networks and traditional practices despite overall agreement on goals.[24] Regional policies and agreements on water, however, often manage to find a good fit between water needs and local values, both within and between countries. The United States does not have an overall national policy, but there has been considerable success negotiating water rights within and among states. Negotiations on the use of river water shared by a number of countries offer good examples at the global level. These intra- and international efforts at sharing (and nonsharing) of water resources provide a mixed picture, and I will glance at them in turn.

The standard classification in considering the claim that a country (or for that matter a U.S. state) has on river water is threefold: *absolute sovereignty* (the upstream country of a river has an absolute right to use it); *riparian rights* (the right of all countries along a river to have an equal share of water); *prior appropriation—first in time, first in right* (initially used in California during the Gold Rush: the first prospector to make use of a stream or river for panning had the sole right to use it.) Egypt has made use of this principle with respect to the Nile.[25] Turkey has managed to corral some 50% of the water from the Tigris and Euphrates, considerably reducing the water available to Syria and Iraq and often using its control of the upper reaches of those rivers as political (and sometimes) military bargaining chips. Along with China, Turkey was one of the few countries to vote against the 1997 UN Watercourses Convention affirming riparian rights for all nations.[26]

China, for its part, has built dams mainstream on the Mekong River without consulting Laos, Thailand, and Cambodia. Yet India and Pakistan, despite decades of hostility, agreed to share water from the Indus River, an agreement that was honored even during the 1971 war between those two countries. There is also agreement among India, Nepal, and Bangladesh on sharing the Ganges River. In California agreement was reached after a years-long struggle over moving water from northern California (with 75% of the state's water) to desert-like Southern California (with 75% of the state's population). Affected states along the Colorado River, which has water drawn from it starting in California and down through the southwestern United States, did work out agreements for water sharing, although Mexico, at the bottom of the river has been a loser.

## TECHNOLOGY: GAINS, LOSSES, DILEMMAS

Technology plays a role in the strategies used to deal with water problems. Historically, as Steven Solomon richly lays out in his book, *Water: The Epic Struggle for Wealth, Power, and Civilization*, over the centuries water needs and problems have been the making and breaking of civilizations. The Mayan civilization's collapse after 800 C.E., resulting in a loss of 90% of its population, was brought about by a terrible long-term drought that eventually overcame irrigation canals and underground cisterns that collected water (both of which I would broadly characterize as technologies). The Agricultural Revolution some 5,000 years ago and the Industrial Revolution more recently advanced because of the development of effective means of managing and exploiting the control and use of water. Irrigation and dams have played a central role in economic development and population survival and vitality for much of human history. Those same techniques are still in use, enhanced by ever more helpful innovations but also accompanied by some troubling dilemmas and political and scientific struggles.

I will begin with some of the less controversial water innovations, a sampler only.

*Leaking pipes.* Leaking water pipes are a problem in almost all countries, often the result of outdated water infrastructures of a kind expensive to modernize. One estimate is that the United States loses around 7 billion gallons of clean drinking water a day. And in many cases the leakage from domestic sewer and industrial waste is a source of $CO_2$ delivered into the atmosphere.[27]

*Rain harvesting.* The collection of rainwater in ponds, containers, and cisterns has always been a part of human history. Efforts are under way to find more effective and inexpensive ways to collect rainwater, particularly in developing countries.

*Recycling water.* Efforts to improve and spread the recycling of polluted sewage and industrial waste are being pursued in many places, notably in California and Australia. There are two products of such efforts: clean drinking water and water suitable only for agricultural use (called "gray water"). In Australia, purple-colored pipes carry the latter.

*Tree barriers to combat desertification.* In Africa, what is called the Great Green Wall plan, supported by eleven African countries, aspires to plant a column of trees 9 miles wide and 5,000 miles long, aiming to halt growing desertification and make use of inexpensive technological devices to direct scant rainwater directly to the roots of the trees.[28]

*Desalinization.* In various parts of the world, especially the Middle East, desalinization plants are being used to transform seawater into usable drinking and agricultural water.[29] While not particularly controversial, desalinization efforts are still expensive. Even as the costs decline, the economics of their use is likely to limit them.

*Center-pivot irrigation system.* This system makes use of water sprinklers, with circulating nozzles delivering carefully monitored water quantities based on soil moisture and other variables. It is expensive and not likely to be affordable in most developing countries.

Beyond those technologies, already in use or close to it, there are more that fall into the realm of speculation, but not unrealistically. They include genetically modified crops that require less water; progress with micro-irrigation; remote-sensing technologies; less costly desalinization; and small water turbines to make use of fast-moving streams and generate renewable electricity, thus eradicating the need for dams.[30]

A number of technologies, however, carry with them a combination of benefits and costs. Much of the water for California's Central Valley, one of the most fertile areas in the world, is taken from its own aquifer, reducing its level; and water from the Ogallala, the midwestern aquifer, is at times being taken out ten times faster than it is being replenished. Smaller and cheaper pumps have made water removal in India even more common, also removing much more water from aquifers than can be replaced by rainfall. Additionally, Indian government subsidies for digging the wells and paying for the pumps, together with a lack of effective regulation, have created a dire situation for the future availability of aquifer water. It is a classic instance of Garrett Hardin's "tragedy of the commons,"[31] with unregulated individual withdrawal eventually depriving everyone of the resource. At the same time it is equally true, as noted earlier, that this water largesse has been an enormous help to Indian agriculture and small farmers.

Preserving aquifers while also continuing to grow adequate food for a growing population will become a great cost-benefit dilemma.[32] The Green Revolution, immensely valuable in increasing crop yields, brought millions of people out of poverty and reduced the threat of starvation. But it exacerbated the water problem in the countries that benefited from it. As Fred Pearce notes: "While the new crops were indeed very efficient at delivering more crop per acre, they were often extremely inefficient when measured against water use. They often delivered less 'crop per drop' than the varieties they replaced. . . . [More generally] the world grows twice as much food as it did a generation ago, but it also abstracts three times more water from rivers and underground aquifers to do it."[33]

## Building Dams and Damning Dams

Far more contentious has been the debate for many years on the benefit and utility of dams as a way of managing the availability of drinking and agricultural water and, often enough, simultaneously generating hydroelectric power.[34] The American Hoover Dam was the international pacesetter in creating dams far larger than earlier dams, stimulating a huge burst of dam building all over the world—and almost always with the larger dams creating new problems and threats along the way. One of the most celebrated and politically controversial was the Aswan Dam on the Nile in Egypt, officially opened in 1971. It fulfilled an ambitious aspiration beginning in 1952 of President Gamil Nasser to provide greater food security and economic modernization, and most decisively, to give Egypt control of the Nile. It is an immense dam, 360 feet high, with a 344-mile-long and 8-mile-wide Lake Nasser reservoir. Its construction displaced 100,000 inhabitants and covered many ancient and important monuments. For a time it seemed to work miracles in the production of electricity and irrigation water.

As time went on, however, its unanticipated environmental downside became evident: fertilizing silt, long a treasure of the river, began to be trapped by the dam and built up in the lake behind it; soil salinization reached 30 miles inland; and agricultural acreage declined. Ironically, Ethiopia much more recently began constructing a massive dam on the Blue Nile, a tributary that contributes most of the water flowing in the Nile River, thus affecting Egypt. That effort angered Egypt during the reign of President Mohamed Morsi, who said that "Egyptian blood" would be spent on whatever drop of water was lost to his country. Egypt's current president, however, has drawn back from any direct action but continues to warily watch the dam's progress.[35]

Egypt was not the only country to stake a good deal of its future on a dam. Twenty of the largest countries in the world have massive dams—and no less massive problems, so much so that an international backlash began to emerge in the 1990s. Most of the dams did not have an environmental analysis before their construction, nor did they have the consent of the tens of thousands displaced by their construction. An important retired American government official from an agency that built more dams than any other in the world, Daniel Beard, led the charge against them, once yelling "No more dams!" during a protest against a proposed dam in Japan.[36] Environmentalists joined the movement, and a decision by the World Bank, which had spent $75 billion on large dams in 92 countries, ended by condemning them. Dams usually cost more to build than projected, irrigate less land at an affordable price than expected, and create less hydroelectric power while delivering less water to cities than projected. There are ambitious and expensive plans initiated by the Indian government to rapidly build dams in twenty-eight of thirty-two major river valleys to provide hydropower. It is a plan, however, whose environmental and social impacts have been scantily examined, particularly in light of existing environmental impact laws.[37] The lure of high-tech big bang technological solutions does not easily go away.

In the United States, electricity consumption, spurred by population growth and present-day needs, is expected to rise sharply in the future, with the expectation of a doubling of demand before 2050. The production of electricity requires large amounts of water, drawn from rivers and aquifers, and is now threatened by droughts, water pollution, and lack of necessary water pumping, purification, and treatment systems.[38] Although the installation of more efficient systems to handle water needs has brought consumption per capita down, it is still rising in absolute terms. As an important motive for the building of dams, hydroelectric power is a social and economic need in tension with other water needs.

## WATER AND THE MARKET: SELLING IT AND PRICING IT

If the absence of clean, inexpensive, and available tap water is a desperate crisis in many developing and transitioning countries, there is another side to that coin. It is the rejection of such water by millions who do have excellent access to drinkable water but choose instead to get it from plastic bottles—even when much of that water comes not from pristine spring water but from the same sources our tap water does and costs by far much more. It is an interesting story in its own right but no less a good introduction to the larger issue of the seductive potentiality of water as a market commodity. Even if we put aside the environmental impact of harvesting that water, sometimes superpurifying it, manufacturing the plastic needed for the bottles, shipping it around the country and the world, and then controlling the landfill waste of the empty plastic bottles, it is hard to describe the huge market in such water as meeting a basic human need. It is instead best seen as a consumer luxury sometimes masked as more healthy than ordinary water. About 44% of all such water is known as "purified water," while the rest comes from springs or groundwater.[39] A 2006 study found that 17 million barrels of oil a year are required to produce the plastic bottles, while the cost of tap water production is around a thousandth as much.

Bottled water is an exceedingly profitable item, vigorously advertised, stoutly defended by its manufacturers, and seeming adored by the public. In 2009 Americans spent $10.6 billion[40] to consume 8.4 million gallons of it. It is a strikingly popular item in upscale restaurants, where it is served at 80% markups. A person who drinks the recommended eight glasses a day from New York City tap water (rated as nearly the best in the country) would spend 49 cents a year for that water versus some $1400 for the equivalent amount of bottled water.[41] Americans drink around 29 gallons of bottled water a year per capita and spend almost 50% as much for bottled water as for all other water.[42] While there are a wide range of what can be called "boutique" brands—those whose water is drawn from 3,000 feet below the ocean near a Hawaiian island and advertised as "older than Jesus" and another claiming that "no glaciers were harmed in making this water"—most of the market is dominated by four companies with names famous for their place in international food and beverage sales: PepsiCo, Coca-Cola, Nestlé, and Dannon.[43] The lobbying power of Nestlé was seen in a decision of the Maine Supreme Court in 2009 to support the construction of a trucking facility in Fryeburg against strong community opposition. While recent years have seen a mild environmentalist backlash against bottled water, it is not putting much of a dent in consumption. Profitable products of anything liked by the public are hard to dislodge.

Meanwhile, at a different level, there have been two market-oriented trends of importance, each generating controversy. One of them can broadly be described as the attraction of privatization, treating water as a commodity, and in a milder form, making some features of water a commodity but with some limits. The other trend is to control the use of water by pricing mechanisms.

## Privatization of Water

The privatization attraction can be observed in three dominant forms. One of them is that of the employment of private utility companies, usually with some government oversight and regulation, to manage a community's sale and distribution of water. Another is the buying of water resources by entrepreneurial companies or individuals for profit-making purposes, of which the sale of bottled water cited above is a good entry point. Still another is an aspiration of some economists and corporate interests to see the development of a global market in water, comparable with that of the food and beverage industries. Behind each of them is the correct perception that water, now a half-trillion-dollar industry in the United States, is something people want and need, is often in short supply and unsafe in many places, and is therefore a good target for market mechanisms and policy development. Why not commodify it as is now done with so many other needs of modern life? Even though I put pricing in a category of its own, because it can be used apart from an overall profit-driven market strategy, it is also an important feature of broader market strategies. The commodification of water can thus be seen by wide segments of our society and others as a wholly plausible way to manage water. Whether a use of market mechanisms is ethical or just an acceptable business practice is, of course, challenged, and that clash is no less visible in debates over health care, environmental problems, food, and housing (to which I will return).

Private water companies, initially heavily favoring the affluent, can be dated to the nineteenth century in Europe, and French firms have pursued privatization around the world ever since. In that same century most U.S. water systems were run by private companies, but most were replaced by local government control by the twentieth century. But a variety of problems emerged over time that kept alive the attractions of private sector management in America and many other countries. About 10% of the world water services are provided by private companies. The main argument in their favor is that most users of water, commercial and private, are wasteful and that a market in water fosters efficiency, competitive pricing to foster competition, and improved service.

The fact that the International Monetary Fund, the World Bank, and the Asian Development Bank have a bias toward the market gives the privatization advocates both money and ideological support. In the United States, privatized water systems have had mixed support, with considerable fluctuation in their public acceptance, a shifting in and out of private hands—and a no less considerable resistance to the idea of privatization of water, taken to be a common good—not to be put on the market. Yet the fluctuation of support suggests that local governments can have difficulty managing it as well. On occasion, it also turns out, private water companies have decided that it is not a profitable business. A German company with considerable resources, RWE, entered the U.S market in 2001 with high expectations for the same success it had in forty other countries. But by the end of the decade its stock had sharply dropped, and it decided to phase out its global water business, with its CEO declaring that "water is a very local business" and not suitable for a global enterprise.[44]

Many other companies, however, have stayed in the business, a business subject to political and other fluctuations and uncertainties. Gregory Pierce, noting the debate over private water companies, discerns a new trend, one that shows a reversal of the earlier pattern of increased rejection. Instead, he argues, privatization is slowly in the ascendancy at the moment but moving away from national governments and control by international companies to local and regional decentralized systems. "In other words," he writes, "low- and middle-income countries are creating entirely domestic privatization platforms."[45] He said the trend is toward the French model, "ceding control of service delivery but not physical assets."

## Water as a Commodity

Despite the problems experienced by private water companies, some economists and corporate interests have hankered for a global water market. For one thing, since water can be a commodity, like it or not, a global financialized water market is a plausible goal for those so inclined in a capitalist world. It happened with food and can happen with water. For another, many features of the struggle for water have brought about competition for water, notably in the need for fracking water competing with water needed for agriculture. The requirement of water pipelines in many places and bulk water carriers opens up new markets, with the added possibility of commodity speculation on water prices, an attractive option for investors drawn to that way of making money. There is strong opposition to that movement, led by the Alternative World Water Forum, and the future of a serious global market remains unclear.[46]

"Water"—the legendary tycoon, billionaire, corporate raider, and investor in wind farms, oil, and fracking, T. Boone Pickens, once said— "is the new oil." From a business perspective, it is an ideal commodity (if one wants to make it that): necessary for all humans, not always accessible, and sometimes hard to get. That is particularly true in the dry desert-like environment of the West. If you can get hold of it or work with those who can, then a market can be found if one is clever and ambitious enough. Pickens fits that bill perfectly, good at finding and exploiting natural resources to sell at a profit.

After some large Texas land purchases in the early 1990s, Pickens concluded by 1997 that there was valuable water under that land, the Ogallala aquifer, and that if he could get it to them, it could be sold to the thirsty cities of Dallas, San Antonio, and El Paso. Starting a company called Mesa Water, he then purchased an additional 400,000 acres of land, planning to build a $1.5 billion, 323-mile pipeline to Dallas, which he finally settled upon as his main sales target. All of this was made possible by a Texas law called "pump or perish" that allowed landowners to take unlimited aquifer water, including one's neighbor's water, for whatever purpose.[47] Whether he was aware of it or not, he was following in the footsteps of many entrepreneurs in California, including the wealthy Bass family, trying to find ways of gaining and selling water in a notoriously dry environment. Unlike, say, the IT technology entrepreneurs of recent decades, it takes big money to get into the water world. Pickens also, it might be noted, took an interest in fracking, and that brought him in contact with many water problems associated with that form of extracting natural gas, not only the large amount of water necessary to drill for the gas, but also the dirty water left over by the process. He did not succeed, eventually giving up.

But Pickens does not come across as someone enthralled by environmental protection, any more than oil prospectors worried about such things before water became the new oil. In any case, a distinction can be drawn between wealthy entrepreneurs, investing their money as individuals, and the role of international corporations developing global markets. Companies that sold their services to manage local communities and those seeking other ways of finding a profit in water needs are, for the most part, different in their scope and ambitions. It is thus useful to look at what can be called the global water market, with a number of corporations selling products addressed to different features of the extraction, testing, and delivery of water to whatever customers they can gain. An entry point into that world is the pricing of water.

## THE PRICING OF WATER AND A GLOBAL MARKET

A common theme in much of the research and writing on water is that although it can be costly to collect and deliver it, there is considerable popular resistance to treating it like an ordinary commodity. In many parts of the world water is immediately at hand, in streams, rivers, and lakes, amply falling from the clouds and not far underground, reachable by rope and bucket or an inexpensive pump. But two historical figures, Benjamin Franklin and Adam Smith, noted some special features of water. In *Poor Richard's Almanack*, Franklin noted that "when the well is dry, we learn the worth of water," and Smith observed that "nothing is more useful than water; but it will purchase scarce anything; scarce anything can be had in exchange for it."

Times have changed. Wells still run dry, much more so than in Franklin's day, as do aquifers and, too often, rivers whose water is depleted before they reach the sea. As entrepreneurs discovered, water can be exchanged for money, and even when available its provision requires a wide range of technologies and services to keep it clean and deliver it efficiently. Yet old attitudes linger and the resistance to water pricing remains, so much so that there has been a move to more consciously use prices for two reasons, neither having to do with commercial profit: as an incentive for people to use water more carefully and sparingly to conserve it, and to provide funds to keep water systems in good working order to reduce waste and inefficiency. This is a use of the market too little examined by those who fear the market because of its power to corrupt and monopolize the provision of necessary human goods. Yet it is clear that both strategies are beneficial and in many instances effective in controlling what would otherwise be a profligate use of water. These benefits have been seen as a result of water pricing incentives in western parts of the United States. Of course, there can be a fine line between pricing for profit and pricing as a means of increasing efficiency, but many municipal water systems run and regulated by a community manage to work with that mix.

The idea of promoting a global market for water in the name of profit is another matter. Vigorously rejected by some groups it is no less vigorously pursued by others. A 2011 report by a unit of Citigroup Global

Markets (a division of Citibank, a multinational financial services company), provides a clear picture of the attractions of global water–related investments.[48] The global water market is described as being worth $450 billion. The investment possibilities cover a wide range of activities and rest on a basic distinction between the drinking water and wastewater sectors. Combined, there are some twenty subsectors of the water market, including water and wastewater treatment, industrial water treatment, pumps, valves, infrastructure, engineering and consulting, irrigation, desalination, and water testing. Singling out urbanization as particularly important for future uses of water, the report goes on to identify twelve international companies that work in that area.

While some 70% of current global water use is agricultural, the Citigroup report projected an increase of water needs in urban settings from 20% to 30% between now and 2050, and a decline in the share of agricultural demand to 60% between now and then. The urban population of the world passed the 50% mark in 2012 and is expected to grow to 70% by 2050. By and large, urban areas are more affluent than agricultural areas and will be particularly impacted by the growing demand for meat that goes with a rise in income and "dining out," another common feature of urbanization. Water for household and institutional and infrastructure needs will continue to grow. One projection has the number of indoor flushing toilets increasing by 400 million in the next few years, with a consequent impact on sewer quality and capacity.

The Citigroup report ends with a paper by Willem Buiter, its chief economist. Building on the data provided in the preceding report, he directly takes on the opposition to a global water market. As do many others, he points out that many countries not only price water well below its actual economic value but commonly exacerbate that practice with subsidies (India would be a prime example). He believes the opposition to pricing rests on what he describes as two fallacies. "Fallacy 1: Water is essential for life: therefore it should come free. We disagree. Water is essential for life. That is why it should be priced or physically rationed to reflect its scarcity value. . . . Fallacy 2: Water comes from God: therefore it should come free. We disagree. . . . After all, diamonds come free from God. Should diamonds therefore come free to everyone who wants them? Scarcity and opportunity cost are the drivers of fair and efficient allocation and distribution."[49] Buiter does agree, however, that a safety net will be necessary for those who cannot afford some basic access to water. It would be wrong to price them out of the market—and pricing practices in general will also require regulation.

The Alternative World Water Forum flatly rejects this whole line of argument. "Buiter's essay," they say, is a vision of the future "predicated on his argument that water is a commodity." Moreover, "vigilance is necessary because some of the greediest economic interests in the world are promoting water markets as part of their plan to financialize nature."[50] I will not try at this point to judge that clash of values and visions, but will return to it in part II of this book. I would only emphasize that I do see a difference between the use of pricing as a way of controlling waste and husbanding water resources and the use of pricing as part of an entrepreneurial or industry-based way of making money.

## DEVISING WATER STRATEGIES

"Too many cooks," the old saying has it, "spoils the broth." "Too many broths," an amended aphorism might hold, "can overwhelm the cooks." In the case of the broth that is water, there are any number of pots with many kinds of soup in the global kitchen. One of the pots is technology (high tech and low tech); another is the market; still another is global, regional, and local integration and cooperation; and yet another is more focused on individual behavior changes. Each of the pots has a broth that overlaps with ingredients from some of the others, and some master chef, overseeing the kitchen, would like to find a way to get the best of each of them in one delicious pot.

But as with each of my five horsemen there is no master chef for water either—and yet almost every analyst and commentator sees the need for one, some way of integrating and coordinating a wide range of problems, actors, ideologies, scientific data, and values. Steven Solomon, whose fine book I have quoted extensively, notes that some important principles have been articulated: "Environmentally sustainable use of water; equitable access by the world's poor to fulfill their basic water needs. . . . Efficient use of existing resources, including recognition of water's value as an economic good." But then he goes on to list the myriad ways that the present downward course of water reform is likely to turn out, a victim of disagreement and a chronic failure to follow through on good intentions with effective action: "As the gulf between those with sufficient water and those without deepens as a source of grievance, inquiry and conflict, the new politics of scarcity in mankind's most indispensable resource is becoming an increasingly pivotal fulcrum in shaping the history and environmental destiny of the twenty-first century."[51]

## SOFT PATHS AND HARD PATHS

Is that an overstatement, an excessively bleak view? In many ways it is, and while hardly any analyst is overly optimistic, some see greater possibilities for positive movement. Particularly striking is a belief among some analysts that a plethora of what can be termed "soft path" solutions can be effective where "hard path" approaches have failed. The distinction between those two paths was formulated by Peter Gleich.[52] A "hard path" takes a big-tech engineering approach: dig more and deeper wells, build more large dams and reservoirs, improve transportation systems for water, and move more water by pipelines from places with much water to those with less. A soft path would include "smaller, cheaper, less intrusive means. . . . Low flush toilets and low-flow showerheads. . . . Drip irrigated farming, underground waterbanking, and toilet-to-tap sewage recycling." "Working with nature," not combating it with expensive technologies and structures, is a theme that emerges in the soft path argument. Alex Prud'homme cites Singapore as a city that has managed to flourish despite limited water supply. It uses a combination of low-tech efficient technologies of a kind recommended by Gleich, complemented by high water taxes and tariffs and government exhortations to its citizens to conserve water. Some 40% of its water is brought in by pipe from Malaysia and another 30% provided by desalinated seawater and recycled wastewater.[53]

Singapore is, of course, a relatively affluent, developed country, and some of its solutions (desalinization) would be far too expensive for use in poorer countries, but the soft path approaches appear more feasible. Oddly enough, Steven Solomon's pessimism is not total. He sees possibilities in new, extraordinary technologies coming along, but he says that "just as important . . . is the gradual humdrum accumulation of low-tech and organizational advancements in the productive use of water supply already available . . . [and] a messy, muddling through process of competitive winnowing and trial and error with diverse technologies . . . Uncertainty, multiplicity, and fluidity are likely to characterize the landscape."[54] Solomon also echoes the view of Holger Hoff on the importance of understanding the interaction of water problems with a great range of other social and environmental realities. "Water needs," Hoff writes, "have to be addressed consistently in climate, energy, trade, agriculture, development and other policies at all scales."[55] Food, energy shortages, and climate change are all interactively related.

Education and community involvement are also important. An interesting study of a Mexican community found a distinction between a present-oriented outlook on life and a future-oriented one. The latter correlated with higher education, a belief in the control of external events, and a more active engagement in water conservation efforts.[56] Peter Rogers and Susan Leal make a good case for the power of community education and discussion in enabling the improvement of sewers, normally out of public sight and of little interest.[57] Community education in some Australian communities made it possible to overcome the "yuck" reaction to recycled sewer water.

If the alarm about water shortage is at the highest level, so too is the number of plausible solutions, each relatively small in impact if taken alone but in the end comprehensive if all are pursued with zeal. Julian Cribb has as good a summary of the possibilities as anyone, some twenty-five altogether, broken down into three categories.[58] Only a sample of them can be listed here, and many were touched on earlier: *awareness, education, and behavioral change*: reduce water wastage, think about water as a scarce resource, educate consumers and farmers on good water-saving practices, water thrift for industry, open up food trade between water-rich and water-poor countries; *incentives and penalties*: price water based on its true cost (including environmental and opportunity costs), price food according to its water cost, introduce markets in water, monitor all water extraction and penalize illegal extraction, end government subsidies that price water artificially low; *technical solutions*: improve water efficiency on farms, reduce leaks in water infrastructure (channels, dams, and pipes), expand rain-fed cropping, invest more in water science and technology, recycle all urban water.

While the possibility of war among countries that need to share water is real in many places, the ideological warfare seen in global warming and food solutions is at a lower level of heat and passion with water—save for the debate about a global market. This may well be a function of the common solutions on offer, neutralized perhaps by their large number and small individual impact: few are likely to invoke values and strategies that will come in direct conflict of a high-profile kind. That is a speculation only, and I may have overlooked some serious ideological struggles. Yet it is probably also true that when so many changes and reforms in thinking and practice are needed at so many levels, making it hard to find a defining focal point, progress may be harder to accomplish.

It certainly makes sense, as the 2012 UN *World Water Development Report 4* concludes, that "there is a need to replace the old ways of sector-based decision-making with a wider framework that considers the multiple facets of the development nexus."[59] The report then goes on to enumerate an intimidating list of ways to implement that goal, encompassing national government and international governing processes, with "appropriate financing and regulatory mechanism[s] to ensure the long-term viability, sustainability and efficiency of water services and infrastructure." It is a list of valid generalities and aspirations that staggers the policy imagination, calling into question, as do excessively long lists, its plausibility.

# 4

# CHRONIC ILLNESS

—

## Rich or Poor, Few Escape

hronic illness—usually called noncommunicable disease (NCD) by international agencies—has long been associated with aging in developed societies. But in recent decades it has also become a serious problem for developing nations, still burdened with a high prevalence of infectious diseases. Chronic illness has long been known as the main cause of death in developed countries—especially cancer, stage 2 diabetes, heart disease, and chronic obstructive pulmonary disease (COPD)—but only in recent decades did it attract the attention of those concerned with global health. A 2011 UN meeting was the first major international conference on the topic. It was by most accounts not a very successful meeting, primarily because coping with chronic diseases is expensive and often beset by opposition to price controls from the pharmaceutical and food and beverage industries.[1] In 2011, in the midst of the recession, there was little expectation that money could be found, and not much has changed since then, at least internationally. But it has also been recognized that chronic illness must primarily be dealt with at the national and local level, not by international agencies. And that is happening to some extent, but with little sign of halting the rise of chronic illness, as we shall see.

Chronic illness can be distinguished from acute illness (such as appendicitis) and communicable diseases (such as polio and malaria) by noting that chronic illnesses typically see a gradual onset, are multicausal, and are long term in duration. Acute conditions are marked by rapid onset, usually with a single cause, and are short term. While not yet nearly as common as those four, Alzheimer's is rapidly gaining ground as aging increases. Among the nonlethal chronic conditions are depression and arthritis, sources of misery and dysfunction.

As with most of my horsemen, chronic illness has a longer history behind it than one might imagine. The medical historian George Weisz cites a striking quote from Aretaeus the Cappadocian (second century C.E.): "Of chronic diseases the pain is great, the period of wasting long, and the recovery uncertain; for they are not dispelled at all, or the diseases relapse upon any slight error . . . and also if there be the suffering from a painful system of cures . . . the patients resile as truly preferring death itself."[2] Weisz goes on to list a number of medical studies in the eighteenth and nineteenth centuries and contends that the concept of chronic illness in recent times has primarily American roots, going back to the New York City Department of Health—but coming to global prominence in the late twentieth and early twenty-first centuries.

In essence, the concept became a public health issue, which moved it out of the medical and clinical domain only. That move, he says, was a "logical consequence of changing disease patterns."[3] His book does not examine specific chronic illnesses. It is only the "shifting and fluid concept" and its incorporation into the public health field that is his focus; and that may have an important bearing on the complaints of many that infectious disease reduction efforts have attracted most of the philanthropic money.[4] For my part, however, I will look at some specific chronic illnesses later in this chapter because of their interaction with many causal features related to other horsemen. The most prominent feature of chronic illness now is that it has become a potent danger for developing countries.

## CHRONIC DISEASE: GLOBAL SCOPE

At present, about 60% of deaths worldwide, some 38 million annually, are attributed to chronic disease, with 29% occurring under the age of 60 (figure 4.1).[5] Some 80% of global deaths from chronic disease occur in low-income and middle-level countries (figure 4.2).[6] Sixteen million NCD deaths occur before the age of 70, with 82 percent of what the WHO labels "premature deaths," occurring in low- and middle-income countries. Cardiovascular diseases account for most NCD deaths, some 17.5 million

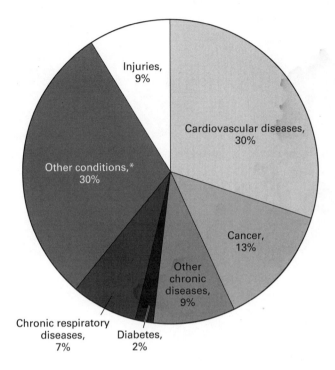

**FIGURE 4.1**

Noncommunicable diseases constitute more than 60% of deaths worldwide. (World Health Organization [WHO], 2005. Preventing chronic diseases: a vital investment. WHO global report. Geneva: World Health Organization.)

annually, cancer for 8.2 million, respiratory disease, 4 million, and diabetes, 1.5 million.[7] Chronic disease now takes more lives globally than infectious disease, but international support for chronic diseases as a group commands only 3% of the $26 billion provided by private donors, international organizations, and foundations. Their focus is on the search for cures or vaccines.

It should be noted, however, that much of the global interest in chronic disease focuses on those under 60, not on those in old age. But of course those that survive or avoid chronic disease at a young age will go into old age certain to incur it and die from it. A UN Fund for Population Activities report projected that there will be 2.6 billion aged in 2050, with well over three-quarters of them over 60, most of whom will die of a chronic

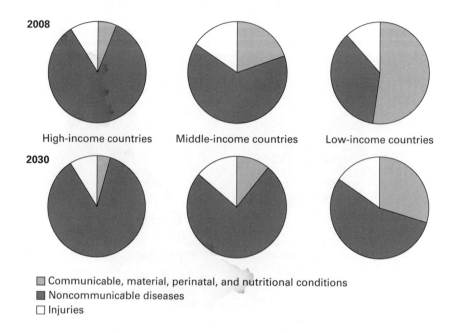

FIGURE 4.2

The increasing burden of chronic noncommunicable disease, 2008 and 2030. (World Health Organization, Projections of Mortality and Burden of Disease, 2004–2030)

disease.[8] The international focus on the developing countries is thus well justified, with the increase going much faster than in the developed countries. Indeed, the low- and moderate-income countries will have a triple threat to deal with: unabated high mortality from infectious disease, continued high incidence of chronic illness in young people, and an increasingly aged population with the chronic disease that accompanies it.

In the United States 70% of deaths each year are from chronic disease, with the numbers skewed heavily toward Medicare patients but including younger patients as well.[9] Most older people will die from one or more chronic diseases, some from as many as four or five. Most of those diseases are preventable in younger persons, that is, for those under 65, though not in the long run. In addition to industry research, those diseases are the leading beneficiaries of federal research money, led by the

National Institutes of Health (NIH). That money goes for research on cancer, heart disease, diabetes, and chronic obstructive pulmonary disease, the leading killers of the old. Even so, infectious disease—once thought all but vanquished some fifty years or so ago with the elimination of smallpox—continues at a high a level.

HIV/AIDS and antibiotic-resistant bacteria account for much of that continued prevalence in developed countries. But just as American chronic care costs continue to increase with more and more baby boomers retiring (now 10,000 a day), so also may the developing countries approach the high projection of elder costs the United States incurs.

But "may" is the operative word, and "probably not" might be more accurate a term. It is almost unimaginable that now-poor countries will in the foreseeable future be able to spend the kind of money the United States and other affluent countries currently spend to cope with those diseases in their aging societies.[10] Meanwhile, the struggle against chronic illness will continue in developed countries. Over the next thirty years, with no improvement in managing costs and new technologies, it is estimated that $30 trillion will be spent globally on chronic illness (what would have been close to 50% of global GDP in 2010). And that is not sustainable.

## Health Impact

The global rise over the years of longer life expectancies, increased affluence, and GDP growth ultimately lie behind the increase of chronic disease.[11] Life expectancy increased from 31 years in 1900 to 67 years in 2009 in developing countries. In developed countries longer lives have meant a sharp increase in the number and proportion of the elderly, with the leading increase in those over 70, few of whom are exempt from one or more such illnesses. In the developing countries many of the victims are children who survived infancy, becoming old enough to incur earlier onset chronic illnesses. There is a painful irony in these instances of progress, with both young and old at greater risk of chronic disease because of increases at both ends of the spectrum of life extension, childhood and old age.

The proximate causes of chronic illness in developing countries are now well known, easily heavily traceable to four leading conditions: alcohol consumption, tobacco use, physical inactivity, and unhealthy diet.[12] The last two increase the likelihood of obesity, which in turn carries a number of health threats. Tobacco kills around half of all those who use it, about 4 million people each year, and is a well-known cause of cancer and heart disease. Some 7% of all deaths can be traced to alcohol use, and it is responsible for various forms of cancer, cirrhosis of the liver, murders, and auto accidents. Of the top ten chronic diseases around the world, the WHO has estimated that a complete elimination of those conditions would lead to an 80% reduction of heart disease, stroke, and type 2 diabetes, and an approximately 40% reduction of all cancers.[13] Since the rising number of the elderly will inevitably contract one or more chronic diseases, it is obvious that chronic disease cannot altogether be eliminated; even if chronic diseases are reduced in younger adults they will eventually appear as those adults age. Nonetheless, to be sure, a considerable amount of money, not to mention human misery, would be saved by reducing chronic illness in the young.

In many ways, chronic illness has a greater impact on children than adults. Children have a better chance of suffering for many years from whatever disease or diseases they may have, and the rate of increased younger victims is considerable. Obesity is a major driver worldwide. Slightly over 10% of children aged 5 to 17 are obese, with the United States the leader, with 35% of its children overweight or obese. China's childhood prevalence of obesity is high, at over 27%, and Mexico follows close behind at 26%. The net result of this global phenomenon is a sharp increase in type 2 diabetes. One in three American children were projected in 2000 to contract type 2 diabetes during their lifetimes.[14] The use of tobacco among children is growing around the world as well, with the poorer countries in the lead—stimulated by aggressive and unregulated advertising. A likely prospect is that gains in life expectancy in the United States will in coming years stall or be reversed.

A notable feature of NCDs is that, unlike most communicable diseases, patients can now live for long times with them (HIV/AIDS is an anomaly). For researchers and philanthropists, the attraction of infectious diseases is that they can be conquered and in many cases eliminated entirely, as was the case with small pox, or dealt with effectively and decisively by vaccinations. NCDs now, however, feature a mixture of declining death rates for most of them together with a high incidence of new cases and a persistent prevalence of them by patients who live with them in ways not possible earlier. Most of us can think of friends or family members with severe breathing problems (COPD), or suffering from heart failure though living on, or patients with diabetes whose condition has been managed successfully by drugs and self-care. As one valuable 2015 global study noted, patients can not only live longer with a disease, but require costly diagnosis and then extended treatment thereafter. The result, the studies show, "is that decreases in mortality rates can occur despite increases in absolute numbers of people dying from disease."[15] While death rates from chronic disease reflect differences among countries in reducing death rates, and in male and female rates (which are generally lower) in doing so, most available information is from wealthy countries and, for many poor countries there is little data at all.

While heart disease, diabetes, and COPD are obviously important chronic diseases, I will expand here only on cancer and on one disease not on the major list, Alzheimer's. Cancer is one of the most rapidly growing NCDs, and Alzheimer's has taken a fearful place in the public eye because of the aging population.

## Cancer

Cancer is a particularly important chronic disease. A 2010 report by the American Cancer Society makes it evident why.[16] It is now a leading global cause of death (figure 4.3), with an estimated 7.6 million deaths in 2008 and 12.4 million diagnosed cases, and an annual and rising cost of $895 billion. As the American Cancer Society noted in a 2011 study, cancer now represents the greatest international economic impact of any disease, some 19% higher than heart disease. Sixty percent of those deaths and 50% of the diagnoses are in developing countries, which the report describes as a "silent pandemic." Smoking is a major cause of the deaths. It is the direct result of an intensified effort of tobacco companies to sell their products in developing countries after the developed countries drove use down by means of regulation and taxation. Cervical cancer, readily diagnosed by routine screening and treatment in rich countries but not in

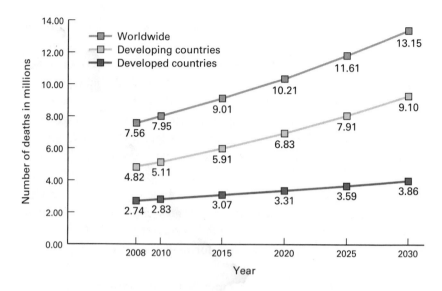

**FIGURE 4.3**

Projected number of cancer deaths worldwide, 2008–2030. (American Cancer Society, http://www.cancer.org/acs/groups/content/@internationalaffairs/documents/document/acspc-029672.pdf)

poor countries, kills approximately 241,000 women in poor countries—out of 274,000 thousand globally. If the relatively inexpensive treatments are unavailable to the poor, it is inconceivable that the expensive new biologics for cancer, running into the thousands of dollars, will be available to the poor in the foreseeable future.

The astute journalist George Johnson, author of *The Cancer Chronicles*, has underscored two points that make researchers and many others uneasy, seemingly putting a powerful damper on the idea of unlimited medical progress and technological innovation: (1) that a person cured of one lethal disease will then inevitably die of another, and that while much more can be done to save and treat those with cancer, (2) cancer may in the end be incurable, the last disease, because of the constant genetic mutation changes that accompany evolution.[17]

I would only add that, in my experience, many health policy analysts balk at accepting the economic corollary of the first point, what I call the longitudinal costs of medical progress. I mean by that the kind of costs in which a cured disease A will be replaced by a later disease B, and that health-care costs will necessarily rise over an individual's lifetime. Self-evidently, the cure of one disease in an individual will not lower costs, as commonly argued. There will be the cost of the next disease, and there always will be one. That is a major reason why elderly people with chronic illness can, over the course of their aging, be very expensive patients: their lives can be saved many times before they die. We will all die of something, if not from one critical event then from the next one. There has been recent speculation, based on computer modeling only, that a delay of aging might be possible, reducing costs.[18]

The worldwide increase in cancer is as good an index as any in revealing not only its potent impact on rich and poor countries alike but also the moral and economic dilemmas it will pose for low- and middle-income countries. Already the global rise in its prevalence and the steady stream of ever more effective, and costly, means of treating it are making it—to use a now commonplace phrase—"unsustainable," even in the richest countries. The new class of drugs, the biologics, have seen the recent advent of super expensive drugs, ranging from $50,000 per course of treatment up to one that is $332,000—but with increased life expectancy of only a few weeks or months at best. The UK Lancet Oncology Commission in 2011 said that it is unacceptable to have an "ethos of very small benefit at whatever cost."[19]

But the affluent countries have not found a good way to pursue that goal. It is everywhere a controversial problem, with even a discussion of it facing two obstacles: the love of and pursuit of medical progress and technological innovation, a deeply imbedded value in affluent societies, and a resistance everywhere to outright uses of rationing.[20] The fact that cancer has a special place as a "dread disease," rightly feared, does not make that discussion any easier. As the British National Health Service (NHS) has found, a decision to limit availability of expensive drugs invariably evokes public and media outcries, and the NHS sometimes capitulates in the face of that response. Clearly, if it is challenging for affluent countries, expensive cancer care is not plausible in poorer countries. The rich and powerful in those countries will get it, the poor will not. But there are two relatively inexpensive drugs that, with outside financial help, could be affordable. One of them is a papillomavirus vaccine to prevent cervical cancer. The other is lifesaving treatments of a kind that save 90% of children with leukemia in developed countries, the lack of which kills 90% of children in poor countries.[21]

## Alzheimer's Disease

In the previous chapter, I suggested that the rise in aging had crept up on those fixed on population growth as the most general threat to future global well-being. The same could be said of dementia as part of aging, and especially Alzheimer's disease, by far its most common kind. Although well known for decades, Alzheimer's is essentially a disease of aging—not true of the other dominant chronic illnesses (even if they now occur most frequently in the elderly). It took large-scale aging populations with longer life spans to bring it to the fore, with the anxiety about it as a notable feature among the elderly themselves. It is plausible now that, at least with the old, Alzheimer's has replaced cancer as the "dread disease." A disease that kills slowly, it also has some notably disturbing characteristics as a chronic disease. It robs its victims of their minds, self-identities, and abilities to function as self-directing person; it can also impose great burdens on caretakers, especially family members, who can have their own health harmed by the pressures of caring for someone with dementia; and it is very costly for society. In poor countries, the movement of young workers from farms to city jobs has left many elders without traditional family care in old age, and in China that has been exacerbated by one-child families. Even in affluent countries, caretaking can be a daunting demand.

The figures on the global rise and prevalence of Alzheimer's and dementia tell the story (table 4.1). In 2010 the estimate was that 35.6 million people had dementia, with the numbers expected to double every twenty years to 65.7 million in 2030 and 115.4 million in 2050. In 2010, 58% of all those with dementia lived in countries with low or middle income; this number is projected to rise to 63% in 2030 and 71% in 2050.[22] China and India could look forward to a 100% increase by 2040.[23] In the

TABLE 4.1 Estimated Global Number of People with Dementia Over Sixty, 2010, 2030, and 2050.

| GBD region | Over 60 population (millions, 2010) | Crude estimated prevalence (%, 2010) | Number of people with dementia (millions) | | | Proportionate increases (%) | | |
|---|---|---|---|---|---|---|---|---|
| | | | 2010 | 2030 | 2050 | 2010–2030 | 2010–2030 | 2010–2050 |
| Asia | 406.55 | 3.9 | 15.94 | 33.04 | 60.92 | 107 | | 282 |
| Australia | 4.82 | 6.4 | 0.31 | 0.53 | 0.79 | 71 | | 157 |
| Asia Pacific | 46.63 | 6.1 | 2.83 | 5.36 | 7.03 | 89 | | 148 |
| Oceania | 0.49 | 4.0 | 0.02 | 0.04 | 0.10 | 100 | | 400 |
| Central Asia | 7.16 | 4.6 | 0.33 | 0.56 | 1.19 | 70 | | 261 |
| East Asia | 171.61 | 3.2 | 5.49 | 11.93 | 22.54 | 117 | | 311 |
| South Asia | 124.61 | 3.6 | 4.48 | 9.31 | 18.12 | 108 | | 304 |
| Southeast Asia | 51.22 | 4.8 | 2.48 | 5.30 | 11.13 | 114 | | 349 |
| Europe | 160.18 | 6.2 | 9.95 | 13.95 | 18.65 | 40 | | 87 |
| Western Europe | 97.27 | 7.2 | 6.98 | 10.03 | 13.44 | 44 | | 93 |
| Central Europe | 23.61 | 4.7 | 1.10 | 1.57 | 2.10 | 43 | | 91 |
| Eastern Europe | 39.30 | 4.8 | 1.87 | 2.36 | 3.10 | 26 | | 66 |
| The Americas | 120.74 | 6.5 | 7.82 | 14.78 | 27.08 | 89 | | 246 |
| North America | 63.67 | 6.9 | 4.38 | 7.13 | 11.01 | 63 | | 151 |
| Carribean | 5.06 | 6.5 | 0.33 | 0.62 | 1.04 | 88 | | 215 |
| Andean LA | 4.51 | 5.6 | 0.25 | 0.59 | 1.29 | 136 | | 416 |
| Central LA | 19.54 | 6.1 | 1.19 | 2.79 | 6.37 | 134 | | 435 |
| Southern LA | 8.74 | 7.0 | 0.61 | 1.08 | 1.83 | 77 | | 200 |
| Tropical LA | 19.23 | 5.5 | 1.05 | 2.58 | 5.54 | 146 | | 428 |
| Africa | 71.07 | 2.6 | 1.86 | 3.92 | 8.74 | 111 | | 370 |
| North Africa/Middle East | 31.11 | 3.7 | 1.15 | 2.59 | 6.19 | 125 | | 438 |
| Central SSA | 3.93 | 1.8 | 0.07 | 0.12 | 0.24 | 71 | | 243 |
| East SSA | 16.03 | 2.3 | 0.36 | 0.69 | 1.38 | 92 | | 283 |
| Southern SSA | 4.66 | 2.1 | 0.10 | 0.17 | 0.20 | 70 | | 100 |
| West SSA | 15.33 | 1.2 | 0.18 | 0.35 | 0.72 | 94 | | 300 |
| World | 758.54 | 4.7 | 35.56 | 65.69 | 115.38 | 85 | | 225 |

*Note:* abbreviations: LA, Latin America; SSA, Sub-Saharan Africa

*Source:* M. Prince, et. al. "The Global Prevalence of Dementia: A Systematic Review and Metaanalysis."
*Alzheimer's & Dementia : The Journal of the Alzheimer's Association* 9, no. 1 (Jan, 2013): 70.

United States the estimate in 2014 was that 469,000 persons had developed the disease, with a projected 59,000 new cases for those between the ages of 65 and 74, 280,000 new cases among those 85 and older, and also increasing for those over 90. The projection is for a doubling of new cases in the United States by 2050.[24]

If it is hard for Americans to imaginatively put themselves in the shoes of those in low-lying foreign islands faced with rising ocean levels, every one of us (particularly if old oneself) is likely to know someone with Alzheimer's—and also their struggling caretakers. My wife was the guardian for ten years of her stepmother in a euphemistically named facility for the "memory impaired." My wife and I recently counseled a friend with a deteriorating wife about his personal dilemma: he is himself beginning to fall apart from the weight of his caretaking but had solemnly promised her years earlier that he would never put her in nursing home. What should he do? In China, there are few nursing homes at all. It is not an exaggeration to say that no country, rich or poor, is in good shape to cope with the increase of Alzheimer's expected over the next few decades. And as usual, the poor countries will be the worst off. No cure is on the horizon, and the few available (and expensive) drugs can hold off Alzheimer's lethal progression for a short time only. If only gradually, medical advances with heart disease, type 2 diabetes, COPD, and cancer are happening. Not so with Alzheimer's. It shares with obesity the obdurate feature of no significant scientific progress to date.

## SOCIAL AND ECONOMIC CONSEQUENCES

The economic consequences of chronic disease are hardly less devastating than their social impact. They include increased labor market costs because of a decline in the workforce and thus reduced productivity, severe impact on families who must care for those afflicted, longer and more devastating impact on individuals in the company of illness, and with declining employment possibilities for them. All of these factors are intensified by urbanization, a leading feature of life over the past century. In the past few years 50% of the world's population has come to live in cities. The cities of developed countries contained 78% of their inhabitants in

2011, while only 47% of the populations of developing countries were city dwellers.[25] The greatest future change in urbanization will come in the poorer countries. Of the projected world population growth between 2011 and 2100, 7% to reach 11 billion, the increase in population in urban areas for that period will be 2.6 billion, moving upward from 3.6 billion to 6.3 billion. Asia, Africa, and Latin American will see most of that increase.

Cities of 1 million or more will account for 47% of the global population by 2035, but the largest number will be in megacities, those with more than 10 million inhabitants.[26] In 1970 there were only two such cities, New York and Tokyo, but there are now thirteen, expected to increase to thirty-seven by 2025. By that time, one out of 13 people on this planet will live in megacities. New York and Tokyo are still among the top ten megacities but are now joined by New Delhi, Shanghai, Mumbai, São Paulo, Dhaka, Beijing, and Karachi, with the populations in those cities of the less developed countries (the majority) ranging from 20 to 33 million.

Cities have long held out the dream of excitement and work, and urbanization in most developed countries has had the same attraction, all the more so as agriculture declines. But the urbanization of poorer countries has come with a high price: increased tobacco use, alcoholism, the availability of cheap processed foods and a rise in beef consumption, increased outdoor air pollution, indoor pollution from cooking fires, crowded housing, dirty water, and poor medical and public health services—all conducive to cancer. China, we recall, has serious air pollution problems, while India has severe urban water shortages. And it is the megacities that have the largest problems. In poorer countries, cities also draw a disproportionate number of single males, often with the hope that they can do well enough financially to bring their families or young brides to join them. Many of the available jobs are low paying, with long hours, and in dangerous factories.[27]

## GOVERNANCE

Many analysts of global chronic illness complain about its low status as a health issue. Despite recognition by the WHO in 1996 and by many national governments thereafter, it does not have the draw of infectious disease, health-care infrastructure development, and medical training. The 3% figure for assistance on chronic illness by private foundations and donors is the starting point for those laments. David Stuckler and colleagues have pointed to a number of likely reasons. By focusing on just about everything other than chronic illness, the influential private donors have failed to note that the disability years in a lifetime caused by chronic illness are considerably greater than for infectious diseases. Infectious disease accounts for a lifetime loss of 264 disability-adjusted-life-years (known as DALYs), but 646 such years are lost to chronic conditions. Funding per death for infectious disease is $422 versus $18 for chronic disease. There is some $6.32 billion in health development assistance for infectious disease and only $500 million for chronic illness.[28]

Why is that? Stuckler and colleagues offer a variety of reasons. Leading the list is the giving pattern of private donors: "Donor funding misaligns with the burden of chronic disease; factors other than need appear to be driving the agenda."[29] Their perception is that the close relationship between those donors and various corporate interests sways their priorities. The interlocking relationship between food and agriculture companies enhances that relationship. Just about every major company of that kind, such as McDonald's, Coca-Cola, Monsanto, and Johnson & Johnson, can be found in the stock portfolios of the Gates Foundation and Berkshire Hathaway. Still another is the bias of, say, the Bill & Melinda Gates Foundation toward technological innovation (not surprising for the founder of Microsoft), whether in the direction of vaccines for infectious disease or for new kinds of seeds for crops such as corn, rice, and wheat.

Of course, that proclivity can also be seen as a reflection of the American love of technological innovation and the special glory that comes from victory over single-cause infectious diseases. The connection between innovation and industry has long roots: constant innovation is a necessity for industry, and many tempting targets: while there is yet no vaccine for HIV/AIDS, the European Medicines Agency in 2015 approved a malaria vaccine, developed by GlaxoSmithKlein, ending a long and expensive search. Technological innovators go where the pickings look most promising. There is, however, some industry support from NGOs, such as the Oxford Health Alliance, supported by PepsiCo and Novo Nordisk, but only in the $5-million-dollar range, far below the billions spent by the Gates Foundation on its health agendas. That trust, I should add, has increased support in the past few years for work on chronic disease. With the exception of the WHO and occasional glances by the United Nations, it seems fair to say there are few international agencies focused on chronic illness, and few that even consider it. The World Bank and the International Monetary Fund have only issued occasional reports on the topic. The United Nations specifically and deliberately left the topic out of its Millennium Development Goals because of its focus on gaps between rich and poor countries in the areas of infectious disease, malnutrition, and unsafe childbirth.[30] Nor is there much greater interest at national agencies, even though Canada, Sweden, and the United States have made some contributions. The Fogarty Center at the NIH provides money for research and training on chronic illness in poor countries, although this is measured in millions rather than billions of dollars. In the United States the Centers for Disease Control and Prevention published a study of chronic illness in 2010, as did the U.S. Department of Health and Human Services and the quasi-federal Institute of Medicine.[31]

While explanations for the weak support of NCD research and policy work are many, perhaps the most plausible is that advanced by George Weisz and Etienne Vignola-Gagne. Citing the dominant attention given to communicable disease, they comment that for many countries, rich and poor, these may "be more compelling and frightening than slowly evolving chronic conditions that primarily affect older people, are difficult to deal with, and for which there is little potential for quick technological fixes…transformation of risky individual behavior requires that healthy alternatives be easily available."[32]

Yet a distinction is in order here between chronic illness as a class of diseases and the individual chronic diseases. In every developed country there is an interest in specific chronic illnesses, notably cancer, diabetes 2, and heart disease. Considerable research on chronic illness is in that sense occurring. In the United States the NIH has various subunits devoted to all the leading chronic conditions, lethal and nonlethal. A number of health ministries around the world have divisions interested in chronic disease, but many are starved for money, and some have none at all even listing such an interest. There are a few NGOs that are focused on chronic disease: the Wellcome Trust in the United Kingdom, a Global

Alliance for Chronic Disease (founded in 2009), and the NageorfCD Alliance. Stuckler and colleagues end their description of and lament about the low priority for chronic disease by saying that it is a reflection of politics, not economics. That conclusion is plausible but not fully borne out by their analysis, although politics is surely a part of it. But then setting priorities always is a matter of politics, but sometimes it is a genuine difference of judgment of importance, and there is a bit of that here as well.

## INDUSTRY

The role of industry in NCD is variegated. First, every one of the chronic diseases makes use of technology in one way or another: surgery for cancer, assorted devices for heart disease, and drugs for both, as is the case with type 2 diabetes. Second, all of them are multibillion-dollar industries, each of which has cadres of lawyers to protect them from government regulation, multiple public relations personnel and salesmen to sell their products, and a public eager to buy them.[33] Third, their greatest profits come from the treatment of chronic diseases, not their prevention (where the main focus of public health is focused); and even their work on vaccines often requires government or private donor subsidies because of their small profit margin. Fourth, the heaviest costs in health care in all of the developed countries come from treating chronically ill patients, heavily skewed toward the elderly. And, to be sure, those industries want to sell their products to developing countries, where the greatest future market exists and their costs will be even more unsustainable.

I offer two telling instances of the place of industry in chronic illness. During the 2011 UN event on chronic illness, opposition to giving it a larger place in international health efforts came from two quarters. The Gates Trust argued that a heavy emphasis on NCDs would be a distraction from other existing global health efforts.[34] At the 2011 UN conference, industry representatives—invited to the conference as "partners" but not allowed a vote on the final report—placed a number of obstacles in the way of a tough and strong conference result, considerably watering it down. They successfully lobbied against an invocation of the trade agreements that led to the provision of life-saving drugs to some 7 million people with HIV/AIDS as a model for chronic illness; against the use of various international agreements for controlling the costs of cancer and heart medicines; against the use of the word "epidemic" to describe the NCD problem; against compulsory targets (such as reduced salt intake); and opted for a voluntary rather than a regulatory approach to pursuing the report goals.

My other example is in a two-part chapter in the Stuckler and Siegel book, which I have amply quoted in this book. In part 1, a most positive description of the various helpful activities of leading industries (primarily food and beverages) is laid out to display a variety of useful initiatives taken to improve health. As Tara Achrya and colleagues observe, "The food industry can and should play an important role . . . this is done through a uniquely acquired understanding of adaptation to consumer needs, combined with technological capabilities and business acumen."[35] They cite the role in this effort played by "top multinational companies," including General Mills, Kellogg, Nestlé, and PepsiCo. One example is that these companies have agreed to reduce sodium, fat, and calories in their products and to enhance potentially healthy ingredients. A number of other examples are provided, indicating a high level of activity toward health goals. The authors conclude by stating the necessity of "building public/private trust as a prelude to alliances" of the public and private sectors. One of the coauthors of this article, Derek Yach, has worked for both the WHO and PepsiCo and has pushed the importance of this relationship, in great part because of the central role of industry in dealing with chronic illness. How can there be any real improvement, they contend, unless industry is given a central role?

Part 2 of the article, written by William H. Wiist, takes a dramatically different stance.[36] While it is not clear whether it was written in direct response to part 1, it systematically attacks the Yach position and finds nothing good to say about it. The most important theme running through this part of the article is that the industry does not in any meaningful way do what it says it is doing, and is in any case doing nothing but serving its own interests, that is, profit. Industry is of its nature always aiming to gain new customers, to develop new products, to use its influence and power by lobbying and election campaign contributions. It is also a large contributor to the education of dieticians and other educational activities of professionals in the field. The article concludes with forty-one imaginative if often implausible proposals that industry will have to adopt to "gain the trust of potential public health partners, governments and human citizens." The first proposal is that government and NGOs have full access to corporate records on health-related activities—what exactly have you done in the name of health?—and the other forty are no less demanding. Depending on one's viewpoint, they can be seen as minimally necessary or outrageously intrusive. My own view is that while the charges against industry are accurate, nonetheless a public-private relationship is necessary (see chap. 11). The question is how that can be brought about in the face of mutual hostilities.

## LONGER LIFE, GOOD HEALTH, AND A QUICK DEATH?

For many years, beginning in the early 1980s, there was a strong belief that a "compression of morbidity" might medically be possible, that is, a long life in good health followed by a quick death in old age. An important 2011 study concluded, however, that "health may not be improving with each generation. . . . Compression of morbidity may be as illusory as immortality. We do not appear to be moving to a world where we die without experiencing disease, function loss, and disability."[37] There is already some evidence that obesity is beginning to lower life expectancies.[38] The demographer Jay Olshansky has predicted that life expectancy will not indefinitely continue to rise or go beyond an eighty-five-year average (the life expectancy of Japanese women). "While bodies are not designed to fail," Olshansky has tersely written, "neither are they designed for extended operation."[39]

As with so many global problems, the contrast between rich and poor countries is striking. In the case of chronic illness the developed countries now have to deal with its expensive technology-driven costs, most of which are generated in the arena of elder care. In the developing countries chronic illness problems and costs are showing up much earlier in life, and these countries often lack the capacity to pay for even the least expensive technologies. It is unlikely that they will anytime soon (even measured in decades) have the money to bring in costly life-extending technologies. And during the past few years, there has in fact been a change in emphasis in international efforts to take on chronic illness, with the focus primarily on prevention and the reduction of premature deaths, and younger people as the highest priority. There is no serious, even speculative discussion of introducing expensive high-technology treatments for either young or old.

## AVAILABLE RESOURCES: DEVELOPED COUNTRIES

An unavoidable question in light of the enormous future demand for the management of chronic disease centers on the availability of economic and social resources. There is no doubt that developed countries have the resources both to carry out research and to provide good treatment. In many ways, their problem is just the opposite of the developing countries: balancing the high expense of treating chronic illness, particularly in the old, against many other national needs. The greatest emphasis in developing countries is now on prevention, focused on those background socioeconomic conditions that cause or exacerbate it. At the same time it is recognized that there are a number of relatively low-cost diagnostics and treatments that, with adequate funding, could make a difference in developing countries. If, as a public health strategy, prevention is a dominant theme, there is a strong push as well for an emphasis on primary care in those countries.

The United States has the most difficult challenge, now spending a large and expensive portion of health care on a small percentage of its population. In 2009, $1,223 billion was spent on health care. Of that amount $275 million was spent on the top 1% of patients, $623 million on the top 5%, and $821 million on the top 10%—and $36 billion on the bottom 50%. Much of that money at the top goes to the elderly and their chronic illnesses and to intensive care units and follow-up care, where costs of $500,000 and up for a small number of patientsfor a single health crisis are not uncommon.[40] The United States is notorious for its expensive health care, with costs some 25% to 35% higher than other developed countries and with worse health-care outcomes. At stake most directly is the Medicare program for the elderly, with projections of unsustainable costs in the near future.

The projected growth rate of the elderly population in the United States makes clear just why: between 2010 and 2030 their numbers will grow from 39.4 million to 69 million, a 75% increase; and from 2030 to 2050 to 79 million, a 14% increase. Between 2010 and 2050 the number of those over 85 will grow from 5.6 million to 18.2 million.[41] Chronic illness will grow with the elderly; some in their sixties and seventies may be lucky enough to avoid it, but that becomes increasingly unlikely beyond seventy. As with other countries, rich and poor, there are significant economic differences in death from chronic illness in the United States. Our poorest state, Mississippi, has a chronic illness death rate much higher than the U.S. average, while Vermont, one of the most affluent, is below the average; Mississippi has one of the largest proportions of minorities and Vermont one of the lowest.[42] But the other developed countries, helped by universal health care and strong government controls on spending, are also feeling the pinch as their aging populations grow. In Europe there are significant differences between northern and southern countries in chronic illness deaths along the same economic lines as the United States.

But the problem of chronic illness in developing countries is far more difficult. If they will be spared, by dint of poverty, the kinds of pressure our expensive system creates, they will not be spared having too little money and too little technology. If we in America put too little emphasis on prevention because we can afford to pay for treatment, the poor countries seem to realize that is a road they cannot go down. They must choose prevention. That emphasis is obvious in WHO's detailed 2013 report, *Global Action Plan for the Prevention and Control of Noncommunicable Diseases*.[43] The report describes "three pillars" for future efforts: surveillance, prevention, and strengthened health systems, a strategy directed at premature mortality and improving quality of life. It alsoand notes complementary UN and WHO efforts focused on tobacco control, diet and physical activity, and the harmful use of alcohol (each earlier identified as important in the effort to control premature chronic illness). The general goal of the global effort is "to reduce the burden of preventable morbidity and disability and avoidable mortality due to noncommunicable diseases."[44]

## Modest Goals

More specifically, the global action plan sets the following goals to be achieved by 2025:

- A 25% reduction in overall mortality from the leading chronic diseases
- A 10% "relative reduction" in the harmful use of alcohol, "as appropriate, within the national context"
- A 10% relative reduction in the prevalence of "insufficient physical activity"
- A 30% reduction in "prevalence of current tobacco use in persons age 15+ years"
- At least 50% of eligible persons to receive drug therapy and counseling to avoid heart disease and strokes
- An 80% availability "of the affordable basic technologies and essential medicines . . . required to treat major noncommunicable diseases."

What I have just cited is only a small portion of a report that has well over fifty general and specific goals. But it seems to me, with the exception of the last recommendation, it is comparatively modest and in principle achievable—but will make small overall progress. That last goal will, moreover, require the cooperation of relevant industries manufacturing those technologies and financial support from national and international organizations. No major campaign is on the way to bring that about.

The main impediments to achieving even those modest goals are the weakness of the medical infrastructures and policies to pursue them. The contrast between rich and poor countries is striking. The regulation of tobacco and alcohol is exceedingly weak and practically nonexistent in poor countries. It is due not simply to a lack of money to develop educational programs but more importantly to the active and effective campaign of industry to advertise and sell tobacco and alcohol and the ease with which even light control can be evaded.

The 2010 global survey *Assessing National Capacity for the Prevention and Control of Noncommunicable Diseases* lays out the health-care gaps.[45]Only 20% of low-income countries or in the African region have radiotherapy available. Some eighty-one high-income countries have therapies to prevent blindness versus 7% for low-income countries. Slightly more than 33% of those countries provide support for self-care, with only 27% providing support for home-based care, but well over 50% of affluent countries do so. Primary preventive and health promotion are less than half those provided in Europe and America. The same pattern prevails on the availability of tests and procedures for early detection, diagnosis, and monitoring of NCDs. The report does, however, report considerable prog-ress over the past decade in most of its categories. Even so, the modesty of the goals in the 2013 WHO report shows a difficult road ahead. While it would obviously be expensive, a good case can be made for a strengthen-ing of primary care in low income countries. Evidence from the European Union showed a strong correlation between strong primary care avail-ability and good health, particularly important for those suffering from two or more chronic conditions.[46] In lieu of money for chronic illness, an African study found that trained community health care workers other than doctors and nurses, using paper- or cell phone–based screening tools are highly effective at a comparatively low cost.[47] There is no mention at all of possible adaptations, and the mitigations are not soon likely with either obesity or chronic illness. Their rise may be slowed (and even this will not be easily accomplished) but not stopped. The future good of the body may be even more difficult than the good of the planet. Evolution in great part controls the body and its aging and eventual demise; the way poor people live their lives, usually with little choice, adds to the evolution-ary burden. But there is no comparable process in nature that inexorably dooms us to despoil the planet or pollute the atmosphere. That is left to the way we live our modern industrial lives, which has turned out to have its own (so far) inexorable features. While I am not certain just how much comfort it might provide, the Cambridge public health historian Simon Szreter has informatively written about how the Industrial Revolution of the eighteenth century brought about an increase in bad health, and how the shift from rural to urban life was a part of that, but how it was also a great stimulus for a public health movement that eventually reduced the impact of those earlier destructive forces.[48]

# 5

## OBESITY

---

### The Scourge of Bad Diets and Sedentary Habits

Among the five horsemen obesity has some unique features. While there have always been some obese people and sporadic medical recognition of obesity as a problem, it burst out with a special fury as a global issue in the 1970s. That happened first in developed countries and then, not much later, in poor countries usually noted for their malnutrition. The United States has the dishonor of having the largest proportion of overweight and obese people and, not coincidentally, the world's largest GDP. A colleague of mine, just arriving from her home in New Zealand, noted the first thing that attracted her attention in the airport: "How fat everyone was. You just don't see that in my country." She said that some years ago; New Zealand is now catching up. In 2012–2013 the New Zealand Ministry of Health reported that almost one in three adults (ages fifteen and older) were obese (31%), and 34% were overweight.[1] One in nine children ages two to fourteen are obese (11%), and 22% are overweight.

In the United States some 67% of Americans are overweight or obese, with 35% obese. American children are little better off: 18% of children ages six to eleven are obese, and for those ages twelve to nineteen, the figure is 21% (figure 5.1). Globally, 35% of adults age twenty or older were overweight in 2008, and 11% were obese (figure 5.2). Worldwide obesity rates have doubled since 2008. In that year 1.4 billion were overweight and 200 million men and 300 million women were obese. More than 40 million children under the age of five were overweight or obese. Sixty-five percent of the world's population live in countries where being overweight or obese kills more people than being underweight. In principle, obesity is considered preventable.[2] In practice it continues unabated.

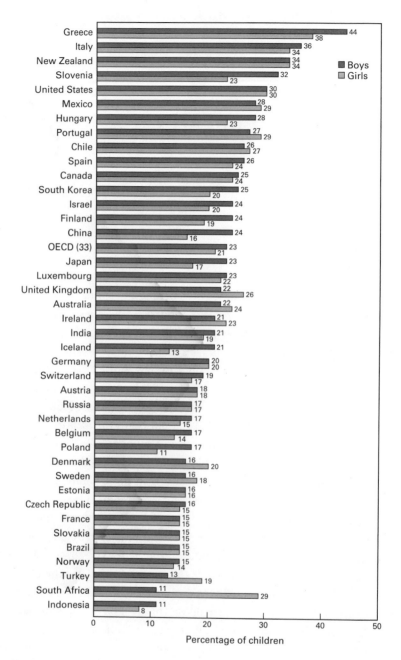

**FIGURE 5.1**

Measured overweight (including obesity) among children aged 5–17, 2010 or nearest year. (International Association for the Study of Obesity, 2013; Bös et al. (2004), Universität Karlsruhe and Ministères de l'Education nationale et de la Santé for Luxembourg; and KNHANES 2011 for Korea.)

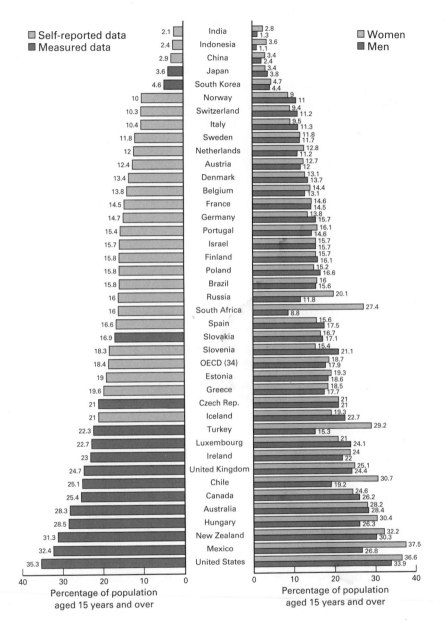

Self-reported data
Measured data

Women
Men

| Country | Self-reported / Measured | Women | Men |
|---|---|---|---|
| India | 2.1 | 2.8 | 1.3 |
| Indonesia | 2.4 | 3.6 | 1.1 |
| China | 2.9 | 3.4 | 2.4 |
| Japan | 3.6 | 3.4 | 3.8 |
| South Korea | 4.6 | 4.7 | 4.4 |
| Norway | 10 | 9 | 11 |
| Switzerland | 10.3 | 9.4 | 11.2 |
| Italy | 10.4 | 9.5 | 11.3 |
| Sweden | 11.8 | 11.8 | 11.7 |
| Netherlands | 12 | 12.8 | 11.2 |
| Austria | 12.4 | 12.7 | 12 |
| Denmark | 13.4 | 13.1 | 13.7 |
| Belgium | 13.8 | 14.4 | 13.1 |
| France | 14.5 | 14.6 | 14.5 |
| Germany | 14.7 | 13.8 | 15.7 |
| Portugal | 15.4 | 16.1 | 14.6 |
| Israel | 15.7 | 15.7 | 15.7 |
| Finland | 15.8 | 15.7 | 16.1 |
| Poland | 15.8 | 15.2 | 16.6 |
| Brazil | 15.8 | 16 | 15.6 |
| Russia | 16 | 20.1 | 11.8 |
| South Africa | 16 | 27.4 | 8.8 |
| Spain | 16.6 | 15.6 | 17.5 |
| Slovakia | 16.9 | 16.7 | 17.1 |
| Slovenia | 18.3 | 15.4 | 21.1 |
| OECD (34) | 18.4 | 18.7 | 17.9 |
| Estonia | 19 | 19.3 | 18.6 |
| Greece | 19.6 | 18.5 | 17.7 |
| Czech Rep. | 21 | 21 | 21 |
| Iceland | 21 | 19.3 | 22.7 |
| Turkey | 22.3 | 29.2 | 15.3 |
| Luxembourg | 22.7 | 21 | 24.1 |
| Ireland | 23 | 24 | 22 |
| United Kingdom | 24.7 | 25.1 | 24.4 |
| Chile | 25.1 | 30.7 | 19.2 |
| Canada | 25.4 | 24.6 | 26.2 |
| Australia | 28.3 | 28.2 | 28.4 |
| Hungary | 28.5 | 30.4 | 26.3 |
| New Zealand | 31.3 | 32.2 | 30.3 |
| Mexico | 32.4 | 37.5 | 26.8 |
| United States | 35.3 | 36.6 | 33.9 |

Percentage of population
aged 15 years and over

Percentage of population
aged 15 years and over

FIGURE 5.2

Obesity among adults, 2012 or nearest year. (Organisation for Economic Co-operation and Development, OECD Analysis of Health Survey Data, 2014)

Most disturbing, the U.S. obesity rate has continued to rise for more than forty years into 2014 with no evidence of decline, despite considerable public health attention—a situation that prevails in most other developed countries as well. Half of all adult populations are overweight or obese in thirty Organisation for Economic Co-operation and Development (OECD) countries (with rising rates in each country), with the United States the leader, but with France, Switzerland, Japan, and Korea at the other end of the scale.[3] The World Health Organization (WHO) uses the phrase "global epidemic," because obesity has begun to affect almost every country and region (save for sub-Saharan Africa, a few countries in Southeast Asia, and some scattered ones elsewhere) (figure 5.3). International projections show 2.16 billion overweight and 1.2 billion obese by 2030.[4]

Just as global warming has the rise of $CO_2$ levels and the warming of the globe as its key consensus measure, obesity has one as well, that of the body mass index (BMI), understood to be a marker of excess bodily fat. Although subject to technical debate, BMI remains the benchmark standard.[5] A person whose weight is 10% to 20% above average is considered overweight,

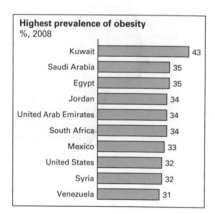
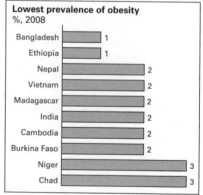

**FIGURE 5.3**

Highest and lowest prevalence of obesity, 2008. (Lucy Hurst, Joshua Grundleger, and Katherine Stewart, Global food security index 2014: An annual measure of the state of global food security. Special report: The burden of obesity: Its relationship with food security, 2014: 4.)

and 20% or more is considered obese. Those who are 100% to 150% above normal are called "morbidly obese." The technical gradients range from a BMI score of 18.5 for underweight, through normal at 18.6-24.9, to 25.0 to 29.9 for overweight, to class I obesity at 30.0 to 34.9, class II at 35.0 to 39.9, and class III at 40 and above. The health risks stemming from obesity begin with "increased" with overweight, are "high" with class I, "very high" with class II, and "extremely high" with class III.[6] Most disturbing, the U.S. prevalence rate has risen steadily despite considerable public health attention, a situation that prevails in most other developed countries as well.

## THE IMPACT OF OBESITY

Save for a few countries and regions, the global impact of obesity is socially pervasive, dangerously unhealthy, economically burdensome, and in many places a source of occupational and medical discrimination. A starting point to grasp those impacts is that of life expectancy. Do the overweight and obese die sooner than others? The answer appears to be *yes*, but with some qualifications. The most thorough study, from the U.S. Centers for Disease Control and Prevention (CDC), found that obesity classes II and III "were both associated with significantly higher all-cause mortality," but not class 1 or being overweight; in fact the latter "was associated with significantly lower all-cause mortality."[7] That phenomenon has been called "the obesity paradox," or it could be said as well that the route from overweight to obesity is a continuum, ranging from the beneficial to the hazardous.

What do the obese die of? Diabetes is the most compelling example. There is a strong correlation between average weight in an adult population and the presence of type 2 diabetes, with a prevalence close to three times higher among the overweight. For a severely obese person that prevalence is ten times higher.[8] Diabetes is just the beginning of an imposing list of other hazards, most with a life-threatening potential: hypertension (with some 78% of cases in men and 65% in women attributed to obesity); metabolic syndromes (abnormal glucose levels, low HDL cholesterol, elevated blood pressure, high triglyceride levels); cardiovascular disease; cancer (at least ten different kinds); gallbladder disease; sleep apnea; and some association with renal failure, blindness, and amputation.

The main hazard for children is that childhood obesity can lead to adult obesity, later type 2 diabetes, and a premature death rate twice that of other children.[9] The economic costs of health care for the adult obese population are estimated to be 5% to 7% of overall U.S. health-care costs, some $100 billion a year.[10] The estimated costs for children are $14 billion.

There is a distinct gap between rich and poor in the adult and child impact. In developed countries, the likelihood of obesity is greater among the poor and less educated than the well-off and better educated. That gap is clearly noticeable in the United States, with those living in our poorest state, Mississippi, having the highest obesity rate. In China, obesity is rising faster among the poor than the rich, a trickle-down feature of its rapid economic growth, allowing greater access to the kinds of food most conducive to obesity. At the same time there is another paradox here: programs in Chile and Mexico aimed at reducing hunger and poverty can carry with them an increase in available food, but often of the wrong kind. In many developing countries families can now have some members who are malnourished and some who are overnourished.[11]

## CAUSES OF OBESITY: INDIVIDUAL AND SOCIAL

There is a large research literature on the causes of obesity, and it can be divided into two categories, one focused on medical and biological causes in individuals, the other on social and cultural causes. At the individual level there is agreement that the rapid increase in obesity since the 1970s occurred much too quickly to be attributed to some striking genetic change. But what caused that sudden change? A good candidate was greatly increased consumption of sugared beverages, with high-fructose calorific sweeteners (HFCS) accounting for 40% of them. An important 2004 study found that "the consumption of HFCS increased 1000% between 1970 and 1990, far exceeding the changes in intake of any other food or food group."[12] This change in sweetener type has implications for obesity, because HFCS is thought to be processed by the body differently than regular glucose, interrupting the signals for the hormones insulin and leptin that signal satiety in the body. A product of treating corn syrup with enzymes, HFCS were introduced in 1957 and refined for industrial production in the late 1960s. In recent years a number of regulations have been adopted to ban HFCS.

## MEDICAL AND BIOLOGICAL PATHWAYS

### Genetics

The effort to understand the biological and medical basis of obesity can be traced back many centuries. Physicians were the first who saw obesity as a threat to health and treated it as a medical malady. The early understanding of it settled on a simple, still lingering, explanation: an excess consumption of calories combined with a deficit of physical activity to burn them off results in obesity. That conclusion was too simple. It seemed to imply that obesity can be avoided by not eating too much or the wrong kind of food and getting regular exercise. Chin Jou described how the history of research unearthed a more complex story: "Weight is clearly far from being within an individual's control. Genetic predispositions, in tandem with the development of food environments that facilitate overeating and built environments require minimal energy expenditure, may help explain why so many Americans are obese today."[13] That summary also succinctly explains global obesity, even with some variations. Yet if it is a complex story to explain the rise of contemporary obesity, it is no less complicated to understand why, once obese, it is exceedingly hard to lose excess weight and keep it off.

Much of the research in recent decades has centered on two important mechanisms of the human anatomy: genetics and the hormone system. Genetics, it is estimated, accounts for some 60% of obesity,[14] but just how that works is still not fully clear. Genes correlating with obesity influence the circulating levels of hormones and the production of the amount and size of adipose (fat storage) tissue. Hormones are produced to signal the brain that sufficient fat storage exists in the body (leptin) or does not exist (ghrelin), influencing appetite and the sensation of satiety. Leptin, made in the adipose cells, signals the body to stop eating and increases the body's metabolic rate. The opposing hormone, ghrelin, is secreted in the stomach and stimulates appetite. The two hormones function through different pathways to regulate eating.[15]

The body's response to and/or production of these hormones is influenced by the obesity genes. One gene in particular, known as the *FTO* gene—"fat mass and obesity-associated gene"—is strongly linked to BMI and a risk of becoming overweight or obese. The significance of genes is demonstrated in weight loss studies, where persons with predisposition appear biochemically to be in a state of starvation after weight loss even with sufficient calories to maintain a healthy weight and lifestyle. This predicts the regain of weight in those achieving weight loss unless radical changes in exercise patterns are undertaken, requiring up to 60 to 90 minutes of daily exercise to maintain lower weight. As a general rule the eating of less food is more effective than exercise for weight loss.

Beyond the *FTO* gene, however, individuals have different genetic responses. A 2011 report said that "recent discoveries in genetics have found that people differ in their perceptions of hunger and satiety on a genetic basis and that predisposed groups of the population may be particularly vulnerable to obesity in 'obesogenic' societies with unlimited access to food."[16] Another study found thirty-two genetic variations associated with obesity.[17] And evidence of familial obesity is strong, as shown in studies of identical twins raised separately.[18]

There has also been some variation among different groups historically. People born before the early 1940s were not at risk of putting on weight if they had the risky variant of FTO. Only subjects born in later years had a greater risk, and the more recently they were born the greater the gene's effect.[19] This suggests that factors in the modern environment influence the expression of the genes, possibly through epigenetic marks that can turn them on and off, affecting gene expression without changing the DNA sequence. These marks are being studied for their role in predetermining obesity, and in particular are linking "early environmental influences [to] epigenetic variation that favors obesity."[20]

## SOCIAL PATHWAYS: THE OBESOGENIC SOCIETY

Yet of all the possible reasons for individual susceptibility to obesity, genetic factors have a dominant place in research, even while the evidence and its scientific meaning remain elusive. As with so many possible genetic determinants of everything human, the relationship between genetics and environmental influences is also difficult to pin down.

Despite efforts to better identify genetic causes of and predispositions to obesity, the main focus of public health analysis has been on environmental factors, those background social influences that together produce an "obesogenic" society. The telltale mark of that kind of society, emerging most clearly in developed countries in the 1970s, was cheaper and more available food. Most notable are processed foods heavy with salt, sugar, and fat; a rise in eating out and expanded restaurant plate and serving sizes; increased consumption of sugared beverages; greater affluence and more women working outside the home (favorable to fast-food restaurants and contributing to a decline in home-cooked meals); massive urbanization and automobile use in developed and developing countries alike; and a decline in physical activity enhanced by commuting by car, sedentary lifestyles (sitting in front of a computer all day and watching TV at night), and a wide range of technological innovations that further reduce the physical movement of the human body (elevators, escalators, remote control devices).

The shift from an industrial society to a service society is also part of that story. Behind those developments lies a plausible biological theory: earlier generations of human beings, often short of food, stored fat when food was available and then—given the nature of agricultural or pastoral societies, entailing considerable physical activity—worked off the fat. Those same evolutionary advantages came to an end with obesogenic societies, whose citizens excessively store fat no longer necessary for survival and do sedentary work that decreases the need for physical labor.[21]

The same new society brought with it the influence of ubiquitous chemicals in the human environment—from fertilizers and pesticides to plastics to fuel—some of which have been shown to interfere with or mimic the hormone actions in the body. These so-called endocrine-disruptors have been shown to adversely influence body weight by interfering with the hormone actions of the body.[22]

If the availability of palatable food increases the probability of obesity, the relationship with other obese people also appears to be a strong determinant of obesity: "a person's chances of becoming obese," an important study found, "increased by 57% . . . if he or she had a friend who became obese. . . . If one spouse became obese, the likelihood that the other spouse would become obese increased by 37%."[23] Closely related to this phenomenon is the widely reported failure of those who are obese to note it in themselves and their children: most everyone they know looks that way. The obesity then comes to seem invisible.

## THE SPREAD OF OBESITY

Obesity has long been understood as a health problem, going back many centuries. The long-standing medical view saw it as a curable condition: simply reduce the intake of food, increase exercise, and, following a doctor's direction, use self-discipline to attain weight loss. But until recently, only in the developed world was obesity seen as a problem. In developing countries, malnutrition was the overwhelming crisis. It was not until the 1970s, however, that national governments and the WHO became interested. By then obesity rates had begun to escalate, and it was understood as well that obesity was an important cause of many medical conditions, something not fully comprehended in the 1940s and 1950s.

Initially, the WHO effort met obstacles. Its initial perception of obesity as a disease only occurring in developed countries meant that obesity fell out of its purview, which was focused primarily on developing countries. Only when presented with evidence that obesity was spreading to the developing countries was the WHO officially able to tackle the problem. But some later opposition surfaced as the WHO undertook careful studies of the medical harms caused by obesity and specified some of the food ingredients that contributed to it. Some of the recommendations presented at a 2003 meeting called by the WHO and the UN Food and Agricultural Organization (FAO) were rejected by 100 ministers of agriculture. They complained about the proposed reduction of sugar intake, calling it a threat to the economies of developing countries. The decision of the American Medical Association to classify obesity as a disease in 2013 aimed to get doctors to take it more seriously. But that move met some resistance from those who saw industry influence at work and an effort to make it mainly a medical and not a public health issue. It has now become a part of both.

Once the increase of obesity in developing countries was noticed and taken seriously, it became clear that obesity was a global phenomenon, with rates of increase in developed countries beginning to be matched in poorer countries. By the end of the first decade of the twenty-first century, international obesity figures rose to astounding heights. In 2014 the estimated total number of overweight and obese individuals was 1.9 billion overweight with 13% obese.[24] Short of unforeseen changes, the total could rise to 2.6 billion overweight and 1.2 billion obese by 2030. The international obesity breakdown for the lower 2030 projection is 22.1 million for established market economies, 21.5 for former socialist economies, 14.6 for Middle Eastern countries, 19.6 for Latin American and Caribbean nations, 4.0 for India, and 4.2 for China. As the authors of this study concluded, "Public health interventions in developing countries are scarce.... Primary prevention of overweight and obesity [is] a more realistic and cost effective alternative for curbing the obesity epidemic."[25] While the obesity prevalence projections for India and China are low, that is not the case with overweight projections: 12.9 million for India and 26.6 million for China.

The peak age for adult obesity is fifty to sixty years in developed countries but is beginning earlier in many developing countries, in the range of forty to fifty.[26] Adult obesity is slightly higher for women in developed countries but much higher in developing countries (where some cultures see it as a sign of affluence). Prevalence figures on overweight children in developed countries show the United States to be the leader, with 39.1% of boys and 36% of girls overweight (2003–2004). The United Kingdom comes in second, with 29% for boys and 29.3% for girls. For developing countries, Bahrain, Russia, and Brazil were high (17%–29% for boys) while China and India (12.9%–14.9%) were lower. Regional differences are common, with France a notable outlier, with a low but still slightly rising prevalence of overweight and obesity. In general, the prevalence is higher in central, eastern, and southern Europe than in western and northern Europe.[27] In the United States, the southern states have much more obesity than northern states, and African Americans and Latinos the highest rates of all. A Danish study of childhood obesity found a pattern similar to other developed countries: an increase from the 1930s until the 1950s, a plateau period between the 1950s and 1960s, and a steep increase thereafter—but with no parallel trends in economic growth.[28]

## DEVISING SOCIAL, POLITICAL, AND BEHAVIORAL STRATEGIES

The doleful reality of my five horsemen is that no one has found a successful consensus strategy to cope with them. Obesity is no exception and has a few notorious features from a policy perspective. One of them is that, once a person is obese, a "success" rate of 10% in losing weight and not regaining it is considered satisfactory.[29] Obese children have a good chance of carrying their obesity into adulthood. It is hard not to be reminded here of the comparable near impossibility of removing $CO_2$ from the atmosphere or replenishing depleted ancient aquifers. The other feature is that there is often strong resistance to interfering with personal behavior. Gentle health education is acceptable, but full-bore intervention using overt social pressure and stigma usually runs into strong opposition from the public health community, and more direct legal and regulatory interventions usually elicit strong industry opposition.

I want now to move on to look at: first, various government strategies being used or proposed to deal with obesity; second, at the role played by industry in advancing its own interests and resisting laws and regulations to curb obesity; and third, at a wide range of efforts aimed at individuals and the obstacles posed by obesogenic societies. If there is any solid axiom about the presence of obesity in a population, it is nicely caught in this statement "The most obvious environmental precondition for a population to develop obesity is sufficient wealth. . . . A degree of economic prosperity is thus an enabler for obesity."[30]

That statement of course recalls to mind the GDP dilemma noted in chapter 2: bringing the poor out of poverty and malnutrition with rising income carries its own hazards, one of which is the likelihood that improved nutrition, which, among other things, entails increased consumption of beef, will facilitate obesity and no less greatly enhance the need for water to raise the cattle (see chap. 3). Not incidentally, rising incomes most often bring with them inequality, and the greater the inequality, the higher the prevalence of obesity. Coping with the pervasiveness of obesity, the economic inequalities of its manifestations, and the number of institutions, public and private, that can influence its future, suggests the need for multiple strategies.

It is impossible here to do justice to the diversity of domestic and international efforts now under way, but a fine study by the Trust for America's Health and supported by the Robert Wood Johnson Foundation—*F as in Fat: How Obesity Threatens America's Future*—provides a good sampler. Here are some of the U.S. initiatives summarized in the chapter on policy approaches.[31]

## State Obesity-related Legislation

- Legislation for healthy schools: national school meal standards, increasing fruit and vegetable consumption at school, reduction of "competitive food" (those items high in calories, fat, sugar, and sodium), and farm-to-school programs
- Physical education, health education, and physical activity programs in schools
- BMI screening and surveillance in schools
- Menu labeling (restaurant menus and menu boards)
- Redesigning streets for all users (walking, cycling, improved urban transit)

## Federal Policies and Programs

- Let's Move (a program launched by Michelle Obama and focused on childhood obesity initiatives)
- Implementation of the Affordable Care Act (a prevention and public health fund, community transformation grants, an interagency national prevention council, coverage of preventive services, menu labeling)
- Strategic realignment of chronic disease programs at the CDC
- Healthy Food Financing Initiative (subsidizing health foods in poor communities)
- National physical activity plan
- Obesity demonstration program
- Support of industry wellness programs

If there are many state and federal programs, there are also a burgeoning number of business efforts, ranging from company wellness efforts offering incentives for employee health improvement and weight-reduction programs—and some instituting penalties for failures to improve health (e.g., higher employee share of health-care insurance). Michelle Obama is the honorary chair of the Partnership for a Healthier America (PHA), which works to bring about changes in the private sector, partners with supermarket chains to bring down the price of healthy foods, and advocates for reducing sodium in restaurants and marketing healthier food. These are hardly dramatic moves, and their growth and extent has been slow. Larry Soler, executive officer of PHA has said that "none of these companies are going to move forward unless they see it's going to be good for their bottom line."[32] Companies that work with incentives to change employee behavior have the diplomatic task of gaining employee support. For such programs to be effective they must find plausible but efficacious ways of pushing their employees in the right direction. They may have to come close to mild coercion to do so, while not overstepping a line that has yet to be determined. A 2013 meta-analysis of industry wellness programs, however, showed that their impact on health (including obesity) was marginal and did little to lower health-care costs.[33]

## Taxing Unhealthy Food and Beverages

Prevention and most government education and programs to combat childhood obesity are relatively uncontroversial, meeting little public or political resistance. That is not so with two other proposed means of obesity management: taxation of unhealthy food and beverages and subsidies for more healthy foods. A public opinion survey with responses from seventy-five countries found the 68% of the respondents favored government regulation of sugar-sweetened beverages. That figure was consistent among the countries, with one exception, the United States. Only 59% of Americans favored taxation, while 84% from other countries did so.[34] New York's Mayor Bloomberg faced full-bore opposition to his (eventually defeated) 2011 effort to tax sugared beverages, and his 2012 attempt to limit serving sizes was met by a massive industry-led campaign and was rejected by 60% of the public in New York City. A state court eventually ruled the proposal illegal. The industry's rhetorical weapon was that of "freedom of choice," a politically effective strategy in individualistic America.

Europeans on the whole, one comprehensive survey found, also appear to be reluctant about both taxation and food subsidies, but mainly as a practical matter, on the grounds that implementing such policies is difficult. But the survey that revealed that proclivity also found some distinct ideological groupings, with representatives of the farming and food-processing industries most opposed and those from the union sector, nutrition experts, and the public health community supportive.[35] Not surprisingly, Finland, Sweden, and Denmark—countries with a long history of using fiscal measures as a means of achieving nutritional goals (price manipulation, retail regulations, nutritional labeling, and public education focused on individuals)—are obviously open to government interventions.

A 2008 economic study by two prominent economists opened with the argument that obesity should be understood as essentially an economic problem, because it reflects two choices people make, calories consumed and calories expended, and because it has social costs.[36] Over time food has become less expensive, while exercise has become more expensive, thus explaining why obesity rises among the poorer groups but decreases among the more affluent. Once obesity is recognized as a public health matter, the study authors argue, government intervention is then commonly embraced, and they say that is acceptable. But such interventions should be judged on the basis of "all the relevant costs and benefits." Altogether missing is any reference to the moral and legal acceptability of government interventions—my right to control my body and what I eat—common in much public discussion.

I have cited this study because it is one of the few that directly challenges some obesogenic remedies much touted but little analyzed. Many have neither strong public support nor a good track record for overcoming legal obstacles (save in a few countries)—and the former will have to be improved if the latter are to be successful. By comparison, one can only look upon Japan with wonder. Already blessed with a good diet (heavy on rice, fish, and vegetables) and the lowest rate of obesity in the developed world, Japan passed a law in 2008 that was designed to combat the metabolic obesity syndrome. It requires the overweight to go to dieting classes. Failure to do so will result in fines for their employers, and in cases in which companies have high percentages of obese employees, the employees themselves are also fined.[37]

## INDUSTRY POWER AND MARKET IDEOLOGY

But Japan is not alone in bringing government regulation to bear on obesity. In the fall of 2011 Hungary passed what was called a "fat tax," specifying taxes on foods with high fat, sugar, and salt content and increased tariffs on sugar-sweetened beverages and alcohol. Finland and Denmark have moved down the same road. Romania considered a similar tax in 2011, but its parliament rejected it, citing rising food costs and the possibility that low-income people would turn to even cheaper food, leading to further deterioration of their diets.[38] Perhaps most arresting is the reality that despite a widespread recognition at the government and public health level of rising obesity rates everywhere in Europe (even with considerable variation), it has been difficult to make use of government regulation and taxation. The posting of calorie counts on menus and menu boards is one of the comparatively few government interventions, save for school programs, that have proved acceptable.

The reason is not hard to find: the enormous political and economic power of industry. It simply resists almost any and all government initiatives to deal with obesity by regulation and taxation. The major companies selling food and beverages are international, highly profitable, and politically well connected. Over the years industry has been highly successful in defending its turf, and nothing so far has much altered its power. There are at least three vantage points for assessing that power. One of them is a common view in the public health community: it is an all-round harmful force in efforts to deal with obesity; there is nothing good to say for it. Another is more ambivalent, particularly groups working to forge government–private sector coalitions. And then there is one more vantage point, one that does not exonerate industry for its bad behavior, but works to understand how it draws upon the support of many interests in the production and delivery of food and beverages, which taken together, play an important economic role.

### Industry's Heavy Hand

Kelly Brownell, former director of the Yale Rudd Center for Food Policy and Obesity, is one of industry's sharpest critics. He has compared the tactics of the food and beverage industries with those employed by the tobacco industry, Big Tobacco.[39] Those tactics consisted of dismissing the science behind a concern with obesity, paying scientists to produce pro-industry research, sowing doubt among the public about the harm of obesity, intense marketing of products to children and adolescents, creating so-called safer products, and promising to regulate itself.

Michael Moss, in his 2013 book *Salt Sugar Fat*, lays out the indictment against industry in great detail. Two companies, one American, the other British, are leaders in pushing the products that fuel the obesity epidemic. The American company, Cargill, was founded in the late nineteenth century and, in its own words, "sells a wide range of food and agricultural products and services." The British company, Tate & Lyle, also dating back to the late 1800s, sells "high quality ingredients and solutions to the food, beverage and other industries." Both are global in scope. As Moss puts it, "Together, the two had the salt, which was processed in dozens of ways to maximize the jolt that taste buds would feel with the very first bites; they had the fats, which delivered the biggest load of calories and worked more subtly in inducing people to overeat; and they had the sugar, whose raw power in exciting the brain made it perhaps the most formidable ingredient of all, dictating the formulations of products from one side of the grocery store to the other."[40] Added to that research agenda was a business strategy, neatly summed up by a chemist at Pillsbury, "taste, convenience, cost." A Kraft executive cited one result of that strategy, "The ubiquity of inexpensive, good tasting, super-sized, energy-dense foods." Moreover, given the number of businesses in the food industry, a former Nabisco executive said, "To be competitive we've got to add fat."

The need for research for technological innovations for my five horsemen, that is, research to reduce the problem, is turned on its head with food research, often aiming to find ways of luring us to unhealthy foods. The use of brain scans to study neurological reactions to some foods, notably sugar, is one strategy. Advertising research and studies to improve grocery store sales by determining the best location for product sales are common practices in the food industry. Advertising research on selling unhealthy food to children is another insidious effort.

In her provocative book *What's Wrong with Fat?*, Abigail Saguy points out the role of psychological and social science research in understanding which strategies are most effective in meeting efforts to reduce obesity. The promotion of "personal responsibility" to dodge the effects of social obesogenic theories about the cause of obesity is one tactic, while another closely related one is the emphasis on the rights of consumers to choose what they want to eat and drink. The Center for Consumer Freedom, with its $9 million budget, is organized to underscore that right.[41] The use of "free choice" arguments is nicely tuned to a value running through American culture, employed by the right and the left for their own purposes and putting a halo of Americanism on commercial values. The dieting industry, often with support from the food and beverage industry, in effect plays both sides of the street. It sells unhealthy fattening products and programs to counteract weight gain.

The contention that obesity is a deadly disease is used to induce the Food and Drug Administration to speed up review of weight-loss drugs while tolerating the risks of using them.[42] Industry also promotes the view that obesity is a medical not societal problem, promoting the view that it is a public health crisis brought on by irresponsible individuals and parental failure.[43]

Marion Nestle has been a tireless critic of the food industry, documenting its efforts over the years to manipulate government policy, the education of nutrition counselors, and national nutrition guidelines. Her book, *Food Politics: How the Food Industry Influences Nutrition and Health*, lays out the indictment in disturbing detail. She begins with an overview of efforts to influence government dietary guidelines, only one use of well-financed lobbyists; then moves on to discuss the industry's involvement with the Food and Drug Administration and the Department of Agriculture, observing the "revolving door" that has seen the movement back and forth of legislators and government officials going into industry and vice versa. Nestle also points out the industry's contributions to election campaigns, using PACS and heavily biased toward Republicans, which aimed to support candidates favorable to the food and agricultural industry. According to Nestle, Senator Richard G. Lugar, the Republican chair of the Committee on Agriculture, Nutrition, and Forestry, received contributions from forty food and agriculture associations in 1997–1998.

The efforts of the industry to influence government is just one of many arrows in a large quiver. Nutrition experts, whose job is to teach people how to eat well and healthily, make a rich target, teaching classes in nutrition, running meetings and large conferences, publishing journals, supporting professional societies, and carrying out research. The food industry has gone after every one of them in overt and quiet ways. Nutritionists need information from the food industry, and it is hardly surprising that there is a revolving door there as well; they need each other. The industry does not need to capture but only to "influence" them. A striking example of how the borderline between control and influence can be crossed came when the University of California, Berkeley, set out to find an industry partner, with the Department of Plant and Microbiology aiming to team up with a corporation in a mutually agreeable private-public partnership. It chose Novartis, a large Swiss agricultural and drug company, in a deal worth some $50 million to the department by way of $25 million for laboratory support and $25 million for research support, an amount equal to the department's annual budget.[44] While the contract with Novartis expired and was not renewed, by 1998 there were 4,800 such university-industry contracts. Another chapter in Nestle's book takes on the beverage industry and its hardly less penetrating efforts with soft drinks, particularly in school systems, and with advertisements in the 1990s and early 2000s. In that case, however, parental and school pressure has brought some needed reform in recent years, still with some way to go. By 2015, however, while little progress had been made with taxation and regulatory efforts, a sharp decline in sugared soft drinks was obvious. It was described as the "single largest change in the American diet in the last decade," particularly notable with children and perhaps a major contributor to evidence of a leveling off of child obesity. The public, it appears, had got the message, even if not ready to accept taxation (save for California).[45]

Marion Nestle's book is 500 pages long, and I have only scratched the surface of its revelations. I have not undertaken a systematic survey of industry lobbying and influences that can be found with the other four of my five horsemen. My impression is that the efforts of the pharmaceutical industry to insinuate itself into the practice of medicine is at least as systematic as with food in its impact on drug research and prescribing, particularly since the leading chronic illnesses are the most lucrative sources of industry income. There is nothing to compare with either of them in the water arena, and similar efforts to slow or stop global-warming reform does not so directly impact individual lives here and now, but will have obvious long-range repercussions.

## Industry and Government Partnerships

The efforts of Michelle Obama and her Let's Move initiative with industry has raised questions about attempts to draw the food and beverage companies into an alliance to combat obesity. The success of that effort after its 2010 start has been doubtful. "Looking back on it," Marion Nestle has been quoted as saying, "It is enough to make you weep. So little has been achieved."[46] As Bridget Huber, who wrote an insightful article trying to diagnose the situation, has put the issue, there is a basic question hanging over the Obama effort: "Should those who seek to address the obesity crisis treat food companies as collaborators or adversaries?"[47] Skepticism about the effort is strong, seeing conciliatory efforts as "a trap," as Brownell has put it, with industry taking "baby steps" but hanging tough in resistance to regulation and taxes—and leaving the Obama administration looking weak and too willing to settle for meager advances.

Marion Nestle does something fairly rare among reform advocates: she details the human price of success. A radical change in American eating and drinking behavior as called for by public health experts would hurt many people. "What might seem a virtue to some people," she has written, "might seem a vice to others—hence, ethical and policy dilemmas."[48] The "others" she is referring to are "cattle ranchers, meat packers, dairy producers, or milk bottlers; oil seed growers, processors or transporters; grain producers (for feeding of cattle); makers of soft drinks, candy bars, and snack foods . . . and, eventually, drug and health care industries likely to lose business if people stay healthy longer." And I have quoted only part of her list. It is intimidating and as good a sign as any of the way the health crisis that is obesity is deeply imbedded in American life, not to mention all countries with an obesity problem. I particularly appreciate her candor in pointing out the extent to which the changes she wants will affect the work and income of those in the affected fields. Such candor has often been missing with global warming, where general statements are made about the social and economic changes that genuine reform will require, but rarely in any detailed ways. And sometimes, as with efforts to stop fracking, the opponents can sound like the global-warming deniers and minimizers who dispute the evidence of possible job loss as little more than concocted fiction.

## MEDICAL AND SOCIAL POSSIBILITIES

### Bariatric Surgery and Medications

There is a range of bariatric surgery procedures, each with a different degree of advantage and disadvantage and no certainty of benefit. None of the procedures are without some risk, in great part because of the nature of the procedures themselves but also because the severely obese are likely to have other medical problems, including the strain on a body as a result of carrying excessive weight and the possibility of postsurgical and nutritional complications.[49] A British study found bariatric surgery generally cost-effective, with some major and minor adverse events depending on the procedure used, and some evidence of long-term gains in keeping lost weight off.[50] While there have for years been efforts to find a pharmaceutical treatment for obesity, there have been some disasters along the way and many failed efforts. I return to the topic of obesity and technology in chapters 7 and 9.

As those two examples suggest, technological hopes and approaches for great breakthroughs have been scant, which is perhaps all the more remarkable given the likelihood that an efficacious and cost-effective technology could find an enormous market. Much more common have been more conventional modes of behavior change, most notably physician counseling and cognitive and behavior therapy. While physicians have a mixed reputation for alerting patients to weight problems, there have been some notable efforts to deal with it in a clinical setting. One study compared a "remote" weight loss strategy (phone, website, and e-mail) with an "in person" strategy with group and individual sessions.[51] It showed a significant advantage with the in person method. Another study, compared two-year counseling efforts, one of brief lifetime counseling, the other long term (called "enhanced counseling").[52] The latter turned out to be superior.

## Public Health Solutions

While I believe there is far more room for working to change individual behavior—and in particular making use of social pressure for prevention and for advising those just beginning to have a weight problem (see chap. 7)—the main line of the public attack on obesity has been to focus on the food environment, the obesogenic culture. Rodney Lyn and his colleagues, making use of public health strategies from other public health areas, look at: the information environment, access and opportunity, economic factors, the legal and regulatory environment, and the social environment.[53] The information environment encompasses *marketing*— making use of educational campaigns and combating misuse of marketing to promote unhealthy foods and beverages; *media campaigns* to advertise improved eating behavior and exercise; and *warning labels, ingredient disclosure, and labeling* aimed at providing more detailed information for making good choices. Access and opportunity focuses on better access to healthy foods and exercise activities and on *the community and school environment*. Economic factors encompass *taxes, behavior, and public health*, as well as *the role of agriculture and farming,* the latter working to deal with potential conflict between public health needs and the interests of farmers; the legal and regulatory environment includes the many *laws, regulations, and litigation* used (if not very successfully) in public health campaigns of one kind or another. The social environment comprises social norms and expectations and particularly receptiveness to collective action, with a change in any or all of these factors bringing about change in each of the other categories.

Two features of all such efforts are worth singling out for attention, one of them using cost-benefit analysis to determine the balance between the costs of implementing policy strategies and their benefits (or lack thereof). The other is the assessment of public support of anti-obesity policies as a guide to their political acceptability. A most useful analysis of cost-benefit research, centered on an Australian study, was sketched in 2011 by Steven A. Gortmaker, Some eight of twenty interventions were found to be health improving and cost saving, while the remainder ranged from likely to meet that standard to those with inconclusive or inadequate evidence.[54] Among the leaders were family-based programs for obese children, diet and exercise, Weight Watchers, and reduction of junk food and beverage advertising directed at children. Of unproven value were after-school activity programs and front-of-pack nutrition labeling (a British initiative using colored symbols for relative healthiness). A general finding was that policy approaches show greater cost-effectiveness than health promotion or clinical activities. Given that judgment, it was perhaps not surprising that the study's proposal for policy strategies provided at best a lukewarm use of efforts to reach individuals, urging only that parents and caretakers of children should "act as role models" and that individuals "need to make healthy food and activity choices."[55]

Two important parallel efforts in the United States and the United Kingdom to sort out public opinion on the causes of obesity and acceptable policies strategies provide some useful insights. The American study employed the concept of metaphors to understand public attitudes.[56] By "metaphors" the researchers mean the way an issue of public importance and debate is articulated. Metaphors, they say, are similar to analogies but not quite the same, being "*partial* comparisons highlighting certain features of a newly identified matter of concern."[57] Obesity characterized as an "epidemic" is the iconic metaphor, one that brought obesity to wider public knowledge and intensified government activity to combat it. In the early 1980s only 2% to 3% of the public considered obesity to be an important health problem, but public opinion changed dramatically in the wake of the study, fueled by increased media attention. Some 90% of Americans now believe that most of us are overweight and consider it a major issue. Reminiscent of the way Abigail Saguy analyzed obesity discourse, the Barry study identifies some metaphors: obesity as sinful behavior, as a form of disability, and as a consequence of choices outside the control of individual (addiction to some forms of commercialized food or distortions that advance the food industry's commercial goals).

## Beliefs About the Causes of Obesity

Most important, individual beliefs about the causes of obesity affect the choice of particular public policies. The researchers distinguish three forms of policies: *redistributive* (e.g., using tax money and government funds to support desirable initiatives), *compensatory* (e.g., food warning labels, prohibition of advertising unhealthy food on media watched by children), and *price raising* (e.g., a grocery surcharge for high-sugar, high-fat foods, a tax on junk food). The obesity metaphors, the study found, had a strong effect on the respondents' beliefs about the causes of obesity and those in turn influenced the policies they would support. In general, *redistributive* (tax-increasing) policies had mixed support, ranging from 68% for healthy food government subsidies for schools to 37% for employer-paid exercise time or gym memberships. *Compensatory* policies ran the gamut, with 63% agreeing on the need for warning labels on food with high sugar or fat content that also indicate those substances may be addictive, to 33.1% supporting antidiscrimination laws to protect the overweight. *Price raising* gained little support, with only 29% supporting grocer surcharges for unhealthy foods, and the lowest support of all, at 24.6%, went to requiring health insurers to charge higher premiums for beneficiaries who are overweight or fail to exercise regularly.

The British study begins by citing what they call "one of the few non-U.S. studies" of public opinion on obesity, one carried out in Germany.[58] The German study found the highest support for policies aimed at reducing childhood obesity and strong support for public health campaigns, but less so for regulation. Also cited was a European study on prevention showing a strong support for parent education and advertising restrictions but not for taxation of unhealthy foods.[59] The British study set out with two hypotheses: (1) situational (structural) explanations of obesity will lead to support of policies aimed at prevalence, and (2) attributing causes of obesity to individuals will reduce support for prevalence-reducing policies (notably government interventions). Those who attribute obesity to individual behavior are likely to identify obesity as caused by the high cost of healthy food or ineffective diets but not to genetics or a lack of willpower.

That belief in turn correlates with a favorable attitude on policies to limit junk food advertising aimed at children and "snack taxes" on unhealthy foods, using the money from the latter to pay for National Health Service costs incurred by obesity. There was little support for giving the obese the same rights accorded the disabled or for a more active government role in protecting the overweight from discrimination. The study concluded that food manufacturers "have been successful in promoting a discourse of personal responsibility" with respect to obesity.[60] Policy interventions have more support from those who believe obesity is beyond public control, but even those who attribute it to individuals support government interventions to promote prevention programs for children. Overall, the authors concluded with the argument that "policies designed around personal responsibility for weight have not been effective; obesity rates have not declined."[61]

Public opinion polls show that the public understands that obesity is a serious prooblem, but they do not want to do much about it. There are no social movements against obesity, no grassroots uprisings, no large marches of concerned citizens. Thin media coverage offers no stimulus for public interest, protests, or publicity of backstage quarrels among experts. Whether that absence explains why there is little action beyond government and foundation reports, and other scattered initiatives, it is hard to think of any comparable major health problem in the United States or elsewhere that incites so little emotion or movement. Only when some health agency or government tries to impose taxes does it get a public reaction, but usually one to stymie any move in that direction. Oddly enough, perhaps, there is considerable media attention paid to food: healthy foods, good recipes, celebrity national and local chefs and outstanding restaurants, and farmer's market—often touching on obesity, but seeming to make little difference to the continuing prevalence of obesity. No doubt with an eye on class and income, my own local upscale restaurants serve well-presented, small servings of expensive menu items. Our low-cost restaurants have no aesthetic aspirations about the presentation of their much larger size servings; and doggie bags for leftovers are routinely offered.

# II

# EXAMINING THE PATHWAYS THROUGH THE THICKETS

———

# 6

## ALWAYS MORE PEOPLE
## AND EVER MORE ELDERLY

———

### Caring and Paying

While exceptions can be found, many of the most prominent writers and researchers on global warming, food, and water put a reduction of population growth rates high on their list of priorities. And why not? More people mean more consumption of everything needed for our collective lives and survival, some of which can never be recaptured or replaced. More people will add to the environmental pollution already being wrought on our planet, further threatening those already alive and jeopardizing generations to come. More children born to mothers hard-pressed with too many children already harms everyone—mothers, children, and fathers.

I propose an alternative way of thinking about population growth, adding to the mainline story a neglected or too readily minimized feature, that of the parallel growth of aging in rich and poor countries alike. That important difference throws a provocative light as well on the rising obesity and chronic illness in developing countries that is coming to be part of that aging process. In turn, that combination—the coming size of populations and the relatively new important additions of aging and new health risks to the elderly—opens the door for a fresh examination of the growth of gross domestic product (GDP). As matters now stand, we might recall, there is an argument between those who think of strong GDP growth as necessary for national and global welfare, and those who believe it to be a significant cause of our global environmental and health problems. In many ways there is considerable overlap among population growth, aging societies, and GDP growth.

## POPULATION

In 2011 the world saw population growth move past the 7 billion mark. In 1960 there were 3 billion people, in 1974 there were 4 billion, and in 1987 there were 5 billion. The number of births in 2013 is estimated to be 86 million.[1] In the late 1960s the annual population global growth rate was 2% but has now declined to 1.14% and is expected to decline still further in the future, to less than 1% by 2020 and less than 0.5% by 2050 (figure 6.1). Even with low birthrates, populations will continue to grow for a time. The United Nations Fund for Population Activities, revising upward earlier global figures, projected 9,725 million people in 2015 and 11,213 in 2100—but with a 2.0 total fertility rate (TFR), with most countries at 1.8.

Asia has 60% of the world's population, 3.2 billion, and is growing at an annual rate of 1.03%, and Africa has 15.5%, some 1.1 billion, but has the fastest annual growth rate at 2.46%. At the other extreme is Europe, with 742 million and 7% of the world's population, and a projected *decrease* of 4% by 2050; and North America with 355 million, some 5% of the global total, and a projected increase of 26% by 2050. China, with 1.385 billion people and India with 1.252 billion have the largest populations, with the United States in third place with 320 million. Projections can, of course, differ, with higher or lower numbers: the future has its own logic, best discerned in hindsight.

Thomas Robert Malthus (1776–1834) was the first to bring population growth to public attention. His *Essay on the Principle of Population*, which went through many editions and emendations, was both influential and controversial. His leading principle was that "population, when unchecked, increases in a geometric ratio," while "subsistence only increases in an arithmetical ratio." As the resources for survival increase, population increases as well, and this combination in turn stimulates increased productivity. But productivity cannot indefinitely keep up with population growth. At some point, Malthus argued, it must stop and be limited. While in general this argument has been rejected, in great part because of technological improvements in productivity, renewed echoes of it can still be found in current discussions of high birthrates in the poorest of nations, notably in sub-Saharan Africa.[2]

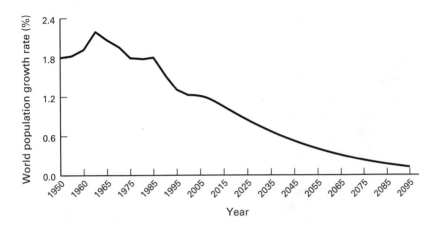

**FIGURE 6.1**

World population growth rate (%). (World Population Clock: Seven Billion People (2013) Worldometers, www.worldometers.info/world population, accessed 10/25/15)

If a shortage of food does not necessarily lead to lower birthrates, there are two variables that have historically been most determinative: a decline in death rates and a rise in income.[3] That fact does not mean that present birthrates are so rigidly determined. Many other causes influence contemporary birthrates, but the original historical momentum, going back some centuries, was a decline in mortality, followed by an increase in income and a rising GDP.

## Stages of Population Growth

Historical demography has made use of a transition model. Stage 1, the premodern phase saw a balance between birthrates and death rates, with a population growth rate that could take 1,000 to 5,000 years to double. Average life expectancy was about thirty years, with an exceedingly high infant mortality contributing heavily to that low average; but even those who survived childhood would be lucky to make it to age fifty. Stage 2 saw a decline in death rates in developed countries (particularly for infants) that began at the end of the eighteenth century but with birthrates remaining high. Poor public health in the premodern era gave way to nineteenth-century improvements in sanitation and clean water supplies, aided by better and more reliably produced food, helping to account for the mortality decline. In 1880 there were only 1 billion people on earth. The numbers would soon start rising sharply.

Stage 3, the industrial era of modern societies (roughly 1870 on), saw a steady decline in death rates and a parallel decline in birthrates. Stage 4, the so-called postindustrial era, our own, has seen a continuation of that decline, with the greatest drop after World War II, and particularly since 1970. The decline of agriculture and the rise of urbanization during stage 3 meant a sharp drop in the need for large families to keep traditional family structures and domestic economies going and a growing need for more education for the young for life in a postagricultural, industrial society. Old age care gradually moved from the family to the state, particularly in northern Europe, although southern Europe saw a continuation of older traditions of family care.

As death rates dropped and family life changed so also did the role of women, greatly stimulated by improved education, a shift that got under way in the late nineteenth century and accelerated after World War II. The greatly improved contraceptive technologies of the twentieth century helped to accelerate the trend to lower global birthrates, already well under way before reliable contraceptives emerged. "The eventual family and sexual mores of an industrial society depended upon controlled fertility," the demographer John C. Caldwell has noted, "but there was no widespread demand for this until mortality, especially infant and child mortality, fell."[4]

If a decline in death rates was the initial driver of a decline in birthrates, industrialization, accompanied by urbanization, was its sustaining engine. Industrialization brought with it a rapid growth in income and affluence. Per capita global income in 1820 was less than double its level in 1520, but had tripled just prior to World War I and by a multiple of fifteen by the end of the twentieth century.[5] As incomes rose, birthrates declined (with the exception of the baby boom era). With affluence and intensified industrialization came the secularization of much of life in the twentieth century, accelerating the decline. What Caldwell has called "agricultural morality," which required high birthrates, passed away, and in its place came the advent of high divorce rates, increased cohabitation, later marriages, and a large proportion of women in the workplace.[6] Beginning in the 1870s or thereabout, fertility gradually declined.

Even though the decline in mortality might be characterized as a macro-shift, of universal importance and influence (along with, as we will see later, affluence), it is also true that the fertility transitions in various countries are both similar and diverse. In the United States, fertility transitions, though obviously influenced by declining mortality, have taken a slightly different route than in western European countries, and are only just now declining, with a total fertility replacement rate (TFR) level of 1.8 children in 2014.

After fertility began to decline around 1870, it steadily dropped through the first half of the twentieth century. By the 1930s, and particularly in 1933 (my wife's birth year), the U.S. birthrate was 1.8 children per woman (approximately what it is now for white women), below, though not strikingly so, the 2.1 replacement level. By the early 1900s, that decline was picking up speed, and in the United States some 25% of women born in 1909 had no children at all (compared with approximately 20% now). The 1930s were the prewar nadir in developed countries—that era when those born just before and after the turn of the century came into their childbearing years. My father's Catholic family in that era had eleven children, nine of whom married. They procreated a total of nine children. Economics trumped theology.

## The Pre–World War II Specter of Population Decline

It did not take long for some widespread anxieties about the dropping birthrates to appear. France has long been noted for its concern about population decline and remains an interesting bellwether country to this day. It was the first country in Europe to take declining birthrates seriously, if for reasons that now seem distant and antique—military might. Wars required large numbers of soldiers; they were the key to victory. In the 1870s, in the aftermath of the French defeat in the 1870–1871 Franco-Prussian War, as Michael S. Teitelbaum and Jay M. Winter put it, "the population question [became] a permanent part of the rhetoric and, on occasion, the substance of French politics."[7] The military focus remained strong for decades and particularly in the years just prior to World War II when Germany was again rising to power. Remarkably enough, despite the hand-wringing, the French birthrates were still high by contemporary European standards. Not until 1975 did the French birthrate drop below the 2.1 replacement rate (and not by much); for many decades after 1870 French TFRs were in the range of 2.4 to 2.8.

Even so, the rhetoric of the anxiety was remarkably similar to that voiced by various American conservatives in recent years in light of much lower birthrates. A Jesuit priest in the 1870s saw the decline as God's punishment for the sin of contraception (which was hardly very effective at that time). Later figures spoke of low natality as a "disease," an "increase in feebleness," an instance of "how a nation dies," a "refusal of life," "demographic bankruptcy," and "biological degeneracy."[8] The more common phrase these days, bemoaning much lower birthrates, is that of national "suicide," but the flavor is identical. Ironically, it is family-friendly France—condemned by many religious conservatives in recent years for its flagrant and official secularity—that has become the one of few developed countries on the Continent to bring its birthrate back to over 2.0 in 2010, although that number had fallen back to 1.99 by 2013.

## BABY BOOM AND BABY BUST

The long decline in the American birthrate, beginning in the late nineteenth century and peaking during the years of the Great Depression, ended with a sudden upswing of births in 1947, dramatically so in the United States and Canada, but strong in other developed countries as well. The U.S. birthrate almost doubled between the 1930s and the 1950s. In Europe during that period it rose by 13% in the United Kingdom, 17% in Italy, 28% in West Germany, and 35% in the Netherlands.[9] Except for the economic security of that period, no theory has fully managed to explain the baby boom period (1946–1964). It was a surprise in its coming, anticipated by neither governments nor demographic experts, and no less surprising in its rapid demise. As the historian Tony Judt noted, "Now—even before post-war growth had translated into secure employment and a consumer economy—the coincidence of peace, security and a measure of state encouragement sufficed to achieve what no amount of pro-natal propaganda before 1940 had been able to bring about."[10] If economic security played a large role, it does not explain the increase in earlier and near-universal marriage and a larger portion of women having two children. Families of three to four increased, but the number of those having five or more actually declined.[11]

If the primary source of fertility decline beginning in the 1870s was a combination of declining mortality and rising affluence, the baby boom era was an anomaly. Once it was over, birthrates resumed their downward course. But why did the baby boom era come to an end? Some leading economists looked for an answer to that question. The economist Gary Becker offers a different account. He contends that time itself has a monetary value, and in the postwar years, rising education levels and income growth meant that time taken out to raise children had an economic cost. That time could otherwise be used to increase earnings.[12] A number of other factors were converging as well. The economy in the 1950s and 1960s in the United States was also becoming more complex, requiring more workers with more education. Women's education was likewise on the rise, not only opening the way for more and better work outside the household but also stimulated by the rise of feminism. Meanwhile, there was a sharp increase during that period in consumption, more people living more affluent lives and buying a whole host of goods, from refrigerators to TV sets to increased air travel. Houses were more plentiful and growing in size; suburbs, with accompanying large shopping malls, were expanding; and two-car families were on the rise.

The demographer John Caldwell adds another twist: "The basic story of the collapse of the baby boom was that women were entering the workforce in larger numbers, that the economy was capable of absorbing them, and that preventing births by contraception, sterilization, or abortion had become easier. This situation was to be one of some permanence."[13] On that last point, by the mid-1950s contraception was almost universal, close to 80% of married couples used it, with the well educated leading the way. If Catholics lagged a bit in the 1950s, they have now caught up.

Even as the baby boom in the United States and other developed countries was well under way in the 1950s, a parallel movement was emerging at the international level. Economic development of poor countries, it was argued, could not take place unless their high birthrates were brought down. Women in those countries often averaged six or seven children. A massive campaign was organized to develop family planning and population limitation programs for poorer countries. It was supported by most national governments in developed countries, the United Nations through its Fund for Population Activities (UNFPA), the World Health Organization, a wide range of private foundations (the Ford and Rockefeller foundations), and nonprofit research organizations (e.g., the Population Council, where I worked for a time). I was hired for a year (1968–1969) to look at the ethical and social problems of trying to lower birthrates and later directed a five-year project for the UNFPA on the same subject, with the focus on developing countries. With the publication in 1968 of Paul Ehrlich's *The Population Bomb*, which sold more than a million copies, population control was riding a large professional and popular wave.[14]

The simple (some would now say simple-minded) premise of all those institutions was (1) that once they understood it was possible, women would choose to have fewer children and (2) that the key to making that possible was education in family planning and available and subsidized contraceptives. In short, the combination of education and technology would do the necessary work.[15] It was, in essence, a technological fix.

## A DECISIVE SHIFT AT THE UNITED NATIONS

Slowly but surely, however, the disappointing fact that birthrates remained high in many countries that implemented that kind of policy led to a shift in thinking. A third intergovernmental world population conference in Cairo in 1994 signaled that change. The first component of it was to move away from a narrow focus on population size and birthrates in favor of a focus on economic growth and sustainable development. The second was to "advocate programs serving goals that people would value for their own good: education, especially of girls; gender equity and equality; reduction of infant, child, and maternal mortality; and universal access to broadly construed reproductive health services."[16]

Although it was not put that way, the Cairo conference was blessing just those social, medical, and economic ingredients that had been an important part of the long-term decline of birthrates in the developed countries since 1870 (lowering infant and maternal mortality) and in the post–baby boom era (particularly gender equity). By the 1990s birthrates were declining rapidly in most developing countries, the fruit mainly of growing economic progress, women's education, and effective family planning. If victory was not declared for the poorest of the poor countries, the emphasis on global population limitation began to decline (at least outside environmentalism circles).

Paul Demeny and Geoffrey McNicoll have nicely summarized the great change in birthrates that emerged by 1970 in the post–baby boom years, when birthrates—initially and mistakenly expected to at least flatten in the aftermath of the baby boom—continued to plunge: "Modernization reduces infant and child mortality, hence necessitated lower fertility in order not to exceed a given family size; made the upbringing of children more costly and diminished children's economic contributions to the family; increased economic opportunities for women."[17]

If the Demeny-McNicoll quote accurately captures the broad demographic patterns, some of the details are no less revealing. One set of them is that of the general trend in population decline: a later age for first marriages, a later age for the birth of a first child, falling birthrates now below replacement levels in many countries, and an increasing old-age dependency ratio of young to old.[18] Three different patterns mark the European countries: those with relatively high birthrates (1.7–1.8), though still below replacement (save for France), most of them in northern Europe (Denmark, Finland, Sweden, and the Netherlands); a number of countries just below them (1.5–1.6 in Germany, Austria, and the United Kingdom); and a number of countries, called the lowest low (under 1.4 and including Italy, Spain, Greece, and Poland). Japan falls into that lowest-low group. Though the line is not all that sharp, the main fault line is between northern and southern Europe.

## COUNTRIES WITH THE LOWEST-LOW BIRTHRATES

While all countries with birthrates of 1.8 or lower face the threat of depopulation, much attention has been focused on those below 1.4 or 1.3 (there is some disagreement about what counts as lowest low). The latter rate will, if extended for some decades, result in radical depopulation, a 20% to 25% decline, and an increased imbalance between the young and the old. A number of questions have dominated the discussion. How low can the lowest low go: Are there any limits? Is the lowest low a function of the delay of childbearing, called the tempo effect, which might eventually be corrected with later pregnancies? Is the lowest low a temporary phenomenon that might be reversed?

Are there limits to the lowest low? Theoretically the lowest low could be pushed lower by a combination of a rising number of childless women and by most families having one child only. For a time, at least, the lowest-low pattern prevailed, with no evidence of upward movement. But the downward trend of TFRs in Europe between 1960 and 1999—which had seen a decline of more than one child, from 2.6 to 1.37 per woman—saw an increase to 1.56 in 2008.[19] It was the first clear evidence of a change, one that had not been predicted. In absolute terms, that last figure may seem slight, but it represents a 20% rise in TFR. Not only was there an overall increase of TFR, but it was experienced across Europe in some fifteen countries, most notably those countries with the lowest-low TFRs, and paralleled by a similar trend in New Zealand, Canada, Australia, and Japan. Germany, however, remained low, moving from 1.24 in 1994 but stabilizing at 1.35 after 2000.

But why did the TFR begin to rise after a long flat period? While pronatalist changes in government policies in some countries may have had an impact, the most likely reason was that a large number of women had simply delayed procreation. Even during the lowest-low period there was persistent evidence that most women wanted more than one child but had put off procreation for economic reasons or to gain an education and establish themselves in the workplace.

## DEVELOPING COUNTRIES

The situation in most developing countries is mixed but with an overall steady decline in overall birthrates. A snapshot from a 2012 UN report throws that pattern into relief:[20]

- Forty-eight percent of the world's population lives in "low-fertility" countries with birthrates below the replacement level (almost all of the European countries, nineteen Asian countries, and seventeen in the Americas).
- Forty-three percent of the world's population lives in "intermediate-fertility" countries, with an average 2.1 to 5 children (Indonesia, Pakistan, Bangladesh, Mexico, and the Philippines).
- The remaining 9% of the world's population lives in "high-fertility" countries, twenty-nine in Africa and two in Asia (Afghanistan and Timor-Leste).
- Rapid population growth in Africa is expected to continue in the near future, achieving 2.1 FTR only by 2095 to 2100; during 2013 to 2100, eight countries are expected to account for half the world's population growth (Nigeria, India, Tanzania, the Congo, Ethiopia, and the United States).
- Countries with high fertility declined from fifty-eight in 1990 to 1995 to thirty-one in 2005 to 2010, and their share of the world population dropped from 13% to 9%, while countries with fertility rates below replacement increased from fourteen to thirty-two.

All in all, population growth remains a serious problem but one that will, with some help, more or less take care itself on its present trajectory. Except for pockets of continued growth, the larger picture is evident. Birthrates have gone down in often dramatic ways and continue in that direction. Yet the experience of those countries with the lowest birthrates, well below the 2.1 parity level, has not on the whole been a good one, and almost all of them are working to raise their birthrates. Too great a disproportionate gap between old and young is good for neither. Yet it is also clear that more intensive family planning programs for those countries with continuing high birthrates are needed, and that will require external international support, which in recent years has been lagging, with the United States often standing in the way because of conservative opposition.

Ironically perhaps, the most recent declines have taken place long after the population bomb era, and "population control" as a rallying call has all but disappeared. Paul Ehrlich might be said to have eventually triumphed, but he lost a famous bet with another economist, Julian Simon, about whether there would be a shortage of basic metals necessary for industry, part of Ehrlich's generally apocalyptic vision of the coming world of environmental and resource decline. Simon had been much more optimistic—presenting the future as exciting and full of redemptive technological innovations—as did his later successor in that pulpit, Bjorn Blomberg. As a recent book on the Ehrlich-Simon battle notes, their ideological differences remain alive and well, with economic growth and innovation in one corner of the ring and a profound cultural change in the other.[21]

An important area of research in recent years has been the effect of population growth on economic progress. As noted earlier, there has been good evidence (countering Malthus) that under the right conditions, population growth can stimulate economic development by stimulating innovation, but under poor conditions, population growth hinders economic development. A recent study of the relationship between population decline and economic growth looks at the other side of the population coin, that of the effects of population reduction on economic growth. Over the years Quamrul H. Ashraf and his colleagues discern three camps: those who think reduction of birthrates is an important variable in population growth; those who consider it to be a necessary condition of economic growth; and those who propose that birthrate reduction is an effect of growth, not a cause. Noting the methodological difficulty of sorting out the population and growth ingredients and judging their variable influence, they do come to some conclusions.

As to whether population reduction would be sufficient to bring poor countries into the ranks of developed countries, Ashraf and colleagues cite a 1986 National Research Council report on the question and decisively say *no*.[22] They go on to say that fertility reduction in itself has few adherents for that kind of leap, but that some analysts give it an "important secondary role" or consider it to be a "necessary condition." On the whole, Ashraf and his colleagues believe that the reduction would have to be considerable to make a significant difference, but that it is also necessary to factor in the costs and welfare impact. Some researchers have noted that the birth of children can have a short-term negative impact but a long-term economic benefit when they grow up and enter the workplace.

## ESTABLISHING A GLOBAL PATTERN

I have gone on at some length on population trends in developed countries for three reasons. One of them is to set the stage for a discussion of the emergence of aging societies and their social and economic impact, in part a consequence of falling birthrates. Another reason is to indicate how trends in birthrates can follow unpredictable and surprising patterns, as illustrated by the temporary but powerful baby boom era and the no less surprising recent increase in TFR in lowest-low developed countries. Be careful what you project. The third reason is that, while they may follow their own historical paths to lower birthrates, developing countries are likely to require many of the same features that have marked the developed countries: rising income, reduced infant mortality rates, and for women, significantly improved education and an enhanced role in the workplace beyond the care of children and families. Many of these factors are beginning to emerge in developing countries as they gradually modernize.

There is considerable agreement that, when speaking of population problems, the greatest difficulties with population growth have been in sub-Saharan Africa, a part of the world with the highest birthrates and the greatest health and economic challenges. Most of the rest of the world has tolerable population growth rates and, as noted, there has been a steady global decline in birthrates. While there may be future surprises, what might be termed the establishment of an international pattern of childbearing has emerged: a rate slightly below the 2.1 child replacement average, with one- and two-child families predominating; a large percentage of unmarried or cohabiting nonprocreating women; and a steady increase of women in the workplace, with greater male-female equity there.

A spate of articles in 2014 suggests that some stability and reasonable projections for the future are possible. With a few exceptions, the ideal of a two-child family has been persistent well over a decade now in Europe, which has witnessed a steady increase in ideal size in formerly lowest-low birthrate countries and a decrease in those desiring more than three.[23] Another study found that, in light of emergent aging patterns, there are two ways of calculating that problem. If the standard of national budget strength is used for forty countries, then a fertility rate well above replacement would be best. If the most beneficial standard of living was sought, however, then a moderately low fertility and population decline would best.[24] Moreover, worries about aging societies would be lessened, by no means entirely but only in terms of the relationship between population growth and aging; the move away from the trends of lowest-low birthrate to just a lower birthrate would make a big difference.[25]

As for sub-Saharan Africa with its high birthrates, has become more evident in recent years that the most important variable in lowering birthrates is not simply a lack of family planning programs or the costs of contraceptives but a surprisingly persistent high mortality rate of children under five years of age. Even so, the important variable is better education of women. Child mortality (ordinarily) declines in countries with higher levels of education for women, and educated women are better prepared to understand and act upon family planning information. That means, to be sure, that family planning programs have to be in place, but also that they will have far better chances of success.[26] Whether a drop in fertility rates for sub-Saharan Africa will result in a "demographic dividend" of a kind that is economically valuable in its first stages is uncertain.

As many of those concerned with population growth recognize, affluent people in developed countries with few or no children—also a trend—and with high spending habits can do far more harm to the environment than large poor families in destitute countries. One need only observe the plentiful ads for resorts and cruises picturing a happy, attractive couple with money enough to enjoy those pleasures and unimpeded by children. The United States, the world's richest country, pollutes the atmosphere per capita with carbon dioxide far more than China, which is the world's leader in overall pollution but has a much poorer overall population. A 1971 formula espoused by Paul Ehrlich and John P. Holdren nicely summarized the causes of that difference—I = f(PATE), designed to calculate the impact on the environment (I) of four important variables: P = population, A = affluence, T = technology, E = ethics.[27] A large affluent population with a high average income, addicted to technologies of a polluting kind and embracing consumerism as a natural god-given right, is the perfect recipe for environmental harm.

## AGING SOCIETIES: AN UNFOLDING GLOBAL FUTURE

The experience of developed countries with the lowering of birthrates serves as a rough template for all countries hoping to achieve the same. It no less serves as a template for the problems brought about by fertility rates below replacement levels in the company of simultaneously aging societies. Population growth and aging societies are both a kind of showcase for the force of technological and health improvements: the reduction of infant mortality (the earliest drag on population growth) has been fueled by the advances of modern medicine; lower birthrates have no less been spurred by effective contraceptives as well as affluence; and medical advances and improved living conditions, also a function of progress, have given us longer life spans. The global aging of societies is no less than the product of technological advances and the socioeconomic benefits of living healthier lives than in the past.

My contention is that the aging problem has emerged as a much greater challenge than that of population growth. The latter is diminishing, the former growing. Yet aging as an issue has been overlooked in the push for lower birthrates. James Gustave Speth, Bill McKibben, and Jeffrey Sachs, all distinguished figures in global-warming research and writing, do not mention the topic. Speth has written, "By any objective standard, U.S. population is a legitimate and serious environmental issue." But the subject of aging is hardly on his environmental agenda. I think it simply snuck up on them, as with many others, especially its emergence in developing countries.[28]

## IS AGING A GIFT?

Bill McKibben made a special name for himself with his 1998 book *Maybe One*, explaining why he and his wife decided to have no more than one child, as should others. He makes a careful and nuanced case, not arguing that no one should have more than one but that proportionately more should make that decision, leaving room for others to have more. But then he also says that "the inevitability of aging" can be transformed "from a disaster into a commonplace, a given, even a kind of gift," and goes on to sketch some highly optimistic ways of gaining that end.[29] The notion of aging as a "gift" will come as news to most of those of us living with it into their eighties. Jeffrey Sachs also minimizes the problem. We should not, he says, think of unemployed, impoverished, young men (many, I would note, blocked from finding jobs held by older workers) as simply victims: "It is to suggest to them, and to us, that reducing the TFR from a very high levels is part of their own security and ours."[30] Are we to believe that the unemployed young or those forced into low-paying jobs, faced with uncertain economic and social futures, will find comfort in the long-term purported security? William Ryerson, president of the influential Population Media Center, makes population limitation the most important need for dealing with environmental problems. But he does not take aging populations seriously. Nor does he offer any alternative to the profligate and harmful ways in which affluent people in developed countries can spend the money they save by not having children.[31]

Here are some basic data about global aging. In 2006 there were almost 500 million people over 65 and by 2030 that number is projected to rise to 1 billion, one in eight of all human beings. The most rapid increases are taking place in the developing countries, which will see a leap to 140% by then (figure 6.2).[32] For the first time in history people over 65 years of age are now a larger portion of global population than those under age 5 (figure 6.3). They also happen to be a far greater comparative burden on social and economic resources than children and working adults. Compare that increase with the decrease of the pace of fertility decline, which now sees forty-four less-developed countries with a fertility rates below the 2.1 replacement rate. Many countries have seen that happen within one generation. But the "compression of aging," as it is called, is occurring rapidly in countries ill prepared or altogether unprepared to cope with such a quick change. Increasing life expectancy is a major reason for the growth in aging, and just how long that will continue is uncertain.

The fastest-growing group are those over seventy-five (figure 6.4). They have the highest rates of disability and chronic illnesses and thus (so far mainly in developed countries) consume a disproportionately large proportion of social resources. By 2030 the projection is that the global health burden of chronic illness and other noncommunicable diseases will generate 54% of all costs, while those of infectious disease will decline

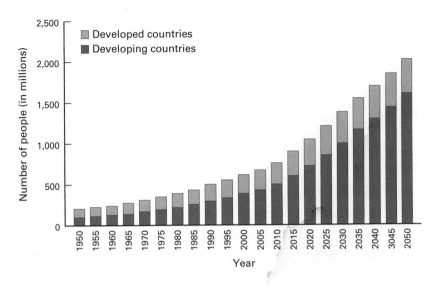

**FIGURE 6.2**

Number of people aged 60 or over: world, developed, and developing countries, 1950–2050. (UNFPA, United Nations Population Fund; www.unfpa.org/pds/ageing.html, accessed 6/8/13)

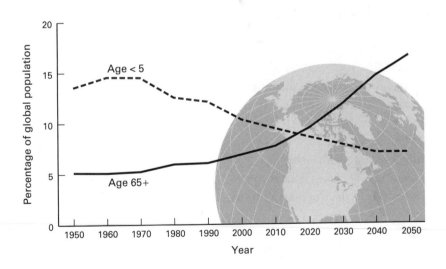

**FIGURE 6.3**

Young children and older people as a percentage of global population, 1950–2050. (United Nations Department of Economic and Social Affairs, Population Division. *World Population Prospects. The 2004 Revision.* New York: United Nations, 2005)

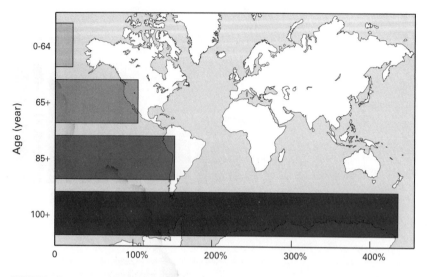

**FIGURE 6.4**

Projected increase in global population between 2005 and 2030, by age. (United Nations Department of Economic and Social Affairs, Population Division. *World Population Prospects. The 2004 Revision.* New York: United Nations, 2005)

to 32%. The combination of declining population growth and increasing aging growth is now a global phenomenon. Aging societies also change family structures, particularly as population decline and aging societies increase the need for caretaking of the elderly population in the face of proportionately fewer young people to care for them. Moreover, what is now a coming crisis of economic support for elderly retirees in developed countries will, along with chronic disease, likely be mirrored even more intensively in developing countries.

## ECONOMIC IMPACT OF AGING

There is a large literature that has accumulated over the years on the economic and social impact of declining population growth, the consequent imbalance between economic needs of aging societies, and the effects as well of changes in the dependency ratio of young workers and retirees.

I have already touched upon the alarm that lowest-low birthrates elicit in countries affected by such ratios. But apart from that imbalance it is worth noting some other effects of aging societies.

How might we counteract the proportionate decline in active workers, ordinarily young, as the proportion of the elderly grows? A 2004 Organisation for Economic Co-operation and Development (OECD) study projected significant declines in the growth (and sometimes the levels) of the labor force in the OECD countries, up to 4 to 5 percentage points on average between 2000 and 2025.[33] That is not a welcome prospect. Many countries are now trying to reverse the decades-old trend toward earlier retirement and, in a compensatory process, to find ways to bring more women, young people, and older people into the workforce. As Gary Burtless has observed, "Increased longevity can change the age profile of the workforce significantly only if it causes people to change the pattern of their labor force participation."[34] That pattern will not be easy to change, he concludes. In most developed countries a high proportion of women are already in the workforce, and in many of those countries high unemployment rates among the youngest workers have been resistant to improvement (Germany, France, Italy); the latter also indicates that in many economies a decline in the workforce by retirement of the elderly does not automatically open up space for the young. A high immigration rate in the United States keeps the birthrate relatively high and thus the dependency ratio fairly level.

## Retirement and Pensions

When most of the present government pension plans were put in place (in 1936 for Social Security in the United States) the benefit level for elders was fairly low, the benefits were greater than the taxes they paid into the system, the longevity of recipients was shorter than at present, the number of tax-paying young workers was growing, and the number and proportion of retirees were smaller. The projected increase on average for developed countries is for a doubling of the number of pensioners from 2000 to 2050.[35] Italy is looking forward shortly to a rise from 35% of earnings to support the elderly to 75% by 2050, an obviously intolerable prospect. Pension spending in Germany and France could go from the (roughly) 25% present level of expenditures to about 50% by 2050.

Government commitments for the future were usually made well before the present economic and demographic problems came into sight and during times of economic growth and confidence in the future. Thus there is a difficult political challenge to reform those programs both in their organization and in their benefit conditions and level. Cutting benefits is notoriously always harder than putting them into place, particularly when the expectations are for rising entitlement benefits, at least equal to cost-of-living increases—and in perpetuity. Some countries face a far more difficult future than others, in great part because of lower retirement ages, heavier dependence upon the government for pension support, and higher levels of benefits. The United States, the United Kingdom, and Canada are comparatively better situated simply because of later retirement, more diversified sources of pension support, and lower benefit levels.

## Are Countries Prepared for Aging Populations?

The Aging Preparedness Index developed by Richard Jackson of the Center for Strategic and International Studies looks at the gap between the rising needs of aging societies and the policies in place to deal with them.[36] The index usefully distinguishes between (1) a *fiscal* side (government preparedness to take on projected costs of government benefits to elders) and (2) an *adequacy* side (bearing on the living standards of the elderly and the strength of government safety nets and family support networks). Strikingly, few countries score well on both counts, and in some there is a often a sharp trade-off between the two. Three of the seven highest-ranking countries on the fiscal sustainability index are among the seven lowest countries on the adequacy index (Mexico, China, and Russia). In contrast four of the seven highest-ranking countries on the income adequacy index (Netherlands, Brazil, Germany, and the United Kingdom) are among the seven lowest on the fiscal sustainability index—not surprisingly, since they are countries with generous but financially troubled welfare and pension programs.

China and Japan offer similar—if perhaps the starkest—stories on the relationship between population growth, or lack thereof, and aging societies. China's official policy of one child per family is known for its draconian (if erratic) implementation, including forced abortions. With a present 1.5 TFR rate, China is well below the replacement rate.[37]

In 1975 there were six children for every older person, but by 2035 there will only two. By 2050 there will be 103 million people over eighty years of age out of 428 million Chinese elderly, some 30% of the elder population. Between 1960 and 2005, average Chinese life expectancy went from 50 years to 73.[38] Moreover, despite China's enormous economic growth, China is still a developing country with a large majority still poor: "While today's developed countries were all affluent societies with mature welfare states by the time they became aging societies, China is aging at a much earlier stage of economic and social development . . . it will have to pay for an age wave of developed country size with just a fraction of the developed world's per capita income and wealth."[39] In 2007 China's per capita income was $5,046 compared with Japan's at $31,607 and $47,488 for Singapore.

Although it has a pension plan for urban and coastal regions (covering only 30% of its overall population), China is still a primarily rural country (figure 6.5). There is no formal retirement plan for most Chinese workers, although one is being devised, and only a tiny minority have personal savings for their later years. The net result is that, given the low birthrate, the ratio of the elderly to young workers is high and steadily rising, providing a weak foundation for a good pension plan. Moreover, the traditional practice of family support for the elderly is eroding in the face of a movement of young workers (predominantly male) from rural to urban areas, not only depriving their parents of caretakers but also exposing them also to all the health-related and social pitfalls of crowded urban life. And China has a shortage of marriageable women (intensified by a long-standing Chinese gender imbalance). In 2015, China announced that it was lifting the one-child rule; many couples, however, do not want a second child.

Japan does not have an official one-child family, but the social structure of Japanese life has resulted in a one of the world's lowest birthrates, 1.4 TFR, which has seen only a slight rise in recent years. For six years in a row up to 2013, death rates exceeded birthrates. The problem for Japan is that the country has been slow to bring women into the workplace and, once there, to provide economic and other support for working mothers (the kind of generous support that has kept birthrates up in some northern European countries). More Japanese women are not getting married now, and those who do marry either do not have children or delay having

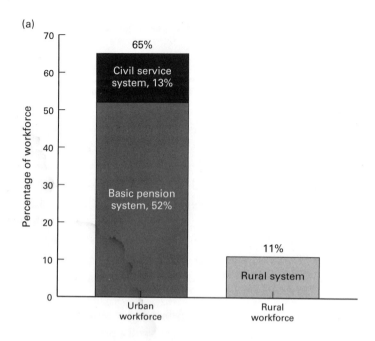

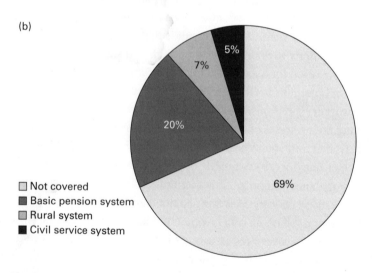

**FIGURE 6.5**

(*a*) Percentage of Chinese urban and rural workforce covered by public pension systems in 2007[*]; (*b*) percentage of total Chinese workforce covered by public pension systems in 2007[*]. (Richard Jackson, et al., *China's Long March to Retirement Reform*. Washington, D.C.: Center for Strategic and International Studies (CSIS), 2009: 11.)

[*]Civil Service System data are for 2006

them—and all this in the context of persistent national economic problems in recent years, conducive to neither marriage nor childbearing. These obstacles have kept marriage rates down and, in the face of recent national economic trouble, have led to those who are married putting off the procreation. The government has responded with efforts to increase birthrates but has been slow to put in place effective policies to do so.

Let me conclude this section with an apt quotation from the fine report *Why Population Aging Matters*: "While global aging presents a triumph of medical, social, and economic advancers over disease . . . [it] strains social insurance and pension systems and challenges existing modes of social support. It affects economic growth, trade, migration, disease patterns and prevalence, and fundamental assumptions about growing older."[40]

While it strays from the main topic of this book, I recently published a paper proposing that the solution to the overlapping issue of chronic illness and aging societies in rich and poor countries alike is to change the underlying model of modern medicine. [41] Most countries and policy makers see the answer to their dual demands as a matter of changes in health and aging policies, that is, a management and organization challenge. I believe that there is no way to deal with those problems other than also changing the value model of modern medicine underlying health-care systems. That value is a full commitment to unending medical progress and technological innovation, a war against death, and an effort to extend life expectancy. What it has given us, however, is neither a reduction of deaths from infectious disease nor the cure of any of the most deadly chronic illnesses. Instead, "progress" has given us more and more ways of expensively extending the lives of the sick, more complex dilemmas about end of life care, and (particularly after age 80) a life not always worth living.

## HOW LONG WILL WE LIVE?

It is fitting to ask that question of potential human life expectancy, The success of medical advances together with other socioeconomic improvements are what have extended human life. But how long will the gain of life years in old age continue and how far can that go? There have been those like myself who want more account taken of the social and individual burdens that go with old age past eighty, and those who would like to see an endless progression of aging, hoping and expecting that further medical advances will help to relieve those burdens. There is an identified group, the transhumanists, who have championed the latter. I will put that debate aside here to say something about the work of demographers in trying to answer the question. And it is a question whose answer will make a great difference in the future of aging, none greater.

There has been considerable debate over the years on that question, commanding the attention of novelists, philosophers, and scientists from many fields. But I will focus my attention on the demographers, because their social science research aims, among other things, at empirically tracking life expectancies and the social implications of that and other changes in human populations. A long-standing debate on how long lives will be advanced in aging has been a feature of the research, and I will single out one struggle in particular, because it encapsulates the central arguments about life extension. I call it the Olshansky-Vaupel debate, named after its two protagonists—S. Jay Olshansky and J. W. Vaupel—both well respected figures in the field.

Vaupel for his part has argued that life extension has been on an upward linear trend for centuries and will likely continue that way. There is no reason to think that the average age of 85, now attained only by Japanese women, will not be exceeded, or that the record age of 122, reached by a Frenchwoman, will not be broken as well. There is no end in sight or conceivable on the limits to human aging.[42] His has been called the optimistic scenario. Olshansky by contrast has been called a pessimist, a label he rejects, identifying himself a realist. His position is that the greatest initial historical gain in human life span came with the reduction of child mortality. While survival after sixty-five has also increased dramatically, it cannot compare with the drop in infant mortality. There is nothing on the aging horizon to suggest there will be anything in aging trends comparable in magnitude with that earlier child mortality decline. He also contends that it is most likely that the eighty-five-year mark will never be passed.

Another prominent demographer, John Bongaarts of the Population Council, has weighed in on that debate, taking a middle position. Noting that life extension can only come from one or more of three variables—reduction of juvenile mortality, background mortality, and senescent mortality—he sides with Vaupel in agreeing that childhood mortality is now out of the picture. If it is unlikely that reductions of senescence can ever equal it in magnitude, he nonetheless takes a more optimistic stance toward the likely continued improvement in old age from advances in medical technology. There is no evidence, he holds, "of approaching limits to longevity," and adds as a final thought that "longevity improvements will be larger and population aging will be more rapid than many governments of high-income countries expect."[43]

If that is correct, the same judgment might be made of aging in developing countries but with perhaps more qualifications based on "background mortality." This debate is a classic instance of the distinction between scientific projections and predictions, a line not sharply drawn here. In any event, I call the debate a draw, one that only the future can resolve, the outcome of which may well exceed my life expectancy at eighty-five—or not. I will note also an unresolved debate, on the impact of obesity on life expectancy, in which the evidence is mixed and contentious. But Neil Mehta and Virginia Chang prudently conclude that the "increased prevalence of obesity among children and adolescents along with an increase in severe obesity among adults, will likely have a substantial impact on future associations between obesity and mortality."[44] While their study was centered on the United States, their findings could likely be duplicated in developing countries (with, of course, some variations).

Demeny sums up the population aging situation persuasively: "If fertility remains below replacement in a given country, the resulting population losses imply not just diminished numbers but a transformation toward a progressively older age structure that would eventually prove socially, economically, and biologically unsustainable. . . . A judiciously mutual mix [of rising birthrates and immigration] is a feasible option."[45]

## ARE THERE, COULD THERE BE, LIMITS TO ECONOMIC GROWTH?

The well-publicized publication of *Limits to Growth* in 1972, focused heavily on resource depletion, opened the door to an important perspective on the environment, and stimulated a debate that has continued into the present. The concept of limits is not itself new: Malthus got there first, but population remains an important part of the environmental equation: more people, fewer resources. In an updated version of the report, subtitled *The 30-Year Update*, some members of the original MIT research team wrote that "in our scenarios the expansion of population and physical capital gradually forces humanity to divert more and more capital to cope with the problems arising from a combination of constraints . . . [such] that it becomes impossible to sustain further growth in industrial output . . . food, services, and other consumption."[46]

Their report made only a passing mention of the implications of a growth of GDP, but a number of economists and others added that ingredient in more recent years. Their thinking became the focal point of more recent debates, responding to the established theories of other earlier economists and policy analysts who came to believe that GDP was the key metric for measuring national well-being. The alternative test should be human happiness and fulfillment (or "capabilities," as Amartya Sen put it). At least some of those economists and others have in their own way also argued for a limit to growth, beginning with an effort to dethrone the idea of constant GDP growth as the key measure of human well-being.

One of the more interesting studies, by Andrew Clark, directly examines the relationship between GDP growth and happiness.[47] It is a controversial economic topic as one might expect, with a good deal of conflicting evidence. Sorting through many studies, Clark's leading question is whether GDP growth will continue to increase happiness in developing countries. As a general proposition "higher income is always associated with higher happiness scores" but with one exception in a debatable area, that of "whether higher national income yields higher well-being." His conclusion, stated in a curious way, is that "it is an idea . . . that cannot be rejected on the basis of available evidence." He later adds that "the relationship between income per capita . . . has diminishing returns over time" but that "there is no consensus on the existence of [a] threshold beyond which the market utility of income falls to zero." A famous study in the United States found that despite a doubling of U.S. GDP over a thirty-year period (1972–2002) "the average happiness of Americans has remained constant."[48] Studies of developing countries have found a strong correlation between a rapid GDP and an increase in happiness when emerging out of poverty—but a flattening after a decade or so of improvement. Table 6.1 presents the global outlook for GDP growth through 2025.

A central variable in both studies is the marginal utility factor, that is, the place of small, incremental gains over time: the poor gain much from growth but those no longer poor may gain little or no more happiness. Some would say there is no marginal utility beyond a certain point of affluence. My own nonexpert interpretation is that it is a mistake to think that even if the impersonal statistical evidence shows no further gains, hope for more plays a large part in human aspirations: perhaps an iPhone 6 will improve my well-being over an iPhone 5, and perhaps a few hundreds or thousands of dollars beyond my already adequate income will being me some marginal gains in happiness.

Some of the high-end businesses make a great profit selling to the already well sated. In consumerist societies, for those with money there is always some marginal utility, if only psychological, beckoning them along. I have long wondered why the very rich, from the Gilded Age on in the United States, have rarely been satisfied with owning only one large house; three to four is more the average, scattered around this country but usually at least one in some other country. William Randolph Hearst built a monster California estate in San Simeon and then constantly bought things to put into it to the point of overflowing. The talk of the real estate industry in recent years has been the flourishing absentee ownership in New York of multi-million-dollar massive apartments forty to fifty stories up in the air, the pieds-à-terre of the 1% who need a place to lay their heads when passing through the city. For such people the concept of marginal utility is a rising challenge, not a declining benefit.

I will note, finally, an absence from the literature of any writing favorable to *Limits to Growth* from any organized political group, nor have I heard of any politician taking it on as a cause. Such a political stance is just about unimaginable in this or any developed country. It is even less imaginable in a developing country. Jean Dreze and Amartya Sen, in their 2013 study of the many problems facing India, take for granted the need for a growing GDP. Sen and another Indian economist, Jagdish Bhagwati, have been at loggerheads for years about the place of the market in India's future but would not disagree on the need for growth, well articulated

**TABLE 6.1** Global Outlook for Growth of GDP, 2010–2025

| | Actual growth, 2010–2013 | Actual growth, 2014 | Forecast growth, 2015 | Projected growth, 2015–2019 | Trend growth, 2020–2025 |
|---|---|---|---|---|---|
| United States | 2.2 | 2.4 | 2.9 | 2.4 | 1.9 |
| Europe* | 0.9 | 1.4 | 1.8 | 2.1 | 1.5 |
| of which: Euro Area | 0.6 | 0.8 | 1.4 | 1.9 | 1.2 |
| Japan | 1.8 | 0.2 | 0.6 | 1.4 | 1.1 |
| Other mature economies† | 3.5 | 3.1 | 3.0 | 2.9 | 2.5 |
| **Mature economies** | **1.8** | **1.9** | **2.3** | **2.3** | **1.8** |
| China | 8.8 | 7.4 | 6.5 | 2.4 | 3.9 |
| India | 6.6 | 5.7 | 5.9 | 2.1 | 5.0 |
| Other developing Asia | 5.2 | 3.9 | 4.6 | 2.4 | 3.9 |
| Latin America | 3.6 | 1.0 | 1.6 | 2.1 | 2.9 |
| of which: Brazil | 3.4 | 0.2 | 0.5 | 2.4 | 3.1 |
| of which: Mexico | 3.5 | 2.3 | 3.5 | 2.1 | 2.8 |
| Middle East and North Africa | 3.4 | 2.9 | 3.4 | 2.4 | 3.2 |
| Sub-Saharan Africa | 4.6 | 3.7 | 4.4 | 2.1 | 5.3 |
| Russia, Central Asia, and Southeast Europe‡ | 4.1 | 0.9 | −1.5 | 2.4 | 1.7 |
| Emerging market and developing economies | 6.2 | 4.7 | 4.4 | 2.4 | 3.7 |
| **World total** | **3.7** | **3.2** | **3.3** | **2.4** | **2.7** |

*Note*: Projections are based on trend growth estimates, which—for the period 2015–2019—are adjusted for remaining output gaps.

*Europe includes twenty-seven members of the European Union (excluding Croatia) as well as Switzerland and Norway.
†Other advanced economies are Australia, Canada, Iceland, Israel, Hong Kong, New Zealand, Singapore, South Korea, and Taiwan Province of China.
‡Southeast Europe includes Albania, Bosnia and Herzegovina, Croatia, Macedonia, Serbia and Montenegro, and Turkey.

*Source*: Conference Board, Global Economic Outlook 2015—Charts & Table, www.conference-board.org/data/globaloutlook/index.cfm?id=27451.

in a book Bhagwati and a colleague wrote with the self-explanatory title *Why Growth Matters: How Economic Growth in India Reduced Poverty and the Lessons for Other Developing Countries*.[49] I hate to say it—since I am drawn to end-growth ideas and a happiness criteria for social well-being—but while it is one utopian idea that is coherent and (maybe) theoretically plausible, it simply is not going to happen politically. Some of its ingredients, however, are attractive.

The overriding need in poor countries is that of reducing poverty, improving health, and increasing economic equality within and among nations—and yet with added expenses of aging, I see no way of achieving those goals without a continuing and steady rise in GDP for both developed and developing countries, but most decisively for the developing countries. I have already made the case for its importance in aging societies, but need now to review the case for a limit to economic growth. I will do that by examining the arguments found in two books that make well-developed and careful cases for reducing GDP growth rate.

## AN END TO ECONOMIC GROWTH? TWO SCENARIOS

Richard Heinberg, a senior fellow of the Post Carbon Institute, offers the boldest contention in his 2011 book *The End of Growth: Adapting to Our New Reality*: the growth of global GDP has already come to an end, signaled by the worldwide recession beginning in 2007–2008.[50] This process will see some countries gain growth for a time, the decline not playing out in an even way, but the general trend will be no growth. "*Economic growth*," he writes, "*is over and done with*."[51] He gives three reasons for this prediction. One of them is "the depletion of important resources including fossil fuels and minerals." Another is "the proliferation of *negative environmental impacts*," both from the extraction and use of resources and the cost of those impacts and efforts to cope with them. Still another is "*the financial disruptions* due to the inability of our existing monetary, banking, and investment systems to adjust to both resource scarcity and soaring environmental costs."[52]

The present mainline position has looked upon economic growth as a necessity for poor countries along with "the creation of a monetary and financial system that *requires* growth. . . . The existing market economy has no 'stable' or 'neutral' setting." In one telling passage, using the criteria of "reality," he concedes that "without growth, we must seriously entertain the possibility that hundreds of millions—perhaps billions—of people will never achieve the consumer lifestyle enjoyed by people in the industrialized nations. Nonetheless, efforts to improve the quality of life in those nations will have to focus much more on factors such as cultural expression, political freedoms, and civil rights and much less on an increase in GDP."[53] A failure to face that reality will lead to no less than a failure to "delay putting in place the support services that can make life in a non-growing economy tolerable." Needed is a "painful but endurable process of adaptation [that] could instead become history's greatest tragedy." I cannot help being impressed by his candid bluntness in creating his own high barrier to making his case.

## A GROSS NATIONAL HAPPINESS?

Heinberg concludes his book by sketching the necessary components of nongrowth, steady-state societies. A redefinition of progress is necessary, and he picks up on the concept of a "gross national happiness," a phrase summing up a number of studies and reports in recent years that sought an alternative to economic growth as the measure of well-being. A key feature of that effort is the citing of many psychological and sociological studies showing that beyond a certain income level, not necessarily very high ($50,000 is often cited), more money does not increase happiness. If a moderate level of income can be assured, a nongrowth society will do good, not harm.

For all the utopian flavor of his book, Heinberg underscores the difficulty of making the transition. He says that while "our problems are resolvable in principle," we will have to work our way through an "if" and a "but clause." "*If we are willing to change our way of life and the fundamental structure of society,*" a solution can be found. "But," he adds, "*our society as a whole is not inclined to do what is required to solve them, even if the consequences of doing so are utterly apocalyptic.*"[54] Only a great crisis, he seems to conclude, can resolve that fundamental tension. Getting through that crisis will require considerable trust and a strong communal bond—which the crisis itself will threaten. Altruism and community are values not easy to preserve in times of upheaval and transition. His optimism about that possibility draws on a number of local and citizen-driven efforts at sustainable ways of life, heavily drawn from British examples. There might be a light at the end of the tunnel, he seems to be saying, but we must squeeze into the tunnel through a narrow (though attractive) opening.

## GIVING PROGRESS A SUSTAINABLE FOUNDATION

In his 2009 book *Prosperity Without Growth: Economics for a Finite Planet*, Tim Jackson, professor of sustainable development at the University of Surrey (UK), makes many of the same points as Richard Heinberg (and Heinberg quotes from parts of Jackson's book). But I put Heinberg first because of his claim that the end of growth it is not only desirable, it is here, a point perhaps more plausible in 2011, during the Great Recession, than in 2009, when the economic downturn was just getting under way. Jackson makes no such claim, or even hints at it. Instead, while making many of the same points in general about the end of GDP growth, Jackson has a few interesting perspectives and comes to a somewhat different conclusion about making it a reality. He emphasizes much more some economic angles, developing the idea of "ecological-macro economics," and also gives a high place to consumerism as a leading feature of the dynamics of GDP growth.

Drawing on the pioneering work begun in the 1970s by the University of Maryland economics professor Herman Daly, Jackson notes that conventional economics sees the GDP as a "measure of the 'busy-ness'" of the economy. All it does really is count up—in three different ways— economic activities of a region or nation: what people or government invest, what people earn, and how much value firms produce.[55] "Lots of things," he writes, "happen outside of markets that result from or impact on economic activity, that don't really contribute additionally to human welfare."[56] "The depreciation of natural capital (finite resources and ecosystem services)," he adds "are not accounted for at all . . . almost perversely, the cost of 'defensive' environmental activities, such as automobile congestion and cleaning up oil spills, are counted; it's economic activity, isn't it?" The solution to such defects is a "new engine of growth." Its aim is to measure a "productive economy" and its "founding concept is the production and sale of de-materialized 'services' rather than material 'products.' . . . The humble broom would be preferred to the 'diabolical' leaf blower."

In the end the aim is "to show that a new ecological macro-economics is not only essential, but possible." The goal of the effort must be as a starting point "to relax the presumption of perpetual consumption growth." He encapsulates the needed strategy in three goals: "establishing the limits" (sustainability), "fixing the economic model" (ecological macroeconomics), and "changing the social logic" (overcoming consumerism and the progrowth logic of capitalism). He does not reject capitalism but aims to "slow down economic growth and thus change its social and economic impact." Toward that goal, three macroeconomic interventions are needed: "structural transition to service-based activities; investment in ecological assets; working-time policy as a stabilizing mechanism."[57] One might call this "growth lite."

## THE BAD TASTE LEFT BY RECESSIONS

It is hard to know how to evaluate the end of growth movement. To be sure, it has a certain plausibility and theoretical solidity: GDP growth can be damaging. I have hardly done justice to the two books I have singled out, which are carefully developed with the arguments well (if not always) defended. And anyone can spot instances of people and small communities of like-minded persons simplifying their lives and note a decline in American young interested in learning how to drive, a decline in credit card debt, and the emergence of urban gardening. But even taking all this together it is impossible to find any strong example of an actual and explicit mass movement toward a lowering of GDP growth, much less the idea of a no-growth society—and that is true some forty years past the limits to growth movement championed by many influential academics, policy researchers, and environmentalists. The research and advocacy literature in favor of it is huge—and unavailing. But the advocates for continuing GDP growth are even more plentiful and influential. The experience of the recession has been to enhance anxieties about sustaining growth, not to reduce them or stimulate a rethinking growth as a value.

A useful debate about growth was an exchange between two economists, Robert J. Gordon and Laurence B. Siegel in 2012–2013, using the United States as a test case. Gordon argued that "the rapid [growth] progress made over the past 250 years may well turn out to be a unique episode in human history."[58] What Gordon calls "headwinds" now working against it are demography, education, inequality, globalization, energy/ environment, and consumer and government debt. Siegel responds by stressing that technological innovation, a key index to GDP growth, continues apace even if there is a characteristic lag in the speed of dissemination. Half the world's population, he notes, has never made a phone call, but some advances, such as cell phones, have been adopted with breakneck speed; with many others are in slower pipelines. He finds a more optimistic reading of the headwinds and sees a continuation of growth in the rest of the world. Gordon, in effect, embraces the ideological combination of growth and technology over the green parties and sustainability.

Beyond that and similar debates, recent data suggest that growth is hardly coming to an end. Heinberg's prediction is turning out to be wrong (though I suppose, to give him a fighting chance, I should add "so far"). A Scotia bank report in August 2013 noted that the U.S. growth rate, in its continued struggle for recovery, was likely to show a 1.5% increase, not high but with good prospects for greater growth.[59] In early 2015, however, the European countries are in trouble and the weaker points globally are the largest developing countries: China, India, Brazil, and Mexico. But then they have been at historical highs of a dramatic kind in recent years: Within the period 2000 to 2011 their growth rates were: China, 9.4; India, 6.6; Brazil, 3.6; and Mexico, 2.2. For 2014 the projections were China, 7.3; India, 6.5; Brazil, 3.5; and Mexico, 4.2. China's 7.3 is hardly a collapse.

As for running out of basic resources, commodity price indexes are a way of discerning whether that is true. There is little doubt that over the past twenty years commodity prices have risen very sharply, with a huge surge after 2005. They persisted after peaking in 2008, but have trended downward since then, with oil prices in particular dropping and with projections that the United States may be already be self-sufficient in oil. A number of efforts to extract oil in Alaska and the Arctic Circle have recently been dropped because of cost and environmental hazards. And if it is appropriate to specify a "so far" proviso for Heinberg's prediction about GDP growth, that is no less necessary with natural resources—and no less necessary as well with the faith that continuing innovation will continue to keep us from going over a cliff.

I suppose one could also say the same for Roger Pielke's "iron law" that GDP growth rates should not be rejected or, put more bluntly, economics trumps just about everything else. But what seems most evident in writing and research on GDP growth is that while developing countries receive a nod along the way, most of the discussion and uplifting examples are from developed countries. For all of the talk about India's and China's gross inequities in income distribution and government and private corruption, and predictions that their present high growth rate is not permanently sustainable, the overpowering reality is that it is bringing them out of poverty in a way nothing else could have. Newfound wealth has given them obesity and resultant rises in early chronic illness, but it has also kept then alive long enough beyond childhood to make that happen. It has not done away with water shortages, undoubtedly intensifying them, or with malnutrition, which might not be happening save for the inequitable distribution of wealth, and they are only now beginning to feel the impact of aging and the chronic illness that goes with it.

But it is better, I believe, to have an aging society with those problems than one with low life expectancies. Developing countries have gained the social maladies that go with affluent countries even if they are not fully there yet; that is the price of progress, which they are finding has its good and bad faces. While efforts to limit GDP growth would not hurt rich countries—assuming they are modest and accompanied by a decline in consumerist and other values—too sharp a drop would do harm to developing countries as well. They are dependent on affluent countries as markets for their natural and agricultural resources and as suppliers of technological innovations.

Even if the poor countries could loosen some of the bonds of globalization and develop sustaining local and regional force, that would not be enough. It is hard to imagine any country these days fully making it on its own. The triple threat of still pervasive infectious disease, the emerging chronic diseases, and all the problems of aging will require money and plenty of it, which can only come with economic growth. As Jean Dreze and Amartya Sen put it, speaking of India, "The need for rapid growth is far from over. . . . India's real income per head is less than most countries in sub-Saharan Africa. . . . Hundreds of millions . . . continue to lack the essential requirements of satisfactory living."[60] The same can be said of China.

# 7

# THE TECHNOLOGY FIX

---

## A Way Out?

The culture of technological progress has a long history, going back to the Industrial Revolution, gathering more speed in the nineteenth and twentieth centuries, with Silicon Valley as a high point going into the twenty-first century. "Innovative technology," the marketing brand of Silicon Valley, is simply the older model of progress in new clothes, but moving faster and faster. To be sure, much of the emphasis is profit and competition driven. iPhone 6 succeeds iPhone 5 which succeeded iPhone 4. The improvements in each are not great, but that does not render them less desirable for millions of customers. What is new is usually more attractive for its own sake in industrial societies than what is old, and that is the way such things are marketed, with buzz and a profusion of apps.

The relatively short history of technology in the management of global warming is far from the explosive exuberance of Silicon Valley—or for that matter is no less great a distance from the innovation that has been the mark of technology for the care of the chronically ill; that is, for the lethal diseases it encompasses. One reason for the latter's energy is that it appeals directly to individual well-being, the saving of life and the reduction of pain and suffering. Everyone can appreciate that, with benefits here and now (putting aside for now the economic burden). Chronic illness has been a splendid, ever-fruitful arena for the pharmaceutical and medical device industries. The diseases it covers get much of their basic research support from the National Institutes of Health (NIH), and beyond that they have easily gained government tax breaks and research subsidies. Best of all, perhaps, they offer enticing markets for private investment. To be sure, there have been innovative technologies for food and water challenges, but many of them have been for niche markets and for relatively low-cost technologies, almost by definition low-profit industries and thus of lesser interest to investors. But efforts to grapple with global warming require major high-cost technologies, expensive R&D investments, and often uncertain markets even if all goes well.

Coping with each of my five horsemen (with one exception—obesity) requires a heavy dependence upon technology, and most of the strategies to manage them make use of mitigation and adaptation. In this chapter I will weave together the themes of energy and technology for four of them, beginning with food and water, then obesity and chronic illness, saving most of my attention for the last one, the most bedeviled: global warming.

## FOOD AND TECHNOLOGY

Just as modern life requires technology for economic growth and the satisfaction of various human needs and desires, it also requires technology for food security. Even in an era in which "natural" food and "organic farming" have become rallying cries in developed countries, the production of food on a global scale requires a number of different ingredients. They include careful national agricultural policies and organization, effective international agencies and NGOs, and technologies both old and innovatively new. The needs of the future to insure that security are daunting.

The Food and Agriculture Organization of the United Nations (FAO) has projected that "global agriculture will need to grow by 70% between 2015–2050 and by almost 100% in developing countries. . . . Doubling the agricultural output of developing countries will require an average annual gross investment of US$209 billion, 50% more than current levels."[1] Marilyn Brown and Benjamin Sovacool in their fine book *Climate Change and Global Energy Security* add a disturbing perspective to the FAO report: "the 20th century saw a 150-fold increase in fossil fuels and electricity used in global cropping, yet only a six-fold increase in yields and a four-fold increase in productivity," and "more livestock production and the raising of animals for food require the conversion of land, and thus [are] connected to massive changes in land use."[2] As if the raising of cattle is not enough of a problem, James Gustave Speth noted that "in 1960 5% of marine fisheries were fished to capacity or overfished. Today [2008] that number is 75%." One projection is at the present rate of fishing "all commercial fisheries could collapse by 2050."[3] Too many cows, too few fish.

Figures and projections of the kind I have just cited are hardly encouraging, but there is considerable optimism about the use of technology to both mitigate and adapt to the challenges to global food security. A number of mitigation efforts have been undertaken, with their starting point the fact that agriculture itself is a major source of global warming, producing an estimated 10% to 12% of greenhouse gases (GHGs). Moreover, a majority of those emissions (some 74%) come from low- and middle-income countries, where small landholders predominate.[4] A fine collection of studies on agricultural mitigation that focus on those landholders note that their collective findings "suggest that the future mitigation will depend more on whether practices are sufficiently attractive to large numbers of farmers rather than whether the mitigation is technically feasible."[5]

There are many ideas and present practices aimed at the use of technology to mitigate threats to food security. They focus on reducing the need for water and on ways to improve efficiencies in the raising of food and to increase crop yields and food quality. I will mention only a few here. Devices to improve efficiency are already available. Growing in use are center-pivot irrigation systems, with large circulating arms that drip water on crops, thus saving water, and electronic soil sensors that measure soil moisture and thus determine with greater precision just where the water is most needed. Available also are commercial services that can determine soil moisture at greater depths (6, 12, 18 inches), and even more specific evapotranspiration (ET) devices that determine the soil moisture of specific plants growing in the field; those same devices can be used to prevent heat or moisture stress.[6]

## Gaining Acceptance for Technology

As with many new technologies, farmers often have to be persuaded to give them a try, especially when, as with those just cited, they can require a significant financial outlay. The use of cropland for ethanol has come under attack both because of the land it takes away from food production and the lack of good evidence that it is an economically effective way to reduce GHGs, even if does reduce to some extent the use of imported oil. In this case the production of ethanol is a technology to reduce dependence upon oil but with a harmful impact on food production. Research is under way to find alternatives to ethanol—wood chips, cellulose, and solid food waste. Far out in the land of speculation one finds climate modeling of the possible impact of solar-radiation geoengineering on food production. One model showed crop yields rising, but even the authors of the study, noting the unknown side effects and risks, did not endorse that solution.[7]

What emerges from a review of various mitigation- and efficiency-directed food technologies is that none of those in use raises serious technical problems. They work. The technologies themselves are not the important obstacles to their use. Two other barriers are of consequence. One of them is inefficiency, poor government, and corruption in many developing countries, which directly affects small landholders.[8] These technologies are simply not taken seriously. The other barrier is a shortage of funds to develop innovative technologies. Most such research has been done or supported by developed countries and international agencies such as the FAO, the World Bank, and foundations. Those funds have decreased in recent years.[9] The worldwide recession, particularly effecting the United States and some of the developed countries, and other competing global needs have had an impact. Countries with millions of small farms need not only good governance but also a favorable international investment climate to assist them. While the earlier Green Revolution provides a model of technological progress, no comparable revolution appears on the horizon. More likely is a gradual improvement in a variety of technological innovations, both in the spread of those already available and in the efficiency refinements possible with them.

## Genetically Modified Organisms

Genetically modified organisms (GMOs) have been a major force in agriculture in recent years, and innovations are still possible. Even though widely used, they are mainly concentrated in the United States, Canada, Argentina, Brazil, and India, and have been met with considerable resistance in Africa, Europe, and some parts of the United States. Their main use has been to introduce varieties of crops resistant to plant diseases caused by insects or with increased tolerance to the use of pesticides. Improved seeds, crops requiring less water use for drought-prone regions, and the control of weeds are part of the research agenda of various multinational food companies, a goal that makes some commentators nervous because of a likely monopoly on patents that will make GMOs more expensive for poor countries.[10]

Despite careful and positive assessments by national and international authorities, public worry about GMOs remains strong if erratic. The strong form of resistance is a refusal to accept any GMOs, while a weaker form is to label all genetically modified food. Traditional foods, genetically modified by long-standing breeding methods, do not meet similar resistance and are not as rigorously assessed. The sources of the continuing resistance to genetically modified foods are fourfold, with the first three being fears of possible allergic reactions, undesirable gene transfer, and outcrossing.[11] The fourth is the concern about toxic effects on organs and biological systems.[12]

Allergic reactions that could come from the transfer of genes from commonly allergenic foods is, however, discouraged, and the World Health Organization (WHO) believes that no allergic reactions have been found with current GMOs. Although the possibility of transferring genetically unhealthy genes is considered to be low, the use of technology without antibiotic-resistant genes has been encouraged by an FAO/WHO panel. Animal feeding trials on genetically modified foods suggests measurable changes to biological systems, but further studies are needed to evaluate the extent of disruptions.[13] Probably the greatest resistance to GMOs comes from fears that the movement of genes from GMOs into other crops (outcrossing) will have an impact on food safety, will affect the susceptibility of such nontarget organisms as beneficial insects, and will cause a loss of biodiversity.

Careful risk-assessment efforts to reduce such possibilities are ongoing and the WHO judgment is that those assessments "have not indicated any risk to human health."[14] While occasionally a wild gene that made it into a foreign plant has been detected, no harms from that quarter have been found either. That kind of endorsement, however, has not mollified critics, whose objections often seem more cultural and ideological than scientific. Fear of a kind that says a failure to "so far" demonstrate harms does not preclude the possibility that they could happen in the future. The suggestion has been made that nothing less than a famine may bring a change of view of the resisters. Resistance has been notably high in African countries.

The most common alternative to banning GMOs has been that of mandatory labeling of food products that have genetically modified ingredients. Bills have been introduced in a number of jurisdictions in the United States and European countries to do that and have met with varying degrees of success.[15] That alternative, a mild form of what might be called soft regulation, has not deterred the production of GMOs nor does it seem seemed to harm the sale of foods with genetically modified ingredients.

## WATER AND TECHNOLOGY

As noted in chapter 3, technology is playing a large part in efforts to develop innovative ways to better manage water shortages and improve existing means of water collection, storage, and transportation. Water-related technologies range from inexpensive pumps to harvest water from aquifers for farmers in developing countries all the way to the construction of massive and expensive dams that aim to use water power to generate electricity, store water, and meet agricultural needs. For the most part (save for dam construction) those efforts are not controversial as technologies—although as in the instance of water pumps they can pose difficult trade-off problems by helping farmers to move out of poverty at the price of drawing excessive and often irreplaceable aquifer water.

If one aim of water management efforts is to manage the quantity and accessibility of available water, the other is water quality. UN general secretary Ban Ki-moon has declared that "without a serious advance in implementing the [UN] water and sanitation agenda, there is little prospect of achieving development for all."[16] That "agenda" is the 2005 UN Millennium Development Goals, and the water effort has the support of an important NGO, the World Water Council. The causes of polluted water are many. Among them are agricultural and industrial runoffs and waste disposal, poor or nonexistent sewer systems, and inadequate water purification efforts. These causes can still be found in developed countries—the Hudson River, next to which I live, is still undergoing cleanup of a massive 1.3-million-pound pumping of polychlorinated biphenyls into the river by General Electric that occurred during the years 1947 to 1977—but by far the greatest damage is being inflicted in the poorest countries.

Lack of effective technologies and knowledge to deal with dirty water have not been the barriers in water quality assurance. For the most part a lack of money and/or assigning a low priority to upgrading water systems have been the obstacles. Putting in a new sewer system or replacing an old one is particularly expensive and can require years of effort. The U.S. Environmental Protection Agency, in its "Technology Fact Sheets" lists about 100 items bearing on water treatments,[17] broken down into a number of categories:

- Combined sewer overflow treatment
- Storm water
- Water disinfection
- Biological treatment (secondary and advanced)
- Water efficiency
- Decentralized systems technology
- Collection systems (sewers)
- Biosolids technology (garbage)
- Wastewater technology (industrial and private)
- Energy conservation and management

Earlier, mention was also made of the role of industry in water matters. It is industry that sells water (e.g., bottled water), manages city water programs, and manufactures and sells the technologies used in water management. Of the 100 subitems under the general categories just listed, most have a technological component and, taken together, amount to a major global industry. As a clever title to a report by GreenBiz (which reports on the business features of environmental issues) puts it: "The Water Industry: A Massive Market Bubbles to the Surface." The article notes that "over the next decade from 2011 [water] demand is projected to increase by 20%–35%. By 2020 37% of the world, developed and developing nations alike, will experience water stress."[18] Not surprisingly, the author reports, the Dow Jones Water [Investment] Index jumped some 80% between 2005 and 2010 and outperformed Standard & Poor's by over 20%.

Industry has noted that growing water stress could cause it serious problems in the future, signs of which are already present. The consulting firm Ernst & Young has singled out water scarcity and poor quality as important for companies and has noted the macro-conditions bringing them about: rising population and rising demand, water quality degradation in environmentally sensitive watersheds, the scale of global deforestation, and extreme weather patterns.[19] It recommends five steps to assess and mitigate the risks of water scarcity:

- Develop a corporate water policy that sets goals and action guidelines
- Understand the current state of water risks at the watershed level
- Understand the business's water footprint
- Engage internally at the facility or corporate level and externally with local stakeholders to evaluate the risks and impact
- Report externally and seek independent assurance; it is important to assure stakeholders that they are receiving accurate information

## OBESITY AND CHRONIC ILLNESS

### Obesity

Obesity is the outlier in the deployment of technology. It plays only a minor role at present in combating obesity. Despite a considerable gain in understanding the biology of obesity, especially the genetic role, that has not translated into a clear pathway for strong technological interventions. The same can be said of efforts to use technology to change obesogenic societies; none have been proposed or pursued. The most prominent initiatives, at least in principle, are for surgical procedures, implantable devices, and drugs. There have been food supplements that do not require government approval, and drugs, devices, and procedures that do. I will put to one side food supplements (running into the hundreds, most of uncertain value), and focus on the latter class of technologies.

The number of bariatric surgeries performed has steadily increased, but the procedure has significant drawbacks. Not all obese patients are medically acceptable, the various procedures have surgical and other risks, some operations simply fail altogether, and all require a demanding postsurgical regimen. It is clearly not going to make a great difference in dealing with population obesity, in which the numbers run well into the millions.

A recent innovation opens up a new technological pathway. Called the Maestro Rechargeable System, it is an appetite-suppressing implant designed to treat obesity by disabling nerves that connect the stomach and the brain.[20] In approving the device, the U.S. Food and Drug Administration (FDA) specified some strict requirements for it use. It is limited to patients eighteen years of age and older who have been unable to lose weight in a supervised weight loss program, have a BMI index of 35 to 40 (in the range of 240 pounds), and one other obesity-related problem, such as diabetes. In addition to those limitations it also has some serious side effects. All in all, this device, like bariatric surgery, has limited possibilities for mass use.

A number of drugs to treat diabetes have emerged in recent years. One of them is the FDA-approved off-label use of a Novo Nordisk injectable type 2 diabetes drug, Victoza® (liraglutide).[21] It is aimed at the overweight who have diabetes, high blood pressure, or high cholesterol—specifications that can cover many thousands of people. The FDA has also approved three weight loss pills in recent years: Qsymia, Belviq, and Comtrave. Contrary to expectations, sales of those drugs have not been great because of limited insurance coverage and high out-of-pocket costs.[22] None, moreover, have been on the market long enough to test their long-term effectiveness.

## Chronic Illness

If obesity relief stands at one end of a spectrum that makes use of technology to solve problems, chronic illness lies at the other end. The principal medical means of diagnosing and treating the leading chronic diseases are almost all technological. They most commonly include the use of diagnostic machines (X-rays, PETs, and MRIs), surgical and treatment devices (stents, organ transplants, artificial hips), and pharmaceuticals of all kinds (cheap aspirin, expensive cancer drugs, and pills for everything from high blood pressure to schizophrenia). Most recently there has emerged a new field, that of the use of "personal health-care systems technology" to manage from a distance nursing home patients and those cared for at home.[23] Information technology allows better record keeping and the easy transfer of patient information. There are few things these days in medicine not touched by technology, with its concentration peaking in intensive care units caring for the critically and chronically ill. The stereotypical picture of a patient with tubes and wires going into and coming out of him closely fits the reality. Close to 60% of American health costs are incurred by the 20% of patients who are chronically ill; and the top 1% of them cost an average of more than $90,000 to treat.[24]

With no exaggeration, medical technology can be called the most blessed of American research efforts and industries.[25] Unlike the situation with technologies to mitigate and adapt to global warming (discussed later in this chapter), medical technology research and innovation have ample government R&D funds (the NIH with its $30 billion annual budget), government subsidies available for innovative work, a profitable pharmaceutical and device industry, a medical community eager to make use of new technologies, and a public long entranced by and supportive of medical technology. So strong is the American cultural support that a frequently voiced concern is that health-care budget cuts could "stifle innovation."[26] The combined annual spending and profits of the global medical device and pharmaceutical industries is $300 billion. At the NIH each of the major chronic diseases has its own research agenda, and in 2013 NIH provided $7,080 billion for cancer research, $3,337 billion for heart disease, $1,067 billion for diabetes, $841 million for obesity, and $331 million for stroke. Much if not all of that money goes for basic research, the main aim of which is technological progress, aiming to assist private industry in exploiting and profiting from the technology. A few years ago—fueled by some congressional anxiety—a new area of research was established by the NIH, translational research, established to enhance the move from the research bench to clinical medicine.

## MEDICAL TECHNOLOGY COSTS AND CHRONIC ILLNESS

But innovation more often increases costs rather than lowering them. It is one of the great sources of economic stress in the provision of health care. The most common estimate is that about 50% of the annual cost increase in American health care can be traced to new technologies and an intensified use of old ones.[27] The advent of PET and MRI scanning devices not only added new costs but did not, as expected, eliminate the use of X-rays. In a 2009 book I likened medical technologies to a beloved dog, one that chews up the furniture, urinates on the floor, and is untrainable—but for all that is a beloved, adored pet.[28] Other developed countries to be sure make use of technology but without the almost religious zeal with which it is embraced in the United States.[29] The overuse of technologies for diagnosis and treatment is regularly lamented by health policy analysts, just as a widespread failure of physicians at the same time to adequately prescribe drugs to patients that are useful is no less bewailed (e.g., for high blood pressure). With the exception of technology costs, there is a considerable writing on the various ways apart from direct medical diagnostic and medical that technology can benefit the chronically ill: distance surveillance and diagnostic devices and various forms of information technology most notably. One of the few sociological studies of those efforts concludes that, while they can bring real patient benefit, they can also bring about a kind of captivity to the technology.[30]

Physicians in the American fee-for-service system have more financial incentives to prescribe or use technologies than to talk with patients. The pharmaceutical industry is both praised for its innovations and condemned for its high prices. A relatively new class of drugs known as biologics now manufactures cancer drugs whose costs can range from $75,000 to one that costs $330,000. No other developed country has technology costs of that magnitude, in great part because of government price control of drugs and devices and fewer economic incentives for specialists to prescribe them. Even so, it is well understood in developing countries that they will not be able to afford the more expensive technologies routinely used in affluent countries, and many of the low-cost technologies are not available either.

## GLOBAL WARMING

The modern world lives on energy, as important for our common life as the blood flowing through our bodies. Energy gives us electricity; runs our industries; fuels our cars, trains, planes, and buses; and warms or cools our buildings and homes. Its generation and dissemination provide thousands of jobs for workers and billions in profit for corporations. That energy comes from the use of coal, oil, natural gas, nuclear power plants, wind farms, dams, and other sources. Some of them have turned out to be dangerous for our planet and our bodies—coal and oil, most notably. Yet one way or another we need energy; it is indispensable. The question is how to find ways to mitigate the harms produced by some forms of energy production, to better adapt those we cannot fully mitigate, or to adopt some combination of both. Technology plays a crucial role in all of those efforts.

The U.S. Energy Information Administration (EIA) prepares an annual "international energy outlook" report.[31] The 2013 report is sobering. It begins by projecting that global energy consumption will grow by 50% between 2010 and 2040. Most of that increase will come from the developing countries, which will account for 90% of the increase, with the Organisation for Economic Co-operation and Development countries (essentially developed countries) contributing only 17%. The fastest-growing sectors of energy production are renewables, increasing by 2.5% a year. That encouraging figure is, however, well overshadowed by the 80% of world energy produced by fossil fuels, with natural gas the fastest growing. The global industrial sector will account for over half the total use of energy by 2040. With China now continuing an ambitious plan to build a large number of coal-fueled plants, that large elephant in the room has been well observed and its dimensions calculated. Between 2010 and 2040 those figures add up as a consequence to a world increase of 31 billion metric tons of $CO_2$ emissions in 2010 to 45 billion tons in 2040, a 46% increase.

Those figures are of course projections, not building in many ideas and schemes for better management of energy between now and then. Technology is seen as the royal road for doing that. Before moving on to survey of technological innovations, I want to set the stage by mentioning first the problem of evaluating their likely benefits. By that I mean improvements in present technologies leading to greater efficiencies and new technologies. I want to recall first the "three rules for technological fixes" by Daniel Sarewitz and Richard Nelson:[32]

1.  The technology must largely embody the cause-effect relationship connecting problem to solution.
2.  The effects of the technology fix must be assessable using relatively unambiguous or uncontroversial criteria.
3.  Research and development is most likely to contribute decisively to solving a social problem when it focuses on improving a standardized technical core that already exists.

I will raise some questions about these useful rules as we move along, but want to add to the above list a rule proposed by Richard Mowery and colleagues (which I number 4):

4.  "Technological solutions to global warming must demonstrate their cost effectiveness, ease of operation and reliability in systems that may be in operation for decades."[33]

I want also to add one of my own and some further discussion of the variables affecting that rule. My rule is meant to take account of the financing challenges of improving or inventing technologies. A wholly feasible and promising technology that no one (government or private sector) will help pay for will not get off the starting block. Since every technological fix must be a accompanied by a financial fix, I add:

5.  Technological solutions must have adequate funding from initiation to implementation, from government, the private sector, or some combination of both.

Two British studies help to elucidate the financial challenges. David Mowery and colleagues, writing for the charitable organization Nesta, reject from the outset the idea of a government program organized along the lines of the World War II U.S. Manhattan Project that produced the atom bomb. It had the special advantage of being a wartime program, with the highest national defense interest behind it, secrecy from the public, the media, and enemies, and the ability to command all the resources it needed. By contrast, civilian technological advances of any magnitude require cooperation between government and industry and that entails a difficult combination of political interests, public support, profit incentives, and various rules and regulations.[34] Only "a sustained and credible policy structure" will do that job, "but the consequences of failure or inaction are even more forbidding."

## TECHNOLOGICAL MITIGATION

Will the technological fixes work and can they be paid for? That is a difficult set of tests, but then mitigation and adaptation contains its own long list of tested and paid-for technologies—and untested schemes and speculative financing ideas. That is the territory the remainder of this chapter will traverse. I note a kind of paradox with much of the literature. Nothing is more common than articles and reports that stress the complexity of the issues, often making the assertion that there are no magic bullets. And then they quickly move on to tout the many present and possible benefits of technology, as if not a magic bullet but a shotgun with many magic pellets is the answer. This paradox may well reflect a recognition that technology offers a way forward that is easier and more promising than other proposed solutions—compared, that is, to getting a globally binding Kyoto-type protocol or persuading China to stop building coal-fired plants. There is an optimism about the technological possibilities often missing in other global-warming strategies. As an important comprehensive 2006 report, *Strategic Plan for the U.S. Climate Change Technology Program*, put it: "accelerated advances in technology have the potential to facilitate progress towards meeting climate change goals. . . . [And] it is expected that the new technologies would create substantial opportunities for economic growth."[35] In other words win-win, the most fertile territory. As we shall see, however, some bumps on the road have occurred since then.

The report goes on to identify six strategic goals:

1.  Reduce emissions from energy end use and infrastructure.
2.  Reduce emissions from energy supply.

3.  Capture and sequester $CO_2$.
4.  Reduce emissions of non-$CO_2$ GHGs.
5.  Improve capabilities to measure and monitor GHG emissions.
6.  Bolster basic science contributions to technology development.

I will not try to provide details for all the items on that list. A number of them are not controversial and are already being widely adopted. But a number of the others are either controversial, far from acceptance or scalable development, untested or still speculative only—or face one or some combination of various obstacles.

## NUCLEAR ENERGY

One of my most memorable real estate horror stories is that of a friend, a professor at the Pennsylvania State Medical School, who owned a house in Hershey directly looking out on Three Mile Island, the scene of a nuclear plant breakdown in one of its reactors in 1979. It was his bad luck to put his house on the market the very day of the disaster. Not a single potential buyer ever came to even look at the house, and it took him years to find one. The Three Mile Island mishap, which did not do great harm, was followed in 1986 by Chernobyl, which did. Most recently the Fukushima event in Japan, caused by a tsunami in 2011, was a real disaster and still had not been cleaned up as of the end of 2014, leading the Japanese government to state that it was going to close down the country's remaining nuclear power plants. In light of the Japanese experience, Germany shut down eight reactors and said the remaining nine would be closed down by 2022.[36] The Indian Point plant, near where I live on the Hudson River, has for years been the target of a number of petitions (so far unsuccessful) to close it down because of safety anxieties and potential harm to fish. I have never worried about the danger and have not signed any of the petitions. There has remained everywhere over the years debate about removing and burying radioactive waste from nuclear plants, one difficulty being finding physically and socially acceptable places to bury it.

Early in 2013 the International Atomic Energy Agency said there were 437 nuclear power reactors in thirty-one countries.[37] France has been the leader, with no meltdowns and nuclear power providing 80% of its electrical energy. The United States has had only one serious problem. In the fearsome context of global warming it is an attractive and tested technology, with increasingly rigorous safety standards and newer, even more efficient reactors coming along. Neither wind nor solar sources, though surely safer, offer the prospect as renewables equaling the effectiveness of nuclear energy. Moreover, nuclear power offers all-weather energy, which wind and solar power do not. For me the dilemma is a clear one, comparing the mitigation of dangerous GHGs with the low-probability harms of occasional nuclear meltdowns. The tiebreaker for me is the imbalance between those two risks, with the GHG evidence coming well ahead.

But there is a serious economic problem with nuclear plants. While not costly to run, they are enormously expensive to build, much more so than coal-fired and gas-fired plants; and there can be resistance in many places to having them located near populated areas. In the past those plants typically received government subsidies for the construction costs, but in recent years those have been more difficult to get, with an additional small decline in funding more recently. Increasing the creation of new plants will be important, but in its favor is the recognition that the sustaining costs of nuclear power are about the same as those for coal-based plants (although not in China) and much more cost-favorable than wind power and solar power.[38] The antinuclear hostility in Vermont reflects the power of pastoral visions of an earlier day but is not viable when more basic human needs, such as energy availability and security of large numbers of people, are at stake. The terror of meltdowns has generated enormous fears; they are dramatic and arresting, but still statistically rare. But in terms of danger they rank well below the more hidden danger of GHGs. And that seems true even if account is taken of the chronic difficulty of finding acceptable places to bury nuclear waste.

## FRACKING

Fracking, the technology for extracting shale oil and gas from shale rock well below the land surface, uses strong water pressure to get at the underground oil and gas contained in the crevasses and small cracks of the rock. The water used contains a variety of chemicals and sand to assist the process. From a global-warming perspective, fracking has the advantage of reducing the use of heavy oil and coal as a source of energy, and shale oil and gas emissions produce half of the GHG emissions of heavy oil. Although its use does not fully eliminate GHG emissions, its 50% reduction compared with coal is hardly insignificant. It has also been the source of thousands of jobs, often in economically poor regions, and is a boon to the industrial oil and energy companies. Taken together, that is an attractive set of benefits to help curb global warming.

The unfortunate rub is that fracking presents its own set of environmental and human health hazards. The combination of the water and the chemicals used to facilitate the fracking can contain hundreds of hazardous chemicals, including radium, lead, and mercury. In addition to requiring large amounts of underground water, fracking involves chemicals that can leak into aquifers and the land around the wells and can also spoil drinking water. Local protests in the United States against the wells have been common and contentious, paralleled by an intensive and well-financed industry effort to combat the resistance.[39] To top a long list of harms, the process also can often leak methane, a much more potent GHG than $CO_2$, thus undercutting some of the gains from fracking. But it is possible to limit the public harms of fracking. While violations of various regulations are now rampant, the harms are reducible, if not entirely so. The targets of the regulation are water and chemical use and the control of methane release.[40] If better regulation does not succeed, then we will be left with an unsavory mix of the good and the bad.

On balance, I would support tightly regulated fracking, with a gradual reduction of its role, aiming for an eventual stop, and that may happen anyway. The UN International Energy Agency (IEA) sees a downturn in shale oil production after 2020 because of rising costs and finite resources, with a concomitant increase in the demand for oil after that point and a return to OPEC's earlier dominance.[41] Ironically, that prospect may allay the fears of many that the rise in shale gas may inadvertently serve as a disincentive to develop renewable resources.

## CARBON TAXES AND CAP AND TRADE

The use of pricing to control the use of an item—whether it be for water consumption (see chap. 3), a tax on sugared beverages, or increased pricing for automobiles with poor gas mileage—is a well-established economic tactic. If other reasons for changing behavior do not work, the slogan "money talks" offers a remedy. Harmful emissions from coal-fired plants travel hundreds or even thousands of miles from these plants, impacting the health of people located both nearby and at a distance. These impacts, called "externalities," are not priced into the costs of running such businesses, a failure of economics to include the harm to the environment and society from business practices (otherwise known as a "market failure"). How can those external social, health, and economic costs be calculated, and how can ways be found to put pressure on the emitters to calculate and be taxed for those costs? I will put to one side the means of cost calculation and focus on efforts to establish effective means of getting the emitters to pay the external costs.

There are two principal approaches. One of them is for government to set a local or regional cap on allowable emissions, called "cap and trade." Some companies will emit at levels higher than the cap and others at levels below the cap. The companies in this system will be allowed to trade with one another: those with low emissions can sell their allowance to emit more to those emitting higher than their cap. The net result will be a gradual net lowering of emissions to stay below the cap. This method has had some success in Europe (though requiring later reforms to provide stronger financial incentives), some regional success in the United States, but a failure at the national level to gain congressional acceptance despite its attractive market ingredients. The energy industries do not want government-imposed regulation.

More attractive, and with strong support, is a carbon tax, with government setting a tax on emissions, the aim of which can be to increase revenue to pay for damage, to invest in development of new technologies, or to fund other valid social needs. The Yale economist William Nordhaus, joining a number of other economists, has been an articulate advocate of that strategy.[42] The specific aim of such a tax is to force a change in behavior toward either mitigating emissions directly or finding technological ways of doing so. "Energy," Nordhaus has written, needs to be priced "at the level necessary to price energy at its social cost . . . [and moreover] environmental taxes have the unique feature of raising revenue, increasing economic efficiency, and improving the public health."[43]

There are two snags. One of them is the complexity of determining the appropriate carbon tax. It needs to take account of the marginal level of environmental damage, say a rise in sea levels or a loss of crops, or a long-range versus short-term impact. To set a price too high invites a political backlash, while too low a price will not cover the cost of the environmental damage. Setting a low price at first, understanding it is too low and gradually raising it, has been proposed as a via media.[44] The other snag is not technical but political: resistance to taxation by energy industries, notably coal, and conservative groups hostile to taxes altogether. Cap and trade had managed to get by in the European Union by virtue of greater openness to taxation—especially in the Scandinavian countries where it first took hold—but also because it had enough market ingredients to appeal to conservative politicians.

Not so in the United States. In 1994, Vice President Al Gore failed in his push for a carbon tax, and in 2009 President Obama supported cap and trade legislation that also failed. Surprisingly, an important political shift appeared late in that year. It was reported that a number of large and important companies—Walmart, General Electric, Exxon-Mobil, some twenty-nine in all—had started to include the economic impact of a carbon tax in their long-range economic planning.[45] Whether that change portends outright political support for carbon tax legislation, a recognition that global warming can be hazardous to their business, or simply a realistic assessment of what the future will bring is not clear. But they are obviously on a different track than the Koch brothers and some other groups that have supported an aggressive campaign against Republican candidates who are supportive of efforts to combat global warning. Global warming is, in any case, now a divisive issue even among Republicans.

Most critically, while the shift of some companies should be helpful, the prospects of a carbon tax bill making it through Congress in the near future are poor. Companies likely to be hit with the taxes will not easily give up, nor are legislators who have made anti–global-warming efforts a core part of their political plank (and their deep-pocketed backers) likely to do so either. Local, state, and regional efforts to curb emissions may have a greater chance.

## CARBON CAPTURE AND STORAGE

Unlike the mitigation strategies already surveyed, carbon capture and storage (CCS) are not particularly contentious. On the contrary, they have been embraced around the world, particularly in the United States and China. But like nuclear power plants, the efforts are expensive, ordinarily requiring large government subsidies. The aim of carbon capture is to prevent large amounts of $CO_2$ emissions from being released into the air, particularly from coal-fired electricity and other industrial sources. There are various technological ways to capture the emissions, which then have to be stored either under- or aboveground, requiring suitable places to be found. All of that entails large amounts of start-up funds, and there is some uncertainty about the financial long-term benefits—CCS is essentially competing against the much cheaper coal-driven emissions that bedevil most technological innovations.

While there is sense of urgency about most possible ameliorations of global warming, sometimes with implausible optimism, in this case the claim of great benefits is plausible. The technologies to carry CCS have been around for some years and have been tested—and pass the three technology fix criteria proposed by Sarewitz and Nelson cited earlier, but not the additional two economic criteria I have suggested. It is now a matter of scaling them up. As the CCS Institute has put it "CCS technology is the one technology available to make deep emission cuts in several industrial sectors (such as iron and steel and cement). Industrial sector emissions account for more than 20% of global $CO_2$."[46] Using an uncommon argument, the IEA has estimated that a *failure* of efforts to move CCS along would increase the costs of mitigation by $2.1 trillion by 2050. The Center for Climate and Energy Solutions has contended that CCS "is the only technology that can reduce $CO_2$ emissions from existing, stationary sources by up to 90%."[47]

Rule 5 of the technology fix criteria—adequate funding from initiation to implementation—is the obstacle now in both developed and developing countries. The IEA projects only sixty carbon capture plants (CCS) globally by 2030.[48] In the United States, that funding got a significant boost from the American Recovery and Reinvestment Act in 2009. It provided $3.38 million for CCS, with the money dedicated to support the development of multiple commercial-scale CCS projects in both the power and industrial sectors.[49] Unfortunately, after the 2009 high point, U.S. federal annual research support steadily declined by nearly 50% by the 2014 budget request. Between 2012 and 2013 there was a decline in CCS projects.[50] Congressional budget cuts have been doing their notable harm in this and other R&D sectors. Particularly unsettling is a proposal to limit federal support to new coal-fueled plants only.[51] That is a good way for the government to shoot itself in the foot, guaranteeing a slow start-up over years as the new plants come online and locking the emissions from the old plants into place for the foreseeable future.

Another twist of the knife here, as Richard Morse has noted, bears on the carbon tax proposals: "The lack of a carbon tax will make it harder to finance some clean-coal technologies. . . . The profitability of CCS technology depends on governments assigning a price to carbon dioxide; otherwise, there is little incentive to capture a gas with almost no value."[52] He then qualifies that dire statement by noting that "slashing emissions from coal does not require a price on carbon, and there is no reason to wait for one." But I will add a twist of the knife of my own. Most of the world's greatest emissions will come from the developing world in the future. Professor Morse comments on the profound effects that "smarter policies" would have in those countries: upgrading existing coal plants and ensuring that "only the cleanest plants" be built. The money would come from "governments, multilateral institutions, and the financial markets of the world." One can hope, I suppose. Even some patients with advanced stage IV cancer now and then have miraculous cures.

## RENEWABLES

As Mizan Khan and S. Timmons Roberts tartly noted a few years ago, "Adaptation is a second-class issue in climate negotiations," neglected in the Kyoto Protocol and failing in recent years to receive much of the promised government subsidies, particularly from the United States (with $30 billion pledged as part of the 2008 Copenhagen Accord).[53] The year 2008 turned out to be an unforeseeably bad one for developed countries to make pledges and for poor countries to begin receiving money. But Khan and Roberts correctly predicted that as mitigation success declined, adaptation pressures would rise. And so it is turning out: the 2013 IPCC report laid much greater stress on adaptation than earlier reports. While mitigation efforts continue, there appears to be a strong strain of realism emerging in climate studies, fueled by a parallel steady rise in temperature, as if global warming were snubbing its nose at mitigation efforts. Or perhaps just magnifying how inadequate most efforts now under way are. It is sober realism that is driving adaptation in many areas threatened by global warming, notably low-lying islands and areas with dense populations in regions even now barely above sea level. They will be the first to feel the effect of rising sea levels. To be sure, those efforts can be expensive, easier for Palm Beach and Manhattan than for Bangladesh.

Renewables are a different matter, a growing industry and a source of optimism. If by no means sufficient to offset carbon emissions, they are a vital secondary level of defense. A number of them are being pursued, each with its own cadre of scientists, innovative efforts, financial supporters, and a matching industry, although with varying strengths. Their natural sources include: tides and rivers, geothermal energy, biomass waste, concentrated solar power, solar panels, wind, and desalinization of salt water. A few comments on some of them will have to suffice.

Desalinization is a growing industry around the world, especially in the United States and the Middle East, with declining prices thanks to innovation. It can, however, have some negative impact on the poor because of inequitable distribution (toward industrial uses) and is itself a $CO_2$ emitter.[54]

Wind power for electricity generation has had a rapid growth, with thousands of wind farms. "Winds," the U.S. Energy Information Agency said in 2011, "have been the fastest growing source of new electric power for several years," with a 28% increase between 2009 and 2011—helped along by federal production and investment tax credits.[55] Its drawbacks are a growing limitation of suitable areas, costly transmission lines from remote areas, and resistance on aesthetic grounds from some communities.

The proliferation of homes and buildings with solar panels and large fields of panels in many desert areas is visible evidence that they have caught on both for sellers and buyers.

Biomass, both wet (food waste) and dry (corn husks), comes from a wide range of sources and has enormous potential; the world is full of trash, waste, and burnable things. As John Houghton has noted, its potential could take care of "75% of current energy consumption" and also has the added benefit that its burning is a "genuinely renewable resource" in that, through photosynthesis, carbon dioxide from that burning 'is turned back into carbon."[56] The greatest potential is in developing countries, but ones with the least money and poor financial and political resources to make anything happen. The use of biomass for household cooking and heating is itself a great hazard in those countries, accounting for millions of deaths from air pollution. That well-recognized problem is being addressed by healthier forms of home stoves and heating systems but has a long way to go and will require continued support from developed countries.

## GEOENGINEERING: THE LAST STAND?

Geoengineering is a utopian idea, and for many it is a disturbing idea as well. What gives it that flavor is that many of the scientists that support the idea themselves consider it as a last, even desperate, defensive strategy. Faced with the plausible possibility of (1) a catastrophic event at anytime, triggered by some sudden massive event (such as the rapid melting of arctic ice), or (2) a far greater rise in the earth's future temperature than commonly projected (with, say, a 10-meter rise of sea levels), creating disasters of all kinds—a more drastic technological strategy may be necessary. Either event would be catastrophic. The need now is for research and testing of climate engineering.

Two types of geoengineering strategies have been envisioned. One of them is solar radiation management, controlling the amount of solar energy to reach the earth, "injecting sulfate particles into the atmosphere, brightening clouds, and building white roads and roofs." It is said to be relatively inexpensive and has demonstrated efficacy with a high probability of scalable success. It has the danger, however, of contributing to ocean acidification.

The other is carbon-cycle management, controlling the amount of heat that gets back into space. There are many approaches to doing this, such as "capturing $CO_2$ relying primarily on biological processes and iron fertilization of the ocean."[57] It has potential hazards as well. The environmental philosopher Dale Jamieson has suggested that there is some confusion in the geoengineering discussion and that it could be clarified by "a four-part division of adaptation, abatement, mitigation, and solar management."[58]

There is a strong debate between those I will call "the anxious worried" and "the optimistic visionaries." A major concern is that individual countries could use those geoengineering methods on their own, risking damage well beyond their own borders. Another is that the efforts might become an alternative to or distraction from carbon mitigation efforts. A variant on that theme is that, out of a false belief that we already have mitigating technologies available, efforts to develop new technologies will lag; and in any case should be seen only as a stopgap measure as the needed new technologies are brought into use. The worst scenario would be a failure to develop new technologies and being forced eventually to embrace geoengineering techniques and the many dangers that go with them, a lose-lose outcome.[59] How, it has been asked, "can you engineer a system you do not understand?"[60]

Even so, its advocates believe that the danger of global warming is so great, and continuing apace, that the risks should be run, research should begin, and even the likely hazards of the research itself must be borne as well.[61] Moreover, as David Keith of the Kennedy School at Harvard, a strong research advocate puts it, the need is "to weigh the risk of doing it against the risk of not doing it."[62] Once again, risk taking is opposed to risk worry. As with nuclear power, I am on the side of taking risks. Everything hangs on how seriously global warming is taken. If the very fate of our common human future and millions of lives are at stake, taking risks to stop that seem imperative.

## CLOSING THE DAUNTING TECHNOLOGY GAP

With that sobering preamble, a closer look at some of the main possibilities and obstacles is necessary. The IEA 2013 energy projections for the next few decades underscore the gap between demand and supply and the further gap between technology advancement and availability.[63] Between 2011 and 2035 global energy demand will increase by one-third. The share of energy supply from fossil fuels will fall from 82% to 76%—hardly great. Renewables could meet some 40% of the increased demand, generating about 50% of electricity demand. Most tellingly, the developing countries will account for 90% of demand growth to 2035. That growth will be led by China, shifting to India and Southeast Asia after 2025. Coal is expected to remain the leading source of energy, even though its share will decline from 41% to 33% by 2035, the decline coming in great part from shale oil (itself of course a source of $CO_2$ emissions, even if only half those emitted by burning coal). The National Oceanic and Atmospheric Administration announced in September 2014 that the "amount of $CO_2$ in the atmosphere reached 396.0 parts per million in 2013. The atmospheric increase of $CO_2$ from 2012 to 2013 was 2.9 parts per million, which is the largest annual increase for the period 1984–2013."[64] In addition, the report also said, "the current rate of ocean acidification appears unprecedented at least over the last 300 million years." The gradual lowering of emission rates in the United States and Europe in recent years has made no difference for total global emissions.

All in all, there are intimidating gaps among (1) the need to rapidly control dangerous emissions in the face of the continuing increase to hazardous levels; (2) the adoption of available technologies and the development of innovative technologies and the pace that is necessary to achieve it, which is hampered by many bureaucratic and political obstacles; (3) the need for more financial support for technology R&D; (4) the much greater need for technology in developing countries and the lack of economic and other resources available to them to move that forward in an effective way; (5) successful dissemination of technologies at scalable and sustainable levels; (6) the continuing recession in industrialized countries, perhaps some donor fatigue, and strong conservative opposition in the United States to the level of foreign aid needed to close the technological gap for countries unable to do so on their own; (7) gaining social acceptance of new technologies in poor countries (and often developed countries as well) even when available because of local customs and practices; and (8) the failure of global climate technology to come anywhere near exciting—to the point of anxiety and agitation—the general public.

Nothing less than this last point is likely to force all the steps necessary for future security.

Popular support comparable to that achieved by social media and information technology is vital for gaining government backing. Will it be possible to fire up that kind of enthusiasm? The history of technological developments has at times been marked by just that spirit, though often only slowly at first: the steam engine, trains, the telephone, air conditioning, the airplane, space exploration. They are useful benchmarks of successful and culture-changing innovation. The question I want to raise at the end of this chapter is whether the available and proposed means of generating that kind of support will be possible—and not just support but enthusiastic support (see also chap. 8 on public opinion).

## DEVELOPING TECHNOLOGY STRATEGIES

The great obstacle facing technology research, production, and dissemination is what Kassia Yanosek has aptly called the "commercialization gap."[65] On the face of it, there should be many incentives to effectively raise money for technological research and implementation. Probably the most important is the fact that the modern world, and especially the United States, has embraced technology not only as one of its singular contributions to its well-being but also as a creator of jobs and wealth. The success of Silicon Valley, touting technological innovation as its product and wealth for its entrepreneurial investors, offers the preeminent model for emulation.

Peter Diamandis, a prominent entrepreneur of innovation technology, has written a book with his colleague Steven Kotler that captures the love affair with technology, *Abundance: The Future Is Better Than You Think*.[66] And Walter Isaacson's book *The Innovators: How a Group of Hackers, Geniuses and Geeks Created the Digital Revolution*[67] showed how the money was raised. Yet that same passion and excitement has not marked innovation and profit for the technologies needed to combat global warming or any of the other horsemen—despite the large number of technological ideas, many with a good track record. Instead, the reality is that raising the needed money is a struggle. Neither the serious reasons for the technological developments nor the potential economic benefits have caught the national imagination.

Michael Grubb has laid out the pathway that successful funding of technological research must follow to move from the beginning, research, to the end, successful implementation and dissemination. He distinguishes between "technology-push" and "technology-pull" views, often seen as opposed to each other. The "push" view assumes a heavy role for government in providing basic research funding. Businesses might not have adequate incentives to undertake expensive research because of uncertain long-term profit or because the dissemination process itself requires even further investment. The opposing "pull" holds that technological advances must originate in the business sector and that this will happen given good economic incentives. But there is a role for government, that of legislating regulatory measures such as GHG emission caps, which will in turn create the economic incentives to create the technologies necessary to stay within those caps. "Profit-seeking business," Grubb writes, "will respond by innovating to produce technologies that will reduce emissions at less cost in order to gain competitive advantage over rivals"[68] Noting that these opposing views represent the different viewpoints of western economics on the role of the state and the market—which may be less true in Asian countries, which are more able to integrate them—Grubb goes on to offer his own ideas on integration. He outlines six stages of technology implementation:[69]

1. Basic research and development
2. Technology-specific research, development and demonstration
3. Market demonstration (to show they work in real-world settings and their potential market)
4. Commercialization (adoption by established firms or creating a firm to do so)
5. Market accumulation (expanding use in niche or protected markets)
6. Diffusion (on a large scale)

Grubb's integration begins with government-financed R&D with "some proof-of-concept" demonstration and, at the other end, with government enforcing a regulatory structure "that can reward innovators" and patent protection. In between that beginning and end the market has its play, most notably bridging the "valley of death," the financial gap between the initial technological innovation and its final diffusion.

The rhetoric, the substance, and the strategy discussed by technology advocates for global-warming relief are far more sober than mark the Diamandis and Kotler book. The latter lay the emphasis on life getting better and better—visions of sugarplums dancing in their heads—while the global-warming analysts are trying to get rid of dire threats now and in the future, losing more and more of what we already have. Hence, the tendency is to focus more on the obstacles and barriers to innovation, which loom larger than the imagined sugarplums. "The barriers," Brown and Sovacool realistically summarize, are "tenacious, interconnected, and deeply embedded in social fabrics, in institutional norms, in regulations and tax codes, and in modes of production."[70] Moreover, they take their analysis a step further: "Clean-energy technologies are often subjected to unfavorable treatment by fiscal, regulatory, and statutory policies, and they are impacted by policy uncertainty that causes market inefficiencies and a reluctance to innovate. Taken together, these barriers create 'carbon lock-in'; that is, they 'lock' societies into carbon intensive modes of energy production and use."[71]

How are these obstacles best dealt with? As with many major issues in this book, a common theme is that of a multilayered approach—one that features a wide range of strategies more or less pursued at the same time—and the creation of a long list of needs (in almost every case too lengthy and detailed to present in this book). The lists include both obstacles and strategies for overcoming them. The absence of magic bullets is notable, with a few exceptions. The use of a carbon tax as a mitigation tactic is one with wide support, and CCS is another—but both with the proviso that, even if adopted, they would by no means solve the problem. Another theme, illustrated by the passage from Brown and Sovacool just cited, is that solutions require the overcoming of deeply embedded values and behavioral patterns and are not simply a matter of changing regulations or financial incentives.

## WHO PAYS FOR IT?

Most proposals for mitigation and adaptation change require forging or greatly enhancing government–private sector alliances, a receptive government with public support, industry incentives to take chances with uncertain long-term profit, and multidisciplinary and integrated systems, among others. Ernest Moniz, nicely summarizes the ingredients in a systematic approach to financing: innovation (R&D), translation (creation of a product), initial adoption, and diffusion. Government is often necessary for the first step but can be helpful with all of them.[72] Often those goals can be more easily achieved at the local or state level, rather than at the federal level, with the global level presenting the greatest challenge because of national sovereignty, which requires coordination of a kind notoriously lacking with anti–global-warming efforts. We should, however, also keep in mind that despite all the investments in renewables, the steady addition of coal-based energy production all but guarantees a continued rise in global warming.[73] The continued availability of comparatively inexpensive coal and natural gas (and the inertia that creates) takes some pressure off the need for innovative technologies. Who should pay for mitigation and adaptation efforts? In answer to that question, arguments from justice are an important variable in the face of inadequate financing in developing countries for implementing renewable energy. At the 2009 Copenhagen climate negotiations, the Sudanese leader of an ad hoc group of 100 from African nations movingly and heatedly said: "ten billion dollars [promised by the European Union] . . . is not enough to buy coffins for everyone who will die because of climate change in Africa. I would rather burn myself than accept those peanuts."[74] Vehement complaints from poor nations on just that point were heard at a 2013 Framework Convention on Climate Change conference in Warsaw meant to pave the way for a larger agreement at the 2015 conference, and they were joined by the voices of a number of representatives from important NGOs. At the center of their complaints was a "loss and damage" agreement on assisting developing countries with mitigation and adaptation costs.[75] The agreement was seen as both too vague and too little likely to gain significant funding. A further criticism of the conference was that it was partially funded by corporate groups, itself raising a red flag of warning for the NGOs.

There can be little doubt that as a rule industry has been self-interested, beholden to its shareholders, resistant to government intervention, its eyes fixed on the bottom line. But that does not mean it cannot be influenced or see ways in which that bottom line can profit in a socially and economically way from working with government on innovative technology. It is open to government subsidies and government-private relationships in moving through the cycle from creation of a technology to its dissemination. That movement is helped along in those cases where a company already has a product foundation from which to fashion an innovation.

## WHAT COUNTS AS "PROMISING"?

Global warming's journey through technology, adaptation, and mitigation has been along wide avenues with many winding side streets. Determining what is promising and what is not is difficult for a variety of reasons: what is technically possible is not necessarily affordable, attractive to the public and industry, or politically possible. Effective and more efficient use of available technologies, even more with plausible new innovative technologies, usually requires that a number of variables come together in ways that overcome the obstacles. A common feature of much research on adaptation and mitigation consists of lengthy lists of strategies to make that happen. My task now is to assess those efforts to determine (1) what difference would those strategies as a whole, if acceptable and successful, likely make in reducing global warming, and (2) what is the likelihood that proposals for particular adaptation and mitigation technologies would be successful? I will get at those questions by using four different technology categories: (1) long-term impact on global warming, (2) short-impact trends, (3) some technologies that may make matters worse, and (4) strategies of hope and glimpses of silver linings now in the clouds. To do this I will make use of the very thorough IEA's 2013 fact sheet, which makes projections up to 2035 based on present trends, and what has been covered on energy in this final section of the chapter.

But I will begin with a category not on my list: apocalyptical warnings of the kind most generally associated with Bill McKibben, but even going further. At the end of 2013 James Hansen and colleagues published a paper on global warming that was wholly alarming.[76] Hansen—often called the world's greatest authority on global warming and recently retired as director of NASA's Goddard Institute for Space Studies—together with some distinguished colleagues was reporting on the results of a number of recent studies. They indicate that not only is the problem of climate control getting steadily worse, it is doing so even faster than anticipated even a few years ago, and the window of opportunity is rapidly closing. "Strong evidence about the dangers of human-made climate change," he writes, "have so far had little effect. Whether governments continue to be so foolhardy as to allow or encourage development of all fossil fuels may determine the fate of humanity. . . . The current global emissions trajectory could within three years guarantee a 2°C rise in global temperatures, in turn triggering irreversible and dangerous amplifying feedbacks." More recently, the Lima meeting revealed a rift on the long-standing belief that 2°C would be the tolerable limit. That standard may itself be too lax, many held. But from a political perspective, setting a lower acceptable level could court rejection or despair. It could be an intolerably heavy and frustrating demand. The resolution of that debate, now uncertain, could make a great difference at the 2015 meeting.

Hansen cites a 2013 IEA projection that "with current policies in place we are locked into a rise of between 2°C and 5.3°C," adding in an interview that 4°C "would be enough to melt all the ice . . . we are now three years away from that point-of-no return." His solution is "a rising carbon fee," and he offers no hope that current mitigation and adaptation technologies could save the day. But then neither does he tell us how to persuade the United States and China, the greatest polluters, to accept a carbon tax that would be so high as to force carbon emitters to wholly desist at once from any and all further emissions. It may be "foolhardy" for them to reject such a tax, but it would take an astonishing overnight conversion experience to bring that about, something so far not on the horizon.

## MAJOR WARMING TRENDS AND MINOR MITIGATIONS

### Major Global-Warming Trends

The IEA has projected a one-third increase in global energy demand between 2011 and 2035, with a small decline in the share of fossil fuel from 83% to 76%.[77] Renewables and nuclear energy will meet about 40% of basic demand during that same period. Fossil fuels will continue to dominate the power sector, with coal as the largest source of energy. The energy sector itself is considered pivotal in $CO_2$ reduction. In developing countries, coal will remain the leading source of electricity, with its share falling from 41% to 33% but still "leaving the world on track for a long term average temperature increase to 3.6°C." Quite apart from Hansen's scenario, these figures underscore the likelihood of a dangerous future, going well past every benchmark of disaster developed by global-warming analysts up to 2013. Here is a final set of figures: "The global cost of fossil-fuel subsidies expanded to \$544 billion in 2012 despite efforts at reform. Financial support to renewable sources of energy totaled \$101 billion." Hansen's comment on "foolhardy" governments is surely apt. Worst of all, whatever benefits mitigation and adaptation may have brought by 2035 are simply wiped out by the larger energy trends.

## Lesser-Impact Technology Trends

It might seem anticlimactic to talk about lower-impact trends when the long-term global-warming outlook is so bleak. I have three reasons to do so. One of them is that we are still a long way from a final environmental disaster, and there are scientific quibbles about the available time frame before all is lost. The second reason is that both Hansen and the IEA leave open some possible ways out—if we are wise enough to pursue them, and with even greater vigor. The third reason is that when so much is at stake, the only choice is to fight, to keep going whatever the poor odds of success. Mitigation and adaptation are the weapons to use in the fight, the only ones available.

It is evident from the figures already presented that investments in renewable energy sources are inadequate, and that those available are being undercut by continuing government subsidies of fossil fuels. That terrible ratio must be reversed. A global carbon tax of a sufficiently heavy kind could reverse that trend if most of the money raised from taxation was applied to migration and adaptation technologies. Will that happen? Given the present resistance of China and the United States to such a policy (and probably India, Canada, and Australia) the odds are heavily against it. Even with the best will in the world and unparalleled global political agreement the three-year time frame presented by Hansen will be hard to meet—and has barely started. What about an enormous and sudden rise in global bad weather or another event of a kind whose effects would impact every country in a severe, inescapable way? That could make a great difference, but there is no reason to believe that it will happen.

The 2013 IEA fact sheet is not supportive of immediate changes. It projects the power generation from renewables to rise from 20% in 2011 to 31% in 2035, with gas to replace coal as the leading source by 2035. Subsidies for renewables would have to increase from $101 billion annually in 2012 to $220 billion in 2035. Solar power would require large subsidies through 2035 to compete with cheaper sources of energy. Shale gas, now increasing at a fast pace, is not expected to continue in availability, and a return to oil is expected in thirty years or so. Wind farms are already finding it hard to expand because of limits on usable sites. While the costs of desalinization are declining they require, of course, proximity to oceans and large-scale pipelines to landlocked regions to make a large difference. Carbon capture has significant promise, awaiting only large government subsidies and commercial incentives. And the word "promising," used so often in the technology research literature and the media, has yet to achieve the public excitement of the communications and data-collecting industry that has made Silicon Valley and its products known and embraced by everyone in both rich and poor countries.

## LOOKING FOR, LONGING FOR THE SILVER LINING

In 2012 the talented journalist David Leonhardt, head of the *New York Times* Washington bureau, wrote an article with the title "There's Still Hope for the Planet."[78] I call this the "silver lining" genre (which we all easily slide into). Not many articles or studies on global warming have a title like that, and I was curious about what kind of case he would make. He noted, aided by two charts that demonstrated his points, that wind and solar energy are getting cheaper globally, dropping for wind power from $500 per megawatt-hours in 1980 to less than $100 in 2010, and for solar power a drop from $7 per megawatt-hour (capital and installation costs only) to little over a projected $1 in 2015. Another chart showed a strong rise in the use of renewable energy. The article itself was filled with similar promising and other hopeful technology policy moves. He did note, however, that wind and solar power would have trouble competing on a "mass scale with existing energy sources."

He then qualified that judgment by noting that one (unidentified) expert group set a target of $25 billion a year for federal research and observing that such a goal was lower than the $30 billion a year or so (now $29) billion that the NIH receives. The NIH is not a great precedent. No other federal agency has ever had such continued congressional bipartisan support over the years. It has rested on a public opinion base that for fifty years has had the kind of enthusiasm for medical research unequaled by anything except recent public support for the technologies coming out of the Silicon Valley technology success story. Moreover, the health, life, and death of the human body is something everyone understands. It is up close and personal.

Leonhardt has another chart, one that decisively shows a decline of federal support for clean-energy expenditures from $44.3 billion in 2009 to a projected 2014 figure of $11 billion. Even so, assuming that the recent shale gas rise, together with stories of storms and other evidence, would motivate more interest, he concluded his article by saying that "there are some reasons for hope—tentative, but full of potential—hiding beneath the surface." I would remind the reader of the wide play given some (often contested) small declines in childhood obesity (about the only hopeful note in that field); the constant refrain of coming cost declines from new medical technologies for treating chronic disease (despite a fifty-year record in that realm of cost increases); and plenty of hopeful ideas out there for food and water shortages (and there are some marginal gains there).

Yet there is one place where hope has been matched by some significant results. The United States has seen an average 2% decrease in carbon intensity per year since the 1970s, and many European countries have also seen decreases to greater or lesser extents. The Obama administration and various groups have determined, however, that the global level requires an annual 4% decrease to meet the goal of no more than a 2°C level, beyond which lies the most dangerous territory. To do that, William Nordhaus has calculated, "would require rapid technological shifts than have not been seen in any industry. This technological fact underlines the daunting challenge posed by climate change."[79] And Nordhaus wrote those words before the 2013 Hansen report was published. A good bit of progress is being achieved at the local level in many countries, so if one is desperate to find some silver lining, it is possible. But that does not come anywhere near doing something about the larger and darker cloud hanging over us. Brought out into the open, "the reasons for hope" look thin.

The comparative success of renewable technologies in recent years—and the kind of hope it inspires in some—in comparison with the continuing, even accelerating, increase of atmospheric $CO_2$, reminds me of a common problem in medical end-of-life care. A patient is dying and the medical prognosis is that he will not live long. His family, however, hopes for some kind of medical miracle, at best a remote possibility. Yet even if his future is inexorably downhill, his day-to-day condition is subject to fluctuations (common with heart failure). On many days, he can seem better and actually be a little better, than the day before. When the family is told that "he is better today," their hope is raised falsely, and even when the physician tells them that there is no change in the downhill course. It is hard to live without hope in the face of threats to life, and fluctuations simply add confusion.

# 8

## A VOLATILE MIX

---

### Public Policy, the Media, and Public Opinion

thread running through this book is that of bringing about large-scale changes in the way people live their lives, particularly when threatened with serious harm. Making progress with each of my five horsemen requires bringing about such changes. I think of change in two ways. One of them I call *social change*, making use of education, science, governance, law and regulation, and finding a good balance between the market and government; that is, focusing change on the various ingredients within the social structure of a society as well as the global relationship among societies.

Then there is *cultural change*. It is made up of a shift in the values held by people, influencing the way they personally behave and live their lives. It encompasses their political ideologies, the histories and traditions of their countries, how they understand their moral relationships to the communities of which they are a part and to the much larger global community.

While the social and the cultural can be differentiated they cannot be disconnected. Cultural values shape and influence social institutions and behavior, while shifts in and experience with social practices can over time often reshape the underlying cultural values. It is a two-way street. The taxation of sugared beverages requires government regulation, but gaining such regulation requires a population willing to let the government attempt to change their private behavior, something Americans resist but the Japanese accept. The culture of some countries or regions encourages local social change, such as opposing nuclear energy (Vermont), more than large countries with a different kind of stake in the issues (France). Genetically modified organisms for agriculture are accepted by some counties (the United States) but rejected by others (some African nations).

This chapter combines three features of importance in the shaping of social and cultural values for coping with the five horsemen. One of them is the relationship between scientific knowledge and the fashioning of public policy. Another is the role of the media as a vehicle for public education, affecting public opinion, and shaping social and cultural values. Still another is that of public opinion as a key component of getting policy supported, enacted, and implemented. Taken together they are the most important influence on devising a response to each of the horsemen (though with varying weights and force). The interplay of these three features makes the difference. Of course, in the case of the media and public opinion they can also be obstacles to change, favoring the status quo, that of providing support for business as usual. The way issues are framed can also influence policy, the media, and public opinion. Mitigation and adaptation play an important role in global-warming policy, but those terms are rarely used as such with the other horsemen, although the ideas behind them can frequently be found, and I make use of them when that seems relevant.

Unlike the previous chapter on energy and technology, organized by working through those issues by examining each of them one horseman at a time, this chapter will be done in the opposite way. I will use the three general topics as my organizing categories and then work through the five horsemen one at a time. As was the case with the previous chapter, global warming is dominant. I will begin with science and policy, move through the role of the media, and conclude with public opinion.

## PUBLIC POLICY: GLOBAL WARMING

The report of the Intergovernmental Panel on Climate Change (IPCC) in late March 2013 was notable from a number of perspectives. It got good media coverage. It gave much stronger scientific support to dangerous trends noted in earlier reports. It emphasized that global warming is already doing considerable harm and is not just a future hazard. While the report stressed that both mitigation and adaptation are necessary, it laid much greater emphasis on adaptation and risk. It paid particular attention to food problems in poor countries. And it touched on the delicate and difficult matter of how much money would be needed from rich countries to help poor countries cope with global warming. The question I want to pose for this part of the chapter, on science and policy, is whether the report—well publicized in the media—will provide the necessary support for stronger national and international efforts. Does it leave any room for lingering doubt, significantly reducing uncertainty? That will set the stage for a closer look at the science and policy issue.

## Science

As with earlier reports, the 2015 assessment casts its findings in terms of the quality and amount of evidence, and agreement among the researchers. The evidence is summed up as limited, medium, or robust, and the agreement as low, medium, or high. Confidence in the validity of a finding puts together the evaluation of evidence and agreement as very low, low, medium, high, and very high. No more than samples of their many assessments can be listed here, but they will serve to indicate the care and nuance that marks their methods. The already present impacts of global warming, for instance, are: In many regions, changing precipitation or melting snow and ice are altering hydrological systems, affecting water resources in terms of quality and quantity (medium confidence). Glaciers continue to shrink (high confidence) affecting runoff and water resources downstream—and causing permafrost warming and thawing in high-latitude and high-elevation regions (high confidence). Based on many studies covering a wide range of regions and crops, there are negative impacts of climate change on crop yields, which are more common than positive impacts (high confidence). Then there are the impacts from recent climate-related extremes, such as heat waves, droughts, floods, cyclones, and wildfires, which reveal significant vulnerability and exposure of some ecosystems and many human systems to current climate variability (very high confidence).

The report identifies with high confidence eight key risks from global warming:

1. Risks of storm surges, coastal flooding, and rising sea levels
2. Risks to urban populations due to inland flooding in some areas
3. Risk of extreme weather events
4. Risks from extreme heat
5. Risk of food insecurity and breakdown of food systems
6. Risk to rural areas and livelihoods because of insufficient drinking and agricultural water
7. Risk of loss of marine and coastal ecosystems
8. Loss of terrestrial and inland ecosystems

In addition the report notes that the magnitude of global warming intensifies these risks. The higher the rise in global warming the greater the risk. Some risks are considerable at 1°C to 2°C (3.6°F) above preindustrial levels but with high or very high risks with an increase beyond 4°C (4.5°C or 8.1°F).

Hardly a week goes by without a news story reporting on research of an emerging event of significance in some part the world. Here is a random spring 2014 potpourri: 1,600 years of accumulated ice in the Andes of Peru have been lost in the past twenty-five years; in recent winters many traditional winter snow and ice events were canceled in Alaska, where temperatures are rising at twice the rate of the lower states with an acceleration of permafrost thawing (which releases methane, a pollutant more dangerous than $CO_2$);[1] the water around New York City has been rising; there is a general global trend to a drier global earth surface; black carbon (soot) is a serious and greater risk than was earlier thought;[2] there has been significant arctic ice loss and loss of land ice in Antarctica; and by 2047 the coldest years may be warmer than hottest years in the past.[3] Each of these stories had scientific support in the 2013 IPCC report.

## CLIMATE SENSITIVITY

To strengthen the overall thrust of the 2013 IPCC report—that the seriousness of the problem continues to increase and greater progress is needed to manage it—some additional important information underscores it. The question of what amount of temperature rise is necessary for a doubling of the $CO_2$ level ("climate sensitivity") has been marked by serious uncertainty for some years (and seized upon by climate skeptics). But the 2015 report removes much of that uncertainty on the rapidity side. It confirms with "high confidence" that a sensitivity range determined in 1979 remains solid, that of 1.5°C to 4°C, but with uncertainty just whether its increase will be on the low side, giving us more time, or faster, giving us disaster. In May 2013, measurements in both Hawaii and Alaska showed that $CO_2$ in Hawaii and Alaska reached an average daily level of 400 ppm, a figure that had never been reached in the previous 3 million years. The figure of 450 ppm has been used by many countries and analysts as a maximum acceptable figure for limiting dangers, a number rapidly being approached. In the late 1950s the level was 315 ppm.[4]

The 2013 figure suggests a move higher in the sensitivity range and thus more rapid climate change than earlier projected. I cite one more important 2013 study, that of what are called "wedges." In 2004 Robert Socolow and Stephen Pacala wrote an influential and much cited paper in which they contended that "humanity can solve the carbon and carbon problem in the first half of this century by scaling up what we already know how to do."[5] Their solution was that a number of "wedges" were needed, each focused on some actual and promising large-scale strategies, fifteen in all: efficiency and conservation; reduced reliance on cars; more efficient buildings; improved power plant efficiency; substituting natural gas for coal; storage of carbon captured in coal or gas in power plants; storage of carbon captured in hydrogen power plant storage of carbon captured in synfuels; nuclear fusion; wind electricity; renewable hydrogen; biofuels; natural sinks; forest management; and agricultural soils management.

In 2013 Steven Davis and colleagues published an article pointing out that an even bolder plan with more wedges, perhaps as many as thirty-one, are needed. For one thing, not one of the earlier proposed strategies has worked.[6] For another, some have moved in the wrong direction (worse, not better), and others languished as a result of ineffective policies. "Insofar as current climate targets accurately reflect the social acceptance of climate change impacts," they wrote, "then solving the carbon and climate problem means not just stabilizing but sharply reducing $CO_2$

emissions over the next 50 years."[7] Nothing in the 2013 IPCC report confutes their judgment.

A number of observers did, however, note some changes in tone and specificity in the 2013 IPCC report.[8] Where specific predictions of dangers in specific regions and locales were made in the earlier 2007 climate impact report, they were removed and replaced by more general warnings of possible events. It had become clear that there was more divergence than convergence of regional changes. The hope that stronger science and computer modeling would make good predictions possible turned out to be wrong in many cases, and the memory of past mistakes and false predictions apparently had a sobering effect this time around.

But far more sobering was the assembly of a wide range of global damage specifications. The climate change dangers are already with us and getting worse. Nothing done so far has made a great difference, and present policies are inadequate or too weak to be effective. If one is to respect the climate research, then the need for immediate and accelerated action is imperative. But that is no longer news and hasn't been for years. Yet it is a welcome exclamation point.

## SCIENCE AND POLICY

It is also not news that, despite growing scientific evidence about climate change, no strategy to inspire effective and decisive global action has been discovered. I use the word "inspire" deliberately. Knowledge is not in short supply. Good scientific evidence and many plausible, rational strategies are available. The public here and abroad now recognize global warming as a problem. They just do not feel enough emotional concern or anxiety to give it a high policy priority—nor are they willing to make much personal sacrifice to bring about a decisive change. Just how can that be brought about? How can everyone be motivated to take it more seriously and thus in turn press their legislators and government to do so as well, adopting policies that could be effective?

The relationship between risk and uncertainty is a crucial feature of the global-warming debate, influencing public acceptance, attendant emotional commitment, and policy implementation. The solution to the emotional and motivation issue lies somewhere in that domain, but as yet not discerned. We know that pictures of devastation and human suffering can often move people more than words. No less effective are pictures and vivid writing that can make people take seriously threats they heard about that made little impression on them earlier: vivid descriptions of enemy troops approaching a border accompanied by pictures of dead bodies in their wake. Some of the most effective motivators in global-warming education have been pictures of melting glaciers and declining sea ice. The nearest many of us come to firsthand experience of global warming are storms and drought; that makes it all real.

Or does it? In his 2013 State of the Union address President Obama said, "We can choose to believe Hurricane Sandy, and the most severe drought in decades, and the worst wildfires some states have ever seen were all just a freak coincidence. Or we can choose to believe in the overwhelming judgment of science—and act before it is too late."[9] But the citing of specific weather events to make a case is just the kind of rhetoric that makes scientists nervous and that led the IPCC to reject future predictions of a specific kind. Single events, such as Hurricane Sandy or droughts in Texas or California, do not prove global warming. Only patterns over time can do that. Ironically, that caution takes from the repertoire of arguments supportive of global warming one of the few that can directly catch the public's attention and feeling—or come to be a "teaching moment" as a *Science* magazine article put it.[10] News stories with titles like this can catch our eye these days with some regularity: "Science Linking Drought to Global Warming Remains Matter of Dispute."[11]

I mention the role of emotions because of an uneasiness about the extremely rationalistic nature of many approaches to policy making. The most notable is the belief that if the public and legislators better understood the science behind the risk of global warming and the significance of the risks uncovered, they would be moved to act. But a question can be raised about that seemingly commonsensical belief: Is it generally true and, if it is, is there any guarantee that knowledge will change their *emotions* enough to move them to act, and with what level of intensity? Daniel Sarewitz has noted in his article that more scientific knowledge will not eliminate different interpretations of that knowledge among scientists.[12] Nor would it, I suppose, unearth the varied emotional responses to the information, a factor that will come into play even more with laypeople and public policy implementation.

While it is too early as of this writing to know what the public opinion impact of the 2013 and 2014 IPCC reports will be, we do know that their earlier reports did seem gradually to make a difference. But the translation of scientific information into changed behavior is always slow. It took years after the connection between smoking and lung cancer was known and publicized to bring about change, about twenty-five years; and not only among smokers but also among the public and legislators trying to devise laws and regulations to put teeth into the education strategy (and 20% of Americans still smoke). More scientific evidence on the harm of obesity has done little to sway lawmakers to enact tough laws and regulation, and medical research has not reduced obesity.

## UNDERSTANDING UNCERTAINTY

Understanding the relationship of risk to policy development is a crucial step. One could in fact say that the main point of the research of the IPCC over the years has been to solidify the science that determines the level of risk from global warming. That knowledge is assumed to be the foundation upon which policy must be based. Uncertainty about that knowledge is then a significant threat. That is not only because wrong projections and predictions can be embarrassing, hurting overall credibility and playing into the hands of skeptics, but no less because it calls into question the importance of science in the debate on global warming—and thus, most of what it has to say about risk itself. Are the purported risks real or not real, and how serious are they? How will they affect my personal life or that of my region or my country?

As an analysis of environmental uncertainty by Laura Maxim and Jeroen van der Sluijs put it this way: "The current practices of uncertainty analysis contribute to increase the perceived precision of scientific knowledge but do not adequately address its lack of socio-political political relevance. . . . The current practices are incomplete and incorrectly focused. . . . [They] must include quality criteria . . . within political, social, and economic context and processes."[13] The IPCC reports surely fall into the "positivistic" models of analysis the authors note as a standard model of risk analysis, appealing to the mind rather than the emotions. One obvious problem with their alternative is that the IPCC was not organized to work with a richer model. The scientific task was an immense and difficult one, and there was no parallel group working to make the connection. That job has in effect been left to other organizations and researchers, widely scattered at that, and unlikely to draw the kind of media and public attention of a well-publicized major UN effort.

The Maxim and van der Sluijs study, however, maps the terrain very effectively, beginning with a list of six objectives of uncertainty analysis:

1. Increase precision and identify knowledge gaps
2. Increase decision makers' confidence in robustness of scientific results
3. Improve stakeholder's and public's confidence in science
4. Increase stakeholders' confidence in decision-making
5. Improve the quality of decisions
6. Highlight the influence of science communication patterns on decision making[14]

Yet for all of its value, the focus is still on scientific knowledge as the crucial lever for action. I cannot help noting again that the improvement of the IPCC reports over the years, accompanied by much collateral research on each of the six goals, has not made much difference in public support for taxes and regulation. What is missing? First, some ambiguity needs to be addressed regarding the kind of science required for dealing with this kind of problem. Second, how best to understand the psychology and cultural influences that shape responses to risk and how they can complicate the policy-shaping process and its public and legislative reception?

What kind of scientific problem is climate control? Sarewitz's argument that more science per se will not necessarily be helpful seems to me to be plausible, even if a refinement of present knowledge can be helpful—as was the choice to eliminate in the 2013 report the specific regional weather and other predictions found in earlier reports. But science, even in the conventional sense, is always changing, gaining new knowledge or rejecting earlier findings; it never stands still. The "consensus" of scientists on research findings is the gold standard. The word "truth" is rarely used. So too with global-warming science.

As Mike Hulme has observed, the model of good science taken to be canonical many years ago was proposed by the sociologist Robert Merton. Three criteria make up the model: universalism, communalism, and disinterestedness. Its place has been taken over, at least for some kinds of research, by a "postnormal science" that aims to be used for "beneficial social outcomes."[15] It is science for issues wherein "facts are uncertain, values in dispute, and decisions urgent . . . [and] decisions have to take into account of future risks in decisions to be made today." Roger Pielke Jr., thinking along similar lines, considers climate change a problem that can only be managed, not solved.[16] That judgment is based in part on the fact that some of the hazards of global warming will continue indefinitely ($CO_2$ now present in the atmosphere will remain for hundreds of years), and in part on the fact that uncertainty affecting policy decisions will remain contentious. He also rejects exaggeration of risks as counterproductive, an issue I will return to shortly.

## RISK AND UNCERTAINTY

Risks play a large role in our lives but is assessed differently by different people and in the different roles they occupy in public life. There is a large psychology literature on why people deny risk and fail to face up to danger: if it is in the future, it can be put aside for now; it evokes fear leading to denial; it evokes false hopes for some not yet available solution; and it can induce a sense of hopelessness and futility. In some cases the extent of a danger can lead to different reactions, some contradictory. Famously, the debate leading up to World War II was marked by bifurcated responses to risk in the 1930s: the persecution of the Jews, the rise in Hitler of a fanatic dictator, and the rearmament of Germany were clear evidence to many of a risk of eventual war. But not to strong anti-interventionists in the United States. They saw no serious threat and did not want any involvement in foreign wars. Only the Japanese attack on Pearl Harbor finally changed their minds; it took an obvious catastrophe to convince them of the danger. At the beginning of the European war, British prime minister Neville

Chamberlain believed he could negotiate with Hitler. He turned out to be terribly wrong. The more hawkish Britons did not believe negotiations with the ruthless Nazi leader were possible, and Chamberlain was quickly labeled an appeaser. Working against the anti-interventionists and Chamberlain were President Franklin D. Roosevelt and Winston Churchill, both with earlier promilitary attitudes and a different set of values concerning the global roles of their countries. The stance taken by Roosevelt and Churchill was eventual proved to be right, but from the very beginning of the 1930s, the governments of both countries had the same information. It is hard also not to forget that there is a debate to this day about why the American government slighted evidence of Japan's military aspirations. Why were these risks discounted?

## Too Often Overlooked: The Risk to Values

I want to add to the discussion of risks and policy making one further thought. There is agreement among most analysts that neither the science nor the policies based on them can be kept free of the values of the scientists in postmodern science, or the values of those fashioning policy, or the values of those who will be affected by the policies. But most of the risk analysis is focused on *physical harms* to the planet; that is, for instance, harms to crops, rivers, forests, the air, and harm to our bodies indirectly as a result of harm to the physical ingredients and resources necessary for our maintenance and survival.

But I want to add a category of risk analysis I do not find clearly distinguished among the "values" said to influence policy making and acceptance. I call it a *risk to values*. Many analysts, to be sure, emphasize the inevitable interplay of science and values, both in the science and in policy. But I am interested in the particularly strong impact of moral and political ideologies on the reception of policy proposals. I earlier distinguished between social and cultural risks. Someone can have no objection to the science behind a policy proposal but may morally and politically reject its implications for social change. The obvious example is that of conservative legislators hostile to increased government regulation, no matter whether applied to global warming or business practices. Biases of that kind can work in both directions: liberals will be more likely to accept a strong government role. It is not surprising that European countries with strong welfare policies are more willing to accept regulation than the United States. It was not skepticism about the science of global warming that kept America from agreeing to the Kyoto Protocol but, as George W. Bush said quite clearly, that it was a threat to jobs. He was likely correct in seeing that possibility. The defeat of an apparently bipartisan "cap and trade" bill in the U.S. Senate in 2010 was defeated by two value ideologies moving in different directions: conservatives saw it as an "economy-killing tax," while liberals viewed the tax as an efficient and acceptable way to move forward to control emissions.[17]

There are also any number of historical and contemporary examples of values obviously trumping risks. U.S. prohibition laws in the 1920s were eventually repealed not because they were failing—they significantly reduced drinking and reduced the deaths and other harms of alcohol consumption—but because they spawned crime and interfered with the freedom of people to choose what they judged good or bad for them. The large number of deaths from guns, and even the Newtown shootings, does not deter those supporting gun control. In the aftermath of Newtown, the National Rifle Association fought every change to tighten gun control. In fact, gun sales increased after Newtown out of a fear that regulation would make it tougher to own guns. The predicted 33,000 deaths from auto accidents in 2015 exceeds the projections of 32,000 deaths from gun violence. Our love of automobiles together with their central place in our culture leads us to accept their deadly toll on health. The Second Amendment right to own firearms, even assault weapons, easily beat out the terrible mortality outcomes of their availability.

Motorcycle helmet laws have been taken off the books in many states despite the well-known fact that they save lives. A reduction of automobile speed laws by 10 mph could eradicate 10% of auto deaths—it happened in the 1970s when such laws were adopted to deal with a gasoline shortage. But that is much too radical a solution for the public. We love our cars. They are worth the risks, to ourselves and others. Seat belts, air bags, and safer autos have been more acceptable adaptations to the dangers. It may well be, although I have not seen it mentioned, that the 2013 IPCC emphasis on adaptation may be a kind of tacit recognition that the more difficult mitigation needs will not be forthcoming because of social and cultural obstacles. There are any number of dangerous sport and recreational activities that are not banned even if lives would be saved. The French have rejected GMOs, even ones known to be safe and cost-effective, on the cultural grounds that GMOs are harmful to their culinary traditions. Good and well-publicized science on standard physical risk grounds is often easily overcome by ideologies and cultural values.

If it is true that global warming will in the end require some basic and radical changes in social organization and the cultural values underlying it—reshaping industrial societies and limiting many of the ways in which live our individual lives—then the examples I just provided are sobering. For many (perhaps half the American population?) that will not be at all acceptable. Scientific knowledge will only sometimes have the power to induce change. But if it runs up against threats to cherished and familiar ways of life, it is often the loser. For many people, the famous statement of Patrick Henry, "Give me liberty or give me death," can be championed well beyond protecting the nation against dangerous enemies. It could be emblazoned on the bare heads of helmetless motorcyclists.

An arresting contrast about fear and the future occurred to me when I read a 2014 article in a leading medical journal on women's response to recent debates on the criteria for undergoing mammography screening.[18] A strong body of evidence has accumulated that below the age of fifty there is a much lower risk of breast cancer detection and a much greater risk or harmful side effects from the screening. The screening can stimulate unnecessary often harmful treatment for up to 40% of those screened. When informed of that risk, however, most women decide to undergo the screening anyway. Yet afterward, even when they had the screening and suffered from the side effects, they were glad they had done it. They preferred the danger of the treatment to the danger of cancer. The evidence with the running of global-warming dangers is just the opposite: better to take the risk of future harm than the dangers to present values and ways of life. I can only surmise that the women were fully aware of the lethal hazards of cancer from seeing it happen to people they knew or heard about, whereas most people's personal sense of global warming is minimal.

## THE MEDIA: RISKS, FEAR, AND RHETORIC

Almost all of the problems of the science and policy issues so far dealt with are eventually filtered through the media, a great force in shaping public opinion and feeding political controversies. Three features of the media can be singled out. There is the political use of the media by advocacy groups to gain favor for their views or disfavor for their opponents. The second is the use and play of rhetoric in the global-warming debate. The third is the role played by journalists, particularly those who are themselves advocates for some position.

Many global-warming researchers argue that a tactic of instilling fear is ineffective and even likely to be counterproductive; analogous, I note, to fears of harmful results from stigmatizing the obese. In both cases, the belief is that fear can paralyze, inducing fatalism and, additionally with obesity, creating public repugnance toward the afflicted by stigmatizing them. Yet a fear of undesirable value and political consequences of various environmental policies has been effective in motivating conservative opponents to rally against global-warming control. Political fear is effective. That effectiveness can be seen in the successful conservative campaign in the United States to reject or minimize global warming, using both the media and lobbying (the latter to be taken up in chap. 9).[19] It has been a well-organized, financed, and focused campaign, beginning with the Reagan administration working against environmental legislation.

Early on, the deniers and minimizers saw that is would be politically prudent to work against that legislation but not against the goal of environmental protection, taken to be less disruptive. By the early 1990s that effort had gained considerable money, backing, and influence. Its main aim was initially to foster skepticism and uncertainty about global warming, later escalating to an outright attack on the science behind it, most specifically the IPCC and the Kyoto Protocol. Much as the tobacco companies were doing to combat the campaign against tobacco use, it enlisted dissenting scientists, heated up the rhetoric of the debate, attacked the scientific peer-review system, exaggerated the cost of change, and belittled scientific institutions and even the ethics of scientists. It gained the support of wealthy individuals (the Koch brothers), conservative foundations (the Heritage Foundation), and the conservative media.

That media includes most importantly the *Wall Street Journal*, a newspaper with the world's largest circulation, and a necessary source of news for the business world. Not only does its editorial page frequently weigh in against global-warming reduction efforts, its numerous op-ed articles echo the same theme. In comparison with some other national newspapers—the *New York Times*, the *Los Angeles Times*, the *Washington Post*, and *USA Today*—it has run over the years far fewer straight news stories (though solid enough) than those other papers by a ratio of five to one in favor of the latter.[20] In its editorial response to the 2013 IPCC report, the *Journal* seized upon some of the changes from and corrections of earlier reports as displaying a "more muted" (and by implication chastised) tone. The other newspapers, as well as many other analysts, saw not a muted and chastened report but good evidence of good science, always correcting itself. Rounding out other conservative journals are the *Weekly Standard*, *National Review*, and the *American Spectator*. Two of the country's most prominent newspaper columnists, George Will and Charles Krauthammer, add their voices. Glen Beck, Sean Hannity, and Bill O'Reilly join in. Some well-known politicians help things along—Oklahoma senator Charles Inhofe once said on the Senate floor that global warming "was the greatest hoax ever perpetrated on the American people."[21] Hardly needing to be spelled out, this whole conservative attack—not looking for negotiation or compromise—has been an important contributor to national polarization. It has been picked up as well by conservative leaders as well in Canada, the United Kingdom, and Australia.

In a *New York Times* article on the use of rhetoric on February 17, 2015, Justin Gillis reported on an effort—with more than 22,000 signatures of laymen and scientists collected by Forecast on Facts—to rid the media and public debate of the term "science skeptics."[22] It should be replaced by "climate deniers," a term often now already used but often treated as the equivalent of science skeptics. At one level, the issue is that of understanding how good science research is conducted: skepticism and constant refinement are basic values, and an important reason why research changes and moves on over time. At another level, the complaint is against the politicizing of global-warming science, shifting the emphasis away from the science, solid and trustworthy, to the ideologies superimposed on them. A small subgroup of climate dissenters are those who do not deny the reality of global warming but contend it is not as serious as the alarmists assert and is thus manageable.

## Invoking Fear

If belittling the science of global warming has been the main thrust, the deniers and minimizers have also played the fear card, emphasizing the role of big government and nasty regulation, a dangerous arrow aimed at the American way of life. The power of this onslaught cannot be minimized. It is a perfect illustration of how a small minority backed by plentiful money, good lawyers and lobbyists, strong media support, and a single-minded goal can have proportionately more influence than a larger group with much weaker assets. The American public opinion polls highlight that influence.

An interesting debate among those in the global-warming research and media community shows both some differences on the invocation of fear as a motivator for action and a strategy to be employed in the media. I have already noted above that fear about a threat to values and ideologies seems to be effective with conservatives. But what about fear as a motive for action among those who agree that global warming is a serious matter but cannot be induced to give it a high priority or to give up much to counter it? As with resistance among obesity experts to the use of stigma in countering obesity, there is a comparable resistance to the use of fear as a rhetorical means of moving emotions about global warming. There have long been complaints about the media's tendency to exaggerate global-warming threats and dangers in ways not justified by science (just as it does with crime and health threats). But a focus on the use of fear presents a different kind of problem. Two well-known environmental researchers, Ted Nordhaus and Michael Schellenberger, wrote an op-ed for the *New York Times* on the use of scare tactics: "More than a decade's research suggest that fear-based appeals about climate change inspire denial, fatalism and polarization."[23]

Another distinguished scientist, Steven A. Cohen, had only a few days earlier criticized both other scientists and the media for simplistic predictions of doom. On the contrary, he held that as they have in the past, "new technologies and new techniques will overcome the problems we now face."[24] That was in essence a strong invocation of hope as the cure for fear. Andrew Revkin did not endorse that rosy picture, but he did not condemn it out of hand in his response; he seems to share the judgment that the use of fear is a poor use of language to engage public attention. David Ropeik, a communication specialist, was drawn into the debate when he took an even more radical stand: "the psychology of risk perception warns against the naïve hope that we can ever achieve that level of concern [of being at war with bombs falling] with effective communication."[25]

That is about as wide a spread as can be imagined in debates about the responsible use of rhetoric. Is there some possible middle ground? Perhaps Ropeik and others (such as James Hansen) are right that we are about out of time to ward off the worst possibilities, and effective communication is no longer feasible. Perhaps Cohen is right that technology will eventually save us. Perhaps Nordhaus and Revkin are right that fear is a bad motivator for change. But if so, that leaves us in a strange situation. No one who believes global warming is a reality, including Revkin, hesitates to call it dangerous. But if something is dangerous, should we not fear it? Or fear it privately but don't talk about it in public? If one is wary of invoking fear, then what is a better way of talking about danger? A 2012 editorial in *Nature Climate Change* contended that "communication should focus on a better society . . . [rather than] on the reality of climate change and averting its risks."[26] This is much the same kind of argument common in obesity research, shifting the language away from an emphasis on obesity, with its harmful connotations, to a positive invocation of good health habits, thus threatening no one. Ropeik may be right that technology may save us in the end, but we will need effective communication to raise the money to pay for it; and in the meantime, and just in case, we will need to at least try to raise fear to a much higher level of intensity to raise that money.

As for Revkin—whose *Dot Earth* blog I follow closely and admiringly for its careful analysis—Matthew C. Nisbet, a professor of communication at American University, wrote a most interesting study of the writing and advocacy career of Bill McKibben, concluding by praising Revkin in comparison. Nisbet's study brings out exceedingly well the assets and liabilities of the use of different kinds of rhetoric used by those who want legislators and the public to take global warming more seriously. "Yet McKibben's line-in-the sand opposition to the Keystone XL oil pipeline," Nisbet wrote, "his skepticism of technology, and his romantic vision consisting of small-scale agrarian communities reflects his own values and priorities rather than a pragmatic set of choices . . . to realistically address the problem of climate change."[27]

Nisbet then goes on to praise McKibben for his advocacy and organizational skills, but then quotes Revkin as saying that he views his role as "interrogatory—exploring questions but not giving you my answers." This stance allows Revkin to say things that advocates might reject. In one of his blogs he summarizes, for instance, a number of studies that show the danger of leaks from fracking to be exaggerated, concluding that "gas production [is] a legitimate successor to coal mining."[28] McKibben does not mind inciting fear or dramatizing hazards, whereas Revkin would. As I have moved along in my research and writing for this book I have gradually moved toward the value of fear as a motivator. Softer approaches work only so far, failing to move emotions far enough along to made action seem self-evidently necessary, banishing doubt and hesitation. I will develop this point further in the final chapter.

## PUBLIC OPINION

While there is a range of judgments, there is considerable agreement among legislators that public opinion is generally influential with them,even if they sometimes ignore it, and who are supportive of its value as well of its importance for democracy.[29] On the skeptical side, a number of public opinion polls indicate a comparatively lukewarm degree of interest in global warming, far below some of the benchmarks I noted earlier. The polls have shown a strong public recognition of global warming as a serious problem. But public opinion seems to fluctuate from year to year, and invariably gives global warming a very low priority in comparison with other social and economic needs. The polls indicate an unwillingness to pay more than nominal tax increases to bring about change or to embrace strong regulatory measures forcing change. Those would be the skeptical findings.

Yet another more optimistic way of reading the polls is reflected in three articles published in the journal *Daedalus* in late 2012/early 2013. Jon Krosnick and Bo MacInnis begin by citing polls that seem less than optimistic, noting a steady decline in support of government efforts to reduce greenhouse gases, about a 15% decline between 2006 and 2012.[30] Another poll they mention looks at policies that would be acceptable to achieve the same end. It found the strongest support for tax breaks for renewable energy sources but considerably less—in the 20% to 30% range—for increased taxes on gasoline and electricity. The polls were relatively static during that period save for the tax breaks for renewable sources, which saw a 10% decline.

Nonetheless, relying on surveys showing public awareness and concern—and despite some of their own data—Krosnick and MacInnis conclude that "for years, most Americans have endorsed a range of policies . . . and have been willing to pay for the implementation of such policies. . . . Willingness to pay appears to be sufficient to fund a great deal of effort."[31] But willing to pay how much? The figures there are not encouraging; technological innovation is expensive. If measured against the need, they do not come close. What may in the end be necessary to jolt the public could be closely repeated weather disasters impinging on a high proportion of a population, inciting fear and anxiety equal in emotional force intensity to the Silicon Valley level of excitement. The low priority given to global warming, despite its acceptance as a serious national issue, is distressing.

Stephen Ansolabehere and David Konisky in a 2012 *Daedalus* issue conclude with a positive slant as well, but for different reasons, one of which is that the control of pollution often finds good support at the local level and is effective over time.[32] Kelly Sims Gallagher, analyzing why and to what extent governments support renewable energy, found a number of important determinants: (1) economic motives; (2) a high endowment of renewable resources and/or a low endowment of nonrenewable resources; and (3) the political system and cultural factors and attitudes. She notes that for many of those reasons the U.S. government has been far more sluggish than other countries in adopting effective policies.[33] At the same time, it should be noted, energy use has been declining in the United

States since 2007, and in 2012 was below the 1999 level, even though between 1999 and 2012 the economy grew by 25%.[34] Something is working.

Another study, complementing hers, compares differences between the European Union and the United States on climate change policy.[35] While one could expect a more open public and legislative willingness in Europe to make strong use of government in setting and implementing policy, they stressed two other important differences. The separation of legislative power in the European Union is less than in the United States, making policy design and negotiations easier. The slowness of policy negotiation in the United States can see timely responses to political developments of a helpful kind left in the dust.[36] Gallagher, noting that thirty-nine U.S. states have initiated renewable energy policies, found that barriers between state and federal government reveal patchwork coordination. Ironically, Gallagher concludes that while the United States shows much less interest in global warming than most other countries, "Americans are very supportive of clean energy on other environmental and public health grounds."[37] American public opinion in tandem with a politically deeply divided electorate and legislatures, moreover, is in a poor position to provide adequate support for expensive technology development.

## PERSISTENTLY LOW PRIORITY

A 2014 Pew Foundation public opinion poll offered no sign of improvement over earlier polls.[38] It found that on a list of twenty national priorities for the president and Congress, climate change was number nineteen. Only 29% would give it a high priority, a slight drop from the 30% in a similar poll in 2009. Some 42% of Democrats gave it a top priority, while only 14% of Republicans and 27% of independents did. By contrast, major deficit reduction came in as a top priority for 63% of those polled, 80% of Republicans and 49% of Democrats. On the solidity of the global-warming science, 88% of Democrats said *yes*, while only 50% of Republicans and 62% of independents did. The partisan gap was equally visible on the question of whether human activity is the main cause: 65% of Democrats, 24% of Republicans, and 43% of independents believe it is. Despite good evidence from this and other polls of weak support of policies requiring strong government activity, another 2014 Pew poll did find a surprising 64% approval rating for emission limits on power plants.[39] The public may not have seen that policy move as one that took money out of their pockets. Aiming to bypass Congress, President Obama decided in early 2014 to use his executive powers to achieve that goal. That strategy of course removes burdens on individuals and shifts it to industry, relieving the former and angering the latter.

Rounding out the 2014 Pew survey was a comparison of the United States and other countries and regions. For all countries, 54% saw global warming as a threat, but only 40% of Americans agreed (Canada, 54%; Europe, 54%; Middle East, 42%; Latin America, 65%; Africa, 54%). An earlier 2009 international survey of twenty-five countries found four countries at the very bottom of a poll on the seriousness of global warming: the United States was number 22, followed by Russia and Poland, with China at the very bottom. Poland has the greatest dependence on coal (for home use and export) of any country in the world.

An insightful 2014 review and summary of public opinion studies around the world (mainly 2000–2009) uncovered many interesting points about regional and national differences on global warming.[40] Surveying a wide range of polls, the authors found that (1) the greater a country's wealth, the lower the concern about global warming (with some scattered counterevidence); (2) countries with the greatest vulnerability to harm (storms, rising water levels) are the most concerned, and those with less vulnerability worried less; (3) willingness to pay taxes to reduce harms showed the United States with 48% willing and China with 68%; (4) on whether people had changed their lifestyles, the United States was lowest with 24% and China the highest at 56%. The authors' overall conclusion: "Our analysis of international attitudes on climate change indicates that rising global awareness and knowledge have yet to be fully translated into greater willingness to bear the cost of climate change mitigation and adaptation."[41] The same, and even more so, can be said of the United States. At present and for at least the near future it is not even imaginable that serious legislation, with real impact, can be passed in Congress. Less certain is whether President Obama's use of his executive power (e.g., enforcing reduction of coal emissions) will make it through congressional opposition.

My earlier discussion of the media and its response to science, the relationship of science and policy, and the invocation of fear and danger, offers some insight into the social and cultural dynamics behind those opinion surveys. On balance they are not the least reassuring. Nor do they offer good clues about the most powerful rhetoric to use or which policy approaches might be most effective. Might it take a global catastrophe of some kind—an "abrupt" change in climate, for instance—a specter long hanging in the background of climate studies that no one could deny or minimize?

## CHRONIC ILLNESS

Chronic illness has a unique position among my five horsemen. Everyone who is an adult will either now or in the future have a chronic disease and will almost certainly die from one or more of them; the same is true of their family and friends. Most of us will have some personal experience with chronic illness or know someone who does. That universality of experience is not true of the other horsemen: few people have so far have had a direct experience of global warming, nor do most people on earth experience water shortages, and not many have known food shortages. Depending on our income, education, and social class we will personally know a few or many obese people—although we all will surely have observed some. With these four, moreover, much depends on where we live and our incomes. Contrast that with chronic illness, which will affect us all, although how early in our lives and with what intensity are uncertain.

Yet in terms of experience, I believe, people do not see "chronic illness" as a single entity, a kind of generic category. That is the way it is perceived in public health, but not the way it is seen in the daily practice of medicine. I do not go to my doctor complaining of "a chronic disease," but because I have COPD breathing symptoms and want that condition to be treated. While there is research on chronic disease as a category, the overwhelming research is devoted to particular diseases. Understanding these differences helps explain why chronic illness as such does not capture the public eye, much less evoke the kind of media attention, full of drama on occasion, that the other horsemen do. The National Institutes of Health has a number of programs that begin with the word "chronic" (e.g., "chronic kidney disease"), but none on "chronic disease." That is left to the Centers for Disease Control and Prevention.

Why have I noted these traits of chronic disease? They help explain, paradoxically in some ways, why a condition in most cases close to our own lives draws less public attention than the other horsemen, which are more remote from our daily life. Why it is harder to raise money for chronic disease (than, say, for cancer), and why does the media show little interest in it? The fine book *Sick Societies*, edited by David Stuckler and Karen Siegel, by far the most comprehensive and useful study of chronic disease, is filled with complaints about the neglect the field suffers and offers some good strategies to gain more support. The nature of chronic disease is that it aggregates a number of specific medical conditions, each with its own etiology, symptoms, treatments, patient impact, and course of illness into a public health category of its own. The rationale for doing so is that, although taken one by one they are all different, taken as a whole they strongly reflect in great part the background social conditions of those with the diseases. These include conditions such as wealth and poverty, the influence of various forms of unhealthy lifestyles and behavior that are causal factors in the chronic diseases (smoking, obesity, alcohol consumption, for instance), and ethnic and racial differences. The traditional goals of public health are health promotion and disease prevention, and those are the main drivers of chronic disease as a group, especially disease prevention.

Perhaps because of the collective nature of chronic illness as understood in the public health community—giving it a kind of remoteness from individual lives in contrast to its vivid meaning in the lives of those with a specific chronic disease—it lacks many of the ingredients of the global-warming debate. No one denies the existence of chronic disease as a problem or minimizes its importance. It has no industry opponents or notable business supporters. I was not able to find a single public opinion poll on it (though there were some on the individual diseases). There seem to be no major struggles on the science of chronic diseases collectively or on the science-policy relationship. Chronic disease has no colorful characters such as a James Hansen or a Bill McKibben, no hostile media figures such as a Glen Beck and a George Will, or proponents or detractors on the editorial pages of major newspapers. The news stories on it are few and far between and almost invariably they report only government studies or those of various scientific or medical committees.

As I noted in the preface to this book, I was drawn in part to the concept of five horsemen to appraise the way individuals and groups argue with one another, and how public policy is formed in conflicted situations requiring basic structural and behavioral change. It is tempestuous territory, with mountains and valleys. Comparatively speaking, chronic illness as a general overarching category is a flat and colorless terrain, but one that hides many more lively debates below the surface on the discrete diseases. Which is why, no doubt, it has little interest to the media and has not seemed of enough public concern to attract the attention of public opinion organizations—or wealthy donors.

## Chronic Illness and the Long-List Syndrome

There is another issue, which I call the "long-list syndrome." While often used in other policy areas, it is notably prevalent in reports on proposed chronic illness policies. By that phrase I mean the development of detailed policy ends and means, in lists often running into twenty to thirty undifferentiated items. Many of the articles and books I have cited contain such lists, and some of them I used in earlier chapters. Most, however, do not propose priorities either in terms of relative importance or, more practically speaking, of easier or more difficult to achieve. The net result is often, at least for me, more bewildering or intimidating than illuminating. It is like standing under a waterfall to get a drink of water. So it is with the many important reports on chronic illness that aim in a comprehensive way to lay out the terrain and chart the way forward. I suppose it might be said that such is the very purpose of a major report, but my reservation is often that it can dull rather than enliven the mind, all the more if one dutifully tries to read more than one. What follows is a shortened and aggregated list of some twenty-six such ends and means for coping with chronic disease, drawn from four such reports by major health organizations: the Robert Wood Johnson Foundation, the World Health Organization/NCD Alliance, the WHO, and the Institute of Medicine:[42]

- Reduce sugar and salt use
- Reduce tobacco use
- Reduce alcohol consumption
- Promote exercise
- Maximize healthy lives at all stages of life
- Reduce the burdens of chronic disease
- Universal health coverage and access
- Improve data collection
- Tobacco-free world
- Improved lifestyles
- Global access to affordable and good-quality medicine and technologies
- Foster health care and public health system changes to improve the health of individuals with multiple chronic conditions
- Maximize the use of proven self-care
- Provide better tools and information
- Facilitate research on reducing blood pressure
- Improve chronic disease surveillance and assessment
- Select a variety of major illnesses for special consideration
- Explore surveillance techniques that are more likely to capture chronic diseases more effectively
- Develop comprehensive population-based strategies
- Greater use of new emerging economic methods
- Improve living with chronic illness
- Evidence-based policy goals
- Incentive programs for all employers to promote heath-promotion programs
- Disseminate information to all levels of government, national and local
- Cost-effectiveness studies of all policies
- Controlling technology costs

Let me voice a dilemma I began to notice as my research and writing moved along. On the one hand, I am convinced that making a difference with any of the horsemen requires some basic changes in well-established social and behavioral patterns, usually with deep cultural roots. Logically, that would entail putting together long and comprehensive lists of needed changes, of the kind I have just presented; that is what big problems require. On the other hand, as with this list, I recoil with skepticism that such a collection of goals has much chance of being successful as a whole; success may be achieved with a few of them but is unlikely with most others and surely will not occur in the near future. Indeed, I suspect that is one reason that chronic illness gains little media coverage, public interest, or the money of the wealthy. The latter, I suspect, are drawn to clearly specified and narrow, plausible goals (the eradication of malaria). A long list with no priorities, or distinctions among different kinds of goals—reducing tobacco use over against achieving universal access to health care—may be interesting but not useful in the face of limited budgets. Some recommendations are for mitigation, others for adaptation, a useful distinction for setting priorities, but also not distinguished. Long and undifferentiated lists can obscure rather than clarify policy options, inducing what I call list fatigue.

## FOOD

With food we enter still another domain. If many of us in the world live in geographical areas not likely to suffer major and lasting agricultural threats or food shortages, others are in dangerous regions, mainly in poor countries. As with chronic illness, media coverage is scant and sporadic, public opinion polls are few. There are no colorful advocates with a large following, no noticeable debates on appropriate rhetoric to talk about the dangers of starvation and food shortages, and the internal struggles within · the food community are strong but obscured, not readily noticeable by the general public.

But as with chronic illness, those working in food policy and advocacy complain about the sporadic attention of the media, often no more than a story about a new report or some particularly startling disaster. Yet the strong emphasis on food in the 2013 IPCC report signaled a heightened interest, and that was picked up by the media. While noting that climate change could be beneficial to some crops, as a global problem it could possibly reduce food production by as much as 2% a decade up to 2100.[43] Meanwhile, food demand is expected to rise as much as 14% each decade. The sudden rise in food prices in 2006–2008 and then again in 2010–2011 was itself a shock, but the fact that prices as of 2014 remain high under-scores the seriousness of the issue.

The IPCC report that tropical countries could be the hardest hit was made clear by a 2013 Red Crescent report. Complaining about a lack of media attention, it noted that 6 million people in southern African coun-tries already face severe food shortages because of either a lack of rain or erratic rainfall, which is exacerbated by poverty.[44] Meanwhile, there also appears to be no great movement away from what Jennifer Clapp called the "dominant response" to food problems, a combination of investor interests and the development goals of international agencies and private foundations (see chap. 2). But those values and practices are not visible to the public, and the food crises have so far been limited to poor countries. That combination does not make them good fare for the media in rich countries; out of sight, out of mind.

A 2008 global public opinion survey (about the only such survey I could locate) found that people in both developed and developing coun-tries were unhappy with high food prices and that a majority in poor countries ate less and ate cheaper foods in response—but not so in rich countries. In any case, those in both rich and poor countries blamed their own governments for the prices, not the underlying "dominant response" just noted or environmental troubles.[45] That finding raises the possibility that thin media coverage has not spotlighted the larger background forces driving food shortages and pricing, leaving an easier and more visible, conventional place to lay blame, the government; in sight, in mind.

An interesting 2011 article by three Belgian analysts focused on fear as a way of raising money for NGOs and various agencies to deal with the food crisis. Its point of departure was the widely held view in 2005 that low food prices were the reason for food shortages. As the UN Food and Agricultural Organization put it, "the long-term downward trend in agricultural commodity prices threatens the food security of hundreds of millions . . . in some of the world's poorest developing countries where the sale of commodities is often the only source of cash."[46] But the price crises of 2008 and 2010 brought a remarkable turnaround. It was not low prices but high prices that were said to be the cause. Oxfam UK said that "high food prices risk hunger for millions of people."[47]

What was going on? The suspicion was a "bias in policy communication." In this instance that meant the intended audience was not the public but donors to NGOs. The aim was to galvanize NGOs via the media; and "events" and "shocks" and "emotionally charged" stories are the way to do it.[48] Both government officials and NGO donors believe that the media news affects public opinion. Various media outlets know their audience and play to it (the CNN factor). Bad news plays best. That article, by the way, was not meant to signal some kind of scandal. It was based on computer modeling, trying to see how organizations craft their message, seeking the most effective way of doing so.

## OBESITY

One of the few American public opinion surveys in recent years on obesity, a 2012 Gallup poll, found that 81% of those polled think obesity is an "extremely" or "very serious" problem, up from 69% in 2005. It was taken to be more serious than smoking or alcoholism. When asked about the role of government involvement, 23% thought that extremely important while 22% thought it very important; that is, when combined, less than 50% favored involvement. That combination showed only 27% of Republicans wanted a government role versus 72% of Democrats.[49]

A 2013 poll of 1,200 readers of the *New England Journal of Medicine* (presumably health-care workers) found that 58% of American respondents favored government regulation of sugar-sweetened beverages, a number lower than in other countries. The Americans opposed cited the interference with personal choice as their reason.[50] A 2013 National Opinion Research Center survey found that obesity was second to cancer as a national problem, that the public understands the causes and impact of obesity, and that it supports school programs and the use of incentives to change behavior—but would not support limits to what could be purchased or taxation of food products.[51]

I have cited polls on obesity in this chapter and in chapter 5. While there has been some upward movement over the past decade, a 2005 poll came out with somewhat different results whose predictions for the future have not been fulfilled.

As the researchers put it, "If the public embraces dominant opinion among experts and agree that that obesity is the result of environmental and genetic factors, we would predict greater support for obesity-related policies in the coming years."[52] That has yet to happen.

The sociologist Abigail C. Saguy has written one of the most illuminating, if sometimes also puzzling, studies of obesity, *What's Wrong with Fat?*[53] Just as the global-warming debate has a small but powerful faction of deniers and minimizers, she speaks out of the context of a small group (the size of which is hard to determine), heavily female and vocal, which holds that (1) the language of obesity as a "disease" and an "epidemic" is overblown, blatantly plays to the media, and distorts the available scientific evidence by exaggerating the harm done by overweight and obesity, failing to note the evidence that many overweight people are in good health and that only those in the worst categories of obesity are in danger; (2) that casting obesity as a medical issue plays into the hands of disciplines that fail to note its social dimensions; (3) that framing obesity as a matter of individual choice, a "sin," ignores those social dimensions; and (4) that all the above lead to a stigmatizing of the obese, particularly women. A number of organizations, heavily female, reflect that perspective: National Association to Advance Fat Acceptance, Association for Size Diversity and Health, and Health at Every Size.

If many of the global-warming minimizers and deniers have been influenced by the opposition of conservatives to government power and by industries with money, friends in the media, and powerful lobbies, there is none of that with Saguy's minimization efforts. It is the power of an aggrieved and burdened group of people to gain greater social acceptance. That is a coherent and plausible movement. What is for me puzzling is that at least over the past few years the public health perspective has gained prestige in its approach. I use the word "obesogenic" as a way of signaling a shift from the language of personal responsibility, a word not used in her book and not common in the research literature in the early to mid- or late 2000s (the period of most of her citations); her book was just on the cusp of that change.

To say that obesity is an obesogenic problem is to say it is caused by the social and cultural mores and practices of society, a force overwhelming enough to shape behavior and beyond individual control. But there is often a confusion here; in fact, there are two. The public health perspective does not deny that the way individuals eat and exercise their bodies is a cause of obesity, but it considers those to be secondary. Instead, the public health perspective views the proximate cause of those individual choices as culturally shaped and often beyond the control of individuals or extremely difficult to control. Those proximate causes are where policy should be focused. But here is the second confusion: the public health approach is not in fact strictly deterministic. In fact, it is self-contradictory. It encourages public education on the risks of obesity, makes copious use of the word "choice" in wanting calories listed and healthy food selected, and wants citizens to fight the power of industry with their votes and condemnation. In short, the message is that we are all the victims of an obesogenic society that robs us of choice—but good choices in our daily lives can empower us to fight back.

In the meantime, there is little good news to report. A 2015 Gallup public opinion survey found that obesity in American adults had crept upward in 2014 to 27.7%. It was higher by 2% since Gallup began obesity polling in 2008 and showing no sign of decline. Noteworthy was the finding that the percentage of those in the overweight category had remained stable since 2013. It was the number of those who had moved from overweight to obesity that stands out, implicitly suggesting a failure of prevention efforts. The survey also found that the obese suffered a loss of "well-being," with lower incomes and long-term unemployment among its manifestations.[54]

## WATER

Very much like food, water draws comparatively little regular media attention. Public opinion polls are few. One of the only global public opinion surveys I was able to discover, from 2009, was surprising.[55] It surveyed 1,000 persons each in fifteen different countries. They included Canada, China, India, Mexico, Russia, the United Kingdom, and the United States. The results showed that those surveyed judged water pollution as the world's top environmental problem, with a shortage of water a close second: 93% saw pollution as a serious or very serious problem, with 92% saying that shortage was also in that category. Government is taken to be most responsible for clean water. In many cases the survey found that worries about water exceed those for global warming. That was a surprising finding and made me regret that other polls are not available at either the national level and international level to see whether their findings would be similar.

What surprised me the most, however, was that unlike global warming water has received comparatively little national media attention—a rare exception being the 2014–2015 California drought crisis. As with food, there are complaints of media disinterest and sporadic attention only. One of the few efforts on the part of journalists to analyze that situation was a 2005 effort by the Nieman Foundation for Journalism at Harvard, as one of its Nieman Reports. They had various contributors from around the world, but with a heavy emphasis on the United States, reporting on different water events and media coverage. I can only mention two, but they nicely capture the flavor of the others.

Stuart Leavenworth, a California reporter, wrote, "I recently finished a stint covering Western water issues for the *Sacramento Bee*. To my chagrin, I had the beat to myself for four years. . . . Papers have tackled problems of water pollution and degradation, but have overlooked fundamental issues of supply—and sustainability. This is curious. . . . It is a fundamental resource for life." The subtitle of his story was "It's the Economy Stupid."[56] A similar kind of judgment, as noted earlier, was made about food coverage, surely up there for human importance with water. A reporter in India noted a great upturn of media interest in water in the early 1900s, but then saw its decline in the late 1990s when the advent of enthusiasm for the market rose. "However," he wrote, "even as environment reporting is languishing, water continues to enjoy media's indulgence, not because rural India is dying of thirst but because the urban middle class is facing an acute water crisis. Even in cities, the water needs of the poor are rarely reported. . . . Indeed, water makes it to the front page only in the summer months, when people in Indian cities start crying hoarse for water."[57]

While it might seem a bit churlish of me to mention it, the Nieman Reports have not repeated the fine 2005 report on water—nor has any other group tried to do so—reflecting perhaps just the kind of sporadic interest the media has in the topic. A 2014 story claimed an increase in

media attention but cited only a scattering of TV programs and interviews.[58] The California Institute for Water Resources reported, however,
that in 2014–2015 there had been a considerable number of news stories
on the California drought,[59] an event hard to ignore.

## DISRUPTORS AND INCREMENTALISTS

Global warming has had considerable media attention, ranging from good
coverage on the Kyoto Protocol, on IPCC reports, and at the country level
with national efforts. People know about global warming and judge it a
serious issue. But "serious" does not mean a willingness to take decisive
steps. The public opinion polls make that clear, fluctuating over the years
on a willingness to accept pain and sacrifice. The net result over many
years is far less than needed, however much is going on and however
many small victories are counted. The same can be said of the response
to chronic illness, water issues, food problems, and obesity. They receive
considerably less attention from the media (save for obesity) and pollsters
but, while global warming has probably had greater success in gaining
media and public attention, it has not been enough to get all the changes
in policy needed. One has to wonder why it has been so hard for all of
them to gain momentum sufficient for significant progress. As noted earlier, it has by no means been impossible for global progress to be made
on difficult fronts, notably with seven of the eight UN Millennium goals,
with global warming the exception. Better education and better "public
communication" may help, but there is something about these problems
that seems to defy the favored nostrums. If all five of my horsemen end in
the same place after many years—stationary at best, perhaps with modest
gains—what is missing, what might be overlooked?

One strong candidate is leadership, often cited as a necessity and generally lacking, especially with global warming but with the other horsemen
as well. In trying to understand the leadership problem I have found it
helpful to think first of all about the tenor of leadership. Al Gore surely
stands out, as the winner of a Nobel Prize for his work on global warming,
as the creator of a major documentary, as the author of a solid and comprehensive book, and as a lecturer around the globe.[60] Less known outside

of the professional field, but strong within it, are James Hansen in science, William Nordhaus in economics, Gus Speth in law and governance, Bill McKibben in organized advocacy, and Andrew Revkin in journalism. I cite them not necessarily because they would get the highest votes if a global survey were conducted, though all would be in the running. Instead, they have distinct differences in their strategic approaches to global warming, each embodying his own form of leadership, what I call their tenor. Most notably, I believe, they fall into two rough groups, the holistic incremental-ists and the fast-moving disruptors aiming at specific targets.

A holistic incrementalist policy is one that calls for very broad, sweep-ing moves to change entire cultures and patterns of social behaviors. This strategy is attractive when the problem to be addressed is caused by a wide range of social practices and behaviors: when there are no "silver bullet" solutions; when the issue is politically and ideologically charged; and when it is believed that an incremental strategy will be more accept-able, less jarring and controversial, than something more immediate and forceful. While it may seem counterintuitive, some of the most radical advocates want massive changes that could take generations to achieve.

A disruptive policy might be called a brute force or blunt instrument strategy. The problem needs to be dealt with as soon as possible, not incre-mentally. It needs to directly achieve its chosen goals in a tough-minded and focused way, and it is willing to put up with a political battle and popu-lar resistance. It is not prone to compromise. There may be no silver bullets, but some of its shotguns will fire large-gauge projectiles, designed to cut through bulletproof vests. I put Revkin and Speth in the tender-minded camp and Hansen, Nordhaus, and McKibben in the tough-minded camp.

Let me bring in first as an example the American health-care reform effort (the Affordable Care Act) and the way its supporters handled the vexed and delicate problem of making benefit cuts to save money. The chosen incrementalist strategy was that of "bending the cost curve" and "getting rid of waste and inefficiency," all of which will take time but are politically palatable. The spreading of large-scale reforms over a num-ber of years in the Affordable Care Act, not all at once, was no less of a holistic strategy. Rationing, an inflammatory word, is unmentionable for the holistic minded. A disruptive approach is impatient with slow incre-mentalism, puts the word "rationing" right out in the open, and likes the idea of regular annual cuts to physician reimbursements in the Medicare program. It applauds the Independent Payment Commission's 2015 requirement of automatic cuts to the Medicare program to control costs, resisted by a Republican Congress.

In the context of the global-warming crisis, incrementalist advocates talk about the "greening" of our modern world, radically changing the way we live our lives and protecting as well the natural world that is the physical context of those lives. Speth is a long-list person with a low-key style and Revkin is as well, moving carefully in taking the issues apart in a calm, often dispassionate way. The disruptive path wants tough action to stop the annual increase of the $CO_2$ level, a reduction of soot, strong and binding international protocols, and a forced $CO_2$ reduction by legislation in the use of coal or the deployment of technology to reduce the harm of coal as a major source of energy. For Nordhaus it is a carbon tax; for McKibben it is an all-court press, making use of fear and cutting through the kinds of rhetorical caution urged by careful scientists (picketing and selective law breaking are acceptable); for Hansen it is the use of fear as well, emphasizing we have a few years left to decisively control emissions. The dangerous outcomes of global warming are already with us! By implication, incrementalism is itself a danger.

Obesity is no less full of tensions between incremental and disruptive policy ideas. The holistic model calls for a full-court press on all fronts—educational, governmental, and commercial. The one thing not allowed, at least in the public health community, is anything that would stigmatize the overweight and the obese. They are simply the victims of hazardous work and leisure patterns, the all-too-effective advertising, marketing, and lobbying of the food and beverage industries, and a lack of regulations and tough guidelines. They are long-list people as well. But by virtue of stressing the full-court press, singling out no one overriding target, it is an incrementalist strategy. Moreover, while there are efforts to put in place some disruptive policies, mainly strong laws and regulations, most of them are defeated by political and industry opposition. New York mayor Michael Bloomberg's push for his department of health first to tax sugared beverages and then to regulate the serving sizes of sugared beverages met a barrage of criticism. He failed both times because of industry opposition and push back by the public.

## URGENCY, PRIORITIES, AND RHETORIC

I should make clear that my distinction between incrementalists and disruptors by no means precludes each side supporting some or much of the other side; it is a matter of urgency, priorities, and the effective rhetoric one believes best for the task at hand. But a decision has to be made about each of them. *Urgency*: How long do we have before it is too late? *Priorities*: Everything cannot be effectively be done at once. *Rhetoric*: How might the need for change be best communicated to the public? It is a demonstrable axiom with the five horsemen that the more social and cultural change that has to be made in a society, the harder it will be. That is why they have made no real progress. Analogously, the urgent threat of war or some equivalent disaster, with imminent loss of life and the most radical social upheaval, may be the only force that can achieve such change. But even there history is full of examples where there was resistance to believe war would happen. The pathology of hope remained too long and realism set in too late.

Yet there is a snag of consequence here. Although as a group they share the almost singular global trait of the (so far) refractory condition of resistance to change, there is no imaginable panacea that would work for all of the horsemen, or any single leader who could speak effectively to and for each of them. Consider some important differences. I take global warming to be the most urgent issue, but I see no way that reducing carbon emissions or mitigation can be achieved rapidly and decisively. At best it will take a long time, putting adaptation high on the priority list; some speed is possible there. In any case, there is already enough $CO_2$ now in the atmosphere that will remain for hundreds of years to give us continuing trouble, regardless of the success of stemming the present rate of increase. Global aging guarantees a rise in chronic illness. The rise in adult obesity everywhere has no cure or serious mitigation prospects on the horizon. Even if one appeared, those now already obese will likely remain so given the poor present treatment prognosis, including a large number who began life as obese children. They will in the process increase the number of the chronically ill on top of this aging momentum. Even with a continuing drop in population growth, including the remaining regions with replacement rates well above the average, the lag time of change makes certain a larger world population in the future. Water and food availability, already in trouble, will likely continue to get worse.

If any great progress is to be made, it will have to come from two sources. One requires a stronger and far more active role for industry. One part of that effort will be to defuse and defang the obstructive role played by so many industries in blocking legislation initiatives, laws, and regulations. Obstructive efforts must be stopped by the power of shame from other industries, their peer group. Positive moves will require joint private-public actions. The other part will be a strengthening of the role of government, the specific aim of which will be to close the gap between the public's understanding of the seriousness of the issues and their unwillingness to give them a high priority or to make much sacrifice to bring a change.

# 9

# LAW AND GOVERNANCE

——

Managing Our Public Planet and Our Private Bodies

ust how much and what kind of government is good for a nation? That is a timeworn question, going back to Aristotle's *Politics* and Plato's *Republic*. Government, I believe, is a necessary and unavoidable part of the management of the five horsemen, even though at different levels and in different proportions. Each of them has both an individual and a social impact, the latter affecting economic welfare, public health, and physical security. The market will have its place if used wisely, but it cannot be given a dominant role. It is congenitally drawn to industry profits and shareholder returns, not the common good; it can become a danger (see chap. 10). In turn, laws and regulation are the principal tools for using the power of governance. The main challenge with the public is to gain sufficient support from it to promote effective governance—both in organizing systematic policies and regulations to make them work.

## PULLING UP DEEP ROOTS

The causes of the five horsemen are deep and penetrating, an invasive plant with pervasive roots. The beginning of those roots can be traced back to the beginning of the Industrial Revolution in the eighteenth century, notably gross domestic product (GDP) growth and technological and manufacturing progress. Later developments spread the progress, accelerating even faster in the post–World War II era, which saw even more rapid population growth and greater affluence, mobility, and globalization. The fastest growth of the horsemen's roots and their increasing ground level visibility has come since the 1970s, spreading throughout the world. A common phrase in global-warming discussions is that of "business as usual," characterizing the dangers that will occur in the future if present human environmental activities go on without any change. Speaking in a cultural sense, a phrase that might be appropriate with the five horsemen is "life as usual": our danger lies in the way we routinely and carelessly pollute the planet and ruin our bodies. Until recently few told us we might do well to do otherwise, and if we cannot see the results it is much harder to discern their roots.

To change life as usual, pulling up the tenacious roots, means asking people to change the ways they live their lives and to accept things often repugnant to them and against their will: to settle for modified industrial progress, to tolerate regulations and laws affecting the food and drink they consume, the weight of their bodies, and behavioral habits that make them waste water or that bring on early or excessive chronic disease. The experience of the last forty years has brutally demonstrated that education and exhortation are insufficient to change life as usual. The roots pay no attention. They have a life of their own and must be pulled up with force and power—and government, assuming public support, is the only known way to do that. People accept that reality in wartime and with natural disasters, but now we must accept it in what can look like, but is not, peacetime. The illusion of surface peace, with seemingly distant clouds, invites denial, evasion, and foot-dragging. And even worse, there is often enough an outright unwillingness to change, even when the dangers are patent and in the open, because of the threat to cherished values: the right to have guns, to eat what I want without interference, to be willing to take my own chances with danger—whatever the harm to myself and others. Those are the acceptable chances that many take in the name of freedom.

I am drawing a harsh and unpleasant picture, and if the problem was only one of the horsemen it might be tolerable, but all five are at stake. My only defense is that no decisive progress has so far been made in decades with any them. It is a history of policy weakness or failure to achieve needed behavioral change. The next steps must of necessity be harder. Part of that demand invites a seductive win-win policy neurosis, which can lull us: a decline in GDP growth can bring us increased happiness and a less consumerist life, renewable climate control technologies can create more jobs; nonmeat diets can be tasty; it is fun and healthy to exercise; for the recalcitrant, carrots can work better than sticks; small is better. There is truth in each of those nostrums. The hazard is to become infatuated with them, as if nothing less will do. A lot of life is not that easy and uprooting life as usual will be hard. There will be real losses.

With that less-than-cheery preamble behind me, I will move to some grittier issues. In the first part of this chapter I will look at various issues of law and governance encountered by efforts to bring about a change with four of the horsemen. The fifth, obesity, has met resistance in many societies, but especially the United States, because some of the regulatory remedies would impinge on personal freedom to do as we choose with our bodies. The professional rejection of stigma and social pressure to reduce obesity reinforces that freedom. After looking at obesity I finish the chapter with a look at a no less strong issue of sovereignty, not of the body but of the nation-state—and I will do so by a comparison of China and the United States in their management of the five horsemen. I want first to distinguish between governance and laws and regulation.

I will mean by "governance," broadly understood, the fashioning of policies, and by "law and regulation" a means of implementing them. Governance has a number of levels, each with different possibilities and limits: international (UN environmental programs), national (emission policies), regional (agreements on sharing river water), states/provinces (school nutrition/obesity policies), cities and towns (pollution and trash control); grassroots and social movements (marches against pipelines and fracking). I will not in this chapter take up the role of NGOs or more informal working relationships among nations, regions, and states, which can be important at each of the levels.

I will mean by "law and regulation" the creation of rules that will be enforced: international (treaties; binding agreements with international agencies on loans); national (clean air regulations); regional (European Union cap and trade regulations); states and regions (emission regulations); cities and towns (regulation of sugared beverages); grassroots and social movements (by definition they have no legal or regulatory authority, although they can force change by picketing, petitions, and boycotts).

## INTERNATIONAL GLOBAL GOVERNANCE: GLOBAL WARMING

If it is true that all roads once led to Rome, it is no less true, as the global-warming debate has underscored, that all roads now lead away from an overarching world government and from an environmental program that has binding regulatory authority. Does that make sense? Assuming that one is not in rigid principle hostile to the idea of top-down governance (and many surely are), the attraction of a single global governmental organization with the power to organize policy and create binding rules and regulations is manifest. But the value of national sovereignty stands squarely in the way, and top-down governance has never gained traction. The net result is that the present efforts to cope with global warming are multiple and varied: national and state governments, international agencies and commissions, NGOs, assorted regimes, and informal grassroots groups. In 1933 the prominent British writer H. G. Wells called for world governance, a new world order, which he called the "only possible solution of the human problem."[1] The League of Nations, founded in the aftermath of World War I, was meant to at least partially play that role but had faded away by the advent of Word War II. Its spirit was taken over by the United Nations, a phrase invented and promoted by President Franklin Delano Roosevelt during World War II and formally employed in 1945. It was not meant to be a world government but at least a powerful force for world peace and human welfare. A more informal nongovernmental grouping, the World Federalist Movement, has held onto the idea of global government, but to no avail.

Yet that idea has been advanced for environmentalism, a World Environmentalism Organization with regulatory force comparable with that of the World Trade Organization (WTO).[2] But the idea has never gained traction, impeded by some daunting international realities, all on display in present global-warming debates: tightly clutched national sovereignty, military and economic competition, global pluralism on values and culture, the balance of market and government, among other obstacles. In that crowd, the United States might well be called the gang leader, flaunting its "exceptionalism," with China, India, Canada, and Australia joining it. Nonetheless, the United States has been a major actor on the international scene, both with government and the private sector.

The international organizations (IOs) working with global-warming issues have little regulatory power. Binding treaties for major global policy initiatives are not possible in that domain now or foreseen for the future. Nonetheless, the United States and other countries, zealous to maintain their sovereignty, do not hesitate to make use of and work with international agencies when it serves their ends. As Kenneth Abbott and Duncan Snidal write, "Formal organizations (IOs) are prominent (if not always successful) participants in many crucial episodes in international politics."[3] There is, they point out, traffic both ways. Sometimes UN agencies carry out the wishes of countries, as when the United States decided to turn to the UN Security Council to stop the Iraqi invasion of Kuwait, but did not when it later decided unilaterally to directly attack Iraq for invading that country. "IO independence," they observe, "is highly constrained: member states, especially the powerful, can limit the autonomy of IOs, interfere with their operations, ignore their dictates . . . but as in many transactions . . . can increase efficiency and affect the legitimacy of individual and collective actions."[4] They want it both ways.

Produced under the auspices of the United Nations, the Intergovernmental Panel on Climate Change (IPCC) reports are a good example of a productive and influential force in the global-warming arena. It might be noted also that another powerful IO, the World Bank, has supported many programs damaging to the environment.[5] After flatly stating in 2015 that no "significant steps have been taken to curb emissions and global warming," William H. Nordhaus has proposed what he calls a "Climate Club." It would be a voluntary grouping of countries that would that would "undertake harmonized but costly emissions reductions"—and would penalize countries that did not join the club by the imposition of tariffs on their imports.[6]

If what has been called "hard law" seems out of bounds for global-warming control, "soft law," that of nonbinding agreements, is common and effective—somewhere between international anarchy and a world government. Joseph Tainter, in his classic 1988 historical study that covers the centuries, *The Collapse of Complex Societies*, alludes to global warming when he writes of a "mutual collapse"—in a vein of what Mike Hulme calls "presaging apocalypse."[7] "That collapse," Tainter writes, "if and when it comes again, will this time be global. No longer can any [powerful] individual nation collapse. World civilization will disintegrate as a whole."[8] In effect, we live in a loose collection of sovereign nations, just on the edge of global chaos, but nonetheless so bound together in formal and informal ways that we could all fall together. But I believe this could only be true of catastrophic global warming, not of the other four horsemen. They bring lots of bad trouble but not on that wide a scale. Of course the top 1% of the world's population, made resource-full with what might be called "wealth as usual," will probably dodge that bullet as well, going underground or being rocketed to the moon. Where there is money there is a way.

## Environmental Law and Regulation: Where It Can (or Could) Work

If national sovereignty making use of strong laws and regulations has been a serious roadblock to a unified global-warming abatement policy internationally, it has not stopped nations from deploying law and regulation within their own boundaries. Roger Pielke Jr. has provided a useful sample survey of those efforts.[9] The United Kingdom enacted the Climate Change Act of 2008, mandating national emission reductions. A Committee on Climate Change will oversee that effort, aiming to cut emissions 80% by 2050. Whether that is a feasible goal is, by the committee's own somewhat skeptical judgment, technically and politically uncertain. In 2005, Japan, already one of the world's most carbon-efficient economies, set a goal for cutting emissions 15% by 2020. Then, in 2009, it pushed the goal further downward from the earlier standard. Japan has been one of the world's leaders in its use of nuclear energy (some 30% of its energy), but in the aftermath of the Fukushima disaster, it decided to gradually shut down its nuclear plants. It then reversed that decision in 2014 but also set less ambitious goals for overall emissions. Australia has had an up and down struggle, voting down a cap and trade proposal but passing a renewable energy packet in 2009—only to see an increase in opposition to emission control efforts when a conservative government came into office in 2013.

The European Union countries have a successful cap and trade program, though not without some bumps in the road and a 2013 rejection of a proposed cut in the pollution contracts.[10] Europe's emissions dropped 10% between 2007 and 2012, most likely due to a sluggish economy and a low birthrate. Germany has an outstanding record of using renewable resources, which now support 25% of its energy, and has set a goal of 40% by 2025. China is the most interesting case and may be even more torn than the United States. Yet it is torn not by political strife but by the dilemma of wanting to continue its strong GDP growth, requiring the continued building of coal-powered plants. Meanwhile it has invested heavily in renewable energy and is working hard to reduce very bad air pollution. It is also trying to have it both ways.

Apart from a most unlikely embrace of a top-down set of global laws and regulations binding all nations, it is at least possible that in December 2015, when the successor to the Kyoto Protocol is fashioned, it will achieve a global agreement by rich and poor countries that will accept targets for reducing emissions. That does not mean, however, that the goals will be enforceable. That is highly unlikely. But it could provide a strong incentive for the rich countries to put up more money to help the poor ones to move to renewable energy, and it could lead to more cap and trade policies.

In June 2014 President Obama, using his executive authority, announced that the U.S. Environmental Protection Agency (EPA) had developed regulations that will set a national limit on carbon emissions from coal-powered plants by up to 20%. It could lead to the shutting down to hundreds of coal-fired plants, and it will force industry to pay for the emissions they create through cap and trade program in states throughout the country. While cap and trade regulation failed in Congress in 2009, moving it to the states can avoid the need to gain congressional approval. The president's action came after a long period of inaction on his part, stymied by Republican legislation and perhaps also by its low priority on public opinion polls. Strong emission-reduction programs making use of cap and trade exist already in some states but are weak to nonexistent in others. They tend to follow liberal-conservative policy lines and the interests of industries that will be hurt by strong policies. California and New York are liberal states, with little fossil fuel (save for gas now) to extract from the earth and few major industrial powers dependent on them. The opposite is true with Texas and Wyoming, with their strong oil and coal resources. Wyoming is the country's leading coal producer, and it has the singular status of a legislature that in 2014 rejected science teaching on global warming in its schools. It was not the science they rejected but, in the candid words of its leading sponsor, that its acceptance "would wreck Wyoming's economy."[11]

But it is necessary to take note of some serious reservations among a few policy researchers on the three approaches that have attracted the most interest, just the three already discussed: carbon tax, cap and trade, and existing carbon emission rules. Of course, standing in the wings, with the support of many economists, is a simple ideal that would cut through global agreements: a carbon tax on all major sources of fossil fuel–based emissions. But I have not heard of, nor would I expect to discover, a computer model of the probability of some kind of international conversion experience among legislators to propose such a tax; or if it were proposed, getting the public to accept it. If there is any global public opinion consensus on the combating of global warming, it is simply this: don't do it by taxation. That matches at the nation-state level a no less simple principle: don't do it by the use of binding treaties. But cap and trade policies have a good track record where used and could gain ground.

## Cap and Trade, Carbon Tax

I was interested to discover some contrarian infighting on the value of both cap and trade and a carbon tax as well, even if they could be achieved. For openers, there is a good chance President Obama's initiative, resting on the use of his executive powers, will succumb to strong congressional resistance, with at the least a number of successful delaying tactics. The latter will likely come from legal challenges in the courts by the EPA's use of the Clean Air Act, originally passed to deal with air pollution. Its use in the carbon emissions context is a stretch, though a plausible one. As Jody Freeman, who served as a counselor for President Obama on energy and climate change, has noted, "Some opponents from states and the utility industry insist that the standards must be based solely on what individual facilities can achieve on-sight with existing technologies. . . . It is entirely possible that some judges would balk at the ambitious approach [that forces the closing of large numbers of old coal-fired plants]."[12] Win or lose, however, real action will not come quickly.

While cap and trade strategies will be stimulated by the Obama initiative, there are other reasons for skepticism about the value of a post-Kyoto
treaty as a goal in 2015—and about both a carbon tax and cap and trade.
Roger Pielke Jr. and signers of the Hartwell Report had claimed earlier
that the Kyoto Protocol was a failure, not to be revived. But other voices
began to be heard even as preparations for the 2015 meeting to define
fresh goals were under way. Former senators Tim Wirth and Thomas A.
Daschle wrote in 2014 that the international community should cease
chasing a binding treaty for the control of carbon emissions.[13] Instead, it
should seek a general agreement that leaves countries free to find their
own solutions to agreed-upon carbon-cutting goals. It is not necessarily ideological resistance to treaties that is the source of opposition (save
for the United States) but more that countries have varied interests and
needs that a treaty might hinder, while individual countries might effectively handle emissions reduction in their own ways. A treaty model is
what Wirth and Daschle criticize as a one size fits all plan and cannot
work. Their own plan, however, puts in its place an ambitious push for
low-carbon solutions and a "race to the top" putting them in place. Even
now, they say, "The world can glimpse the prospect of an economy largely
dependent on inexpensive, affordable clean energy." Al Gore has said
much the same thing.

But they do not lay out the reason for that optimism in the face of a
much wider agreement that even great technological advances will not
be enough to stem the rise of emissions from fossil fuels. Yet if their own
hopes for the future were overly optimistic, they had, as did Pielke and
his colleagues, picked up on rejecting a treaty as the only optimal way
to go. David Shorr, in a 2014 *Foreign Policy* article—giving many of the
same reasons as Daschle and Wirth—noted that giving up on a treaty
attempt would relieve pressure on the countries least likely to sign one,
would create more flexibility, and would allow for the play of peer pressure among countries.[14] That strategy seems to have won in the run-up to
the 2015 meeting.

William Nordhaus, the most prominent advocate for a carbon tax,
has noted some of the drawbacks of a cap and trade policy. Apart from
the political attractiveness of a market-based cap and trade policy, Nordhaus notes the price volatility of the former, creating costly economic

uncertainty (as the European Union has demonstrated).[15] Even so, given political realities, he says both of them would be acceptable but holds on to his argument that a carbon tax would be comparatively more effective. But carbon taxes have some critics. Pioneered by the Nordic countries, beginning with Finland in the early 1990s, the history of carbon taxes in those countries offers a good test of their efficacy. As Anthony Giddens has noted, the level of ambition was at first low, but even so the results were encouraging.[16] Carbon emissions by Finland would have been 2% to 3% higher in 2000 but for those taxes, and even more for Sweden, Norway, and Iceland (3%–4%).

Yet the absolute levels of emissions of those countries continued to rise in the 1990s, although not for Denmark. Why the difference? Only Denmark used the taxes to subsidize energy-reducing taxes. Carbon taxes can also be regressive, with the poor bearing a greater comparative burden. It is hard to choose between a carbon tax and cap and trade, but the political advantages of the latter seem to me compelling. Probably only in the Nordic countries, with strong communitarian values and welfare programs, can a carbon tax make much headway. President Obama's reliance on cap and trade makes sense, assuming his plan can survive the certain war against it that industry and conservatives will wage. But he comes into this effort with a running start from the earlier and ongoing programs in many states.

I have said little here about a less visible but important development in global-warming work. At a more grassroots level, in cities and towns and with informal groups, there is considerable activity on global warming. As James Gustave Speth has put it, "A surprisingly diverse array of local organizations and communities are impatient with international processes. . . . [A] new system is being built from the ground up."[17] They are important for stimulating support for legislative efforts at the state, city, and regional level.

## WATER

The reasons for poor water quality and water shortages are many. They range from economic progress and its attendant increased demand for meat and other water-greedy crops, to increased availability of inexpensive pumps to extract water from shrinking aquifers, to waste and inefficiency, and to the obvious strain placed on water resources by the growing populations who need it. Efforts to manage water problems and shortages require public awareness, incentives or penalties to control its use, and technologies of many kinds.

## International Governance

Water treaties and agreements pervade the globe at the national and regional levels. The United States has more than fifteen treaties and agreements on water sharing among states. A river such as the Colorado, which flows through a number of states, is an outstanding example. Countries in the European Union, Russia, Brazil, and India each have more than fifteen treaties, and dozens of other countries have treaties as well, even if not as many.[18] Perhaps surprisingly—amid long-standing worries and some incidents—international cooperation "is more often the rule than the exception." The United Nations determined that "the last 50 years have seen only 37 acute disputes involving violence compared to 150 treaties that have been signed." The agreements often have to be reviewed and adjusted as water supply and usage shift over time, and sometimes the politics of the negotiation efforts have resulted in inequities (such as imbalance of distribution of Jordan River water between Israel and the Palestinian areas).[19]

The fact that some countries at or near war with one another can make and honor agreements suggests that at some fundamental level the basic need of humans for water should be honored, much like the medical treatment of wounded enemy soldiers. Nonetheless, the negotiations can be difficult and delicate. While below the level of formal agreements or treaties, water management within countries can have many political and equity problems. Political corruption is endemic in many countries, often stemming from contract agreements for operating municipal water systems. Irrigation management often displays an imbalance between large landowners and small, with advantages for the former. In India it is estimated that 25% of irrigation agreements are corrupt, a pervasive reality so often with dams, large-scale engineering projects, and the movement of citizens to make way for dams.

While there is considerable management of water at national, regional, and local levels, that is not the case at the international level. The decision of the United Nations to declare access to clean and available water a basic right, however, was an important signal of its importance for all human existence.[20] In light of the eight UN Millennium Development Goals, the World Water Council plausibly declared that water is a prerequisite for all of the them.[21] But beyond those global declarations the reality of robust governance is decidedly mixed. At the national and local levels, save for the poorest countries, there are almost always laws and regulations concerning water quality. The EPA lists some thirteen federal laws bearing on water, many on water purity, but others on rivers and harbors, watershed protection, flood prevention, and even wild and scenic rivers, many no doubt inspired by Rachel Carson's seminal work on pesticide contamination of water. Americans simply began taking water more seriously, going well beyond dirty rivers and lakes.There are, to be sure, a number of important NGOs, such as the Global Water Council and the Global Water Partnership. But at the global level there are no government regulatory or policy-setting agencies specifically focused on water. Yet there are, in an informal sense, cooperative clusters—usually called regimes—that foster various working partnerships and cooperative relationships.[22] Ken Conca, a political scientist at the University of Maryland, has long been one of the leaders in analysis of water governance and regulation. Conca has tried to characterize those efforts as a group, looking for overt and latent traits they share. He sees "several contradictory forces." They include "sovereign rights and responsibilities, neoliberal structural adjustment (privatization, marketization, commercialization), elitists who push techno-rational norms of management, transnational activism for rights of small communities (human rights, grassroots democracy, and preservation of local cultures and ecosystems)."[23] "But none," he says, "has generated a normative framework governing watershed practices." Even so, despite that diverse range of values there remain some that are dominant and pervasive: "sovereign territoriality, statist authority, and stabilized knowledge [elitist techno-rational management]." He does not try to find a single solution or set of values (though his low opinion of the elites comes through) but opts for "more flexible and hybrid approaches . . . not treaty negotiations."

## Transboundary Water Sharing

That last sentence could have been written by skeptics of treaties for global-warming control, and it is similar to the indirect "obliquity" strategy proposed in the Hartwell Report.[24] In that same vein, Frank A. Ward at the New Mexico State University uses as his point of departure the observation that "few international water-sharing agreements have shown the flexibility to adjust to extended drought; fewer still provide safeguards for adaptation to modern climate variability." Yet current conflicts over the use of transboundary rivers continue and probably always will. The search for negotiated water-sharing arrangements can provide flexibility in the face of change.[25] These arrangements should include, for example, changes in water flow population, climate, technology, and economic activity. He uses examples, however, from two North American agreements and one in South Asia that bring out errors that can occur from that effort.

The United Nations remains on a kind of borderline, usually seeking treaties or conventions among nations but settling on what it can get when that effort fails. Many earlier-initiated but continuing UN efforts on water focused on what is called "transboundary water," that of rivers and lakes impinging on national boundaries. That is an important realm, with some 263 boundary lake and river basins covering nearly half of the earth's land surface. While most are between two countries, many are much higher, with five river basins—the Congo, Niger, Nile, Rhine, and Zambezi—shared between nine to eleven countries. The leader is the Danube, which flows through eighteen nations.[26]

Two recent initiatives show the growing importance of water. One of them is the proposed incorporation of water as a key element in the post–UN Millennium Development Goals period beginning in 2015 under the auspices of the UN Sustainable Development Program. At the center of that effort will be the eradication of poverty, which it calls the most important global challenge, but a wide range of water quantity and quality actions for drinking water and sanitation will also be included. The other and more recent initiative is a "law of transboundary aquifers."[27] This effort is marked by the usual slow development within UN governance activities and some disputes about whether voluntary guidelines rather than a treaty would be most effective. The whole treaty discussion will no doubt reflect some of the contradictions Conca has noted in a mainline techno-rational mode of policy making. Those who call for voluntary bottom lines may reflect the alternative route.

A predictable debate that will emerge will bear on nonsustainable groundwater irrigation, a coming crisis for both drinking water and agriculture. That nasty reality will raise the stakes for some nations as much as and probably more than transnational rivers and lakes.[28] Two articles of importance on nonsustainable groundwater appeared in 2013 in the *Global Water Forum*. One examined the use of the "water footprint" concept, borrowed from its use in climate change analysis of "carbon footprint," thus allowing for a useful distinction among renewable water (rivers and lakes), nonrenewable water (depleted ancient and deep fossil aquifers), and virtual water (the water it takes to produce crops of different kinds). To be a useful metric for calculation of available nonsustainable water and its rate of use, the water footprint concept has been refined to detect unsustainable use of a groundwater aquifer. It is a ratio of that water to the area used for the agriculture that draws upon it. The other article presents data that shows the extent and seriousness of nonrenewable water in different parts of the world. The California Central Valley and parts of India and Pakistan are drawing nonrenewable water. That depletion has also come to include the large Ogallala aquifer in the American Midwest.[29] It is urgent therefore to find ways to limit what has been termed "the overdraft" while not harming food productivity. But so far little progress has been made.

## Water Laws and Regulations

Just as there are myriad transnational treaties and agreements, only a few of which will be touched on here in detail, the same can be said of water laws and regulations for water management within countries: every country has some and usually many. I can only provide a sample of the latter here, just enough to convey the flavor and variety of those efforts; they are drawn from the journal *Water Policy*, which has provided exceptional research and coverage of those efforts.

Here are some examples, first, of citizen participation in water management. One study, comparing water management efforts for a river basin in

Brazil and one in Catalonia, Spain, sought to determine the comparative advantage of two differences between two strategies. One of them tests the success of democratic participation in decision making and the other what is called the "terms of reference" (ToR) management strategy. The Catalonian authorities embraced strong democratic participation while the ToR Brazilian plan relied more on a top-down government effort, setting targets, stages, and activities but involving much less citizen participation. Despite regional differences, each was aiming for the same goal, that of defining water usage rights in a sustainable way.[30] The study found advantages and disadvantages in both. Democratic participation, if truly democratic, invariably elicits a wider and often more contentious debate than government-run programs. It is an ideological question, of course, whether the possible complications of more democratic participation are worth the extra burden on governance. The differences between India and China in managing global warming might be called one of the great case studies of a semichaotic and often corrupt government compared with an autocratic one (also plagued with corruption).

A study in India describes an effort in community-level decision making on the selection of competing choices of technology to supply water in a tribal village. The choice was among six alternatives and required novel strategies embedded in local culture, values, and language. It aimed to give the people of the village a feeling of ownership and control at various stages during the selection process, required for successful implementation and sustainability.[31] Inspired by the 2007 UN Declaration of the Right of Indigenous People, a study in British Columbia examined some untested assumptions of indigenous governance. It turned out that such assumptions can actually harm the process. Changing them could be the basis for enduring policy efforts with indigenous groups.[32]

Moving away from indigenous people, another set of studies examined efforts in Israel to decouple total water supplied and water gained from surface and aquifer sources. One effort involved making use of wastewater and greater efficiency with available water. This is most effective when a region is no longer self-sufficient in water and is generally accepted as true by the affected people. The other effort looked at what happens when a region is now self-sufficient and has the capacity to cope with water shortages. Ironically, those latter efforts can be contentious and political—as are many debates about the avoidance of future problems with competing options.[33]

Another study looked at the special problems of small island developing nations. The country studied was Cape Verde, which has a severe and chronic water shortage. Its finding, making use also of information about countries similarly situated, was the importance of a performance evaluation system and yardstick competition as incentives for those responsible for water management.[34] Competition in the water supply industry was also instituted in Sydney, Australia, during a severe drought in 2000, relieved only by large rainfall and flooding, an endemic fluctuating situation in Australia.[35] A final study I will mention is that of Portuguese municipalities seeking to set tariffs on water use. It found that such efforts can be affordable for average households and incomes, but not for the 20% of the population at the lowest end of the income scale.[36]

## FOOD

### Global Governance

The 2007–2008 food crisis was a jolt to the global agencies that had seemed to be making great and steady progress in improving food security, most strikingly between 1974 and 2000. There was a comfortable belief that food shortages were a thing of the past, eradicated by the Green Revolution. That optimism disappeared in 2008. Widespread food riots in forty countries, the outcome of a sharp rise in food prices, affected developed and developing countries alike, but especially the latter. By early 2009 the number of hungry people in the world was more than 1 billion. The crisis abated a little in 2008 but was still above the 2005 level. In that earlier year the new UN Millennium Development Goals were announced and number one among its eight goals was to "eradicate extreme poverty and hunger."

By 2013 the UN Food and Agricultural Organization (FAO) could say about 842 million people were suffering from chronic hunger, down from 868 million in 2012, about 3%. There had also been a 17% decline since 1990. That was surely progress, but will not come near the Millennium

Development Goal of cutting the number of the hungry in half by 2015. As with poverty reduction there had been real gains in many countries, but sub-Saharan Africa and a number of Asian countries were not among them—despite gains in GDP. In chapter 6 I noted that one reason for persistent high fertility rates in the face of GDP growth in sub-Saharan Africa has been the no less persistent childhood mortality in those countries, thus proving an exception to the general rule that birthrates decline with rising GDP. A lack of adequate food is the main reason for continuing childhood mortality, but there are other reasons. Many of the countries that have improved nutrition have also experienced obesity, that unwanted offspring of "better" diets.

## Achieving Food Security

In addition to the general figures just noted, the *Economist* Intelligence Unit (EIU) has provided a useful breakdown of annual food development in its Global Food Security Index (GFSI), drawing on its own resources as well as those of the FAO and the Global Hunger Index of the International Food Policy Research organization. It defines food security as occurring "when people at all times have physical, social and economic access to sufficient and nutritious food that meets the dietary needs for a healthy and active life."[37] The index goes back to a 1996 formulation developed at the World Food Summit in 1996 and embodies three measures: affordability, availability, and quality. In 2013 the EIU added corruption and urban absorption capacity and obesity as important variables, in addition to earlier efforts to include food safety nets, access to financing for farmers, and protein quality in the average diet.

The organizers of the index are also aware that by measuring what goes on at the national level they will miss developments at the local level and thus "cannot fully capture cultural and political dimensions and risks," an important part of the axiom that all food production is local. But the GFSI makes evident the ubiquitous nature of food security pressures. It cuts through all levels of nations and societies, the effects of which are not easily noticed by most people, who in the end know only what is available for them to eat where they live—or if they are farmers, know only what they can raise and sell. Most people in developed countries get their food from well-stocked grocery stores, where they will mainly feel the impact of food prices. Those of us in America are mostly blind to the estimated 30% wastage of food in the United States, from the food thrown away in restaurants and homes and trucked away from grocery stores at night.

A few further details from the 2014 GFSI are worth noting. Some 70% of countries in the index saw some increased food security in 2014 after a small dip in 2013. What accounts for the improvement, not just in the year but more generally since the 2008 price crisis? Economic growth helped the low-income countries, helped by a decline in wheat and rice prices (e.g., Uganda, Serbia, Sierra Leone, and Portugal). The fifteen countries with the greatest gains benefited from an improved political environment and less risk from urbanization-related food shocks (e.g., the United Kingdom, Singapore, Bulgaria, and Ethiopia).[38] Why did some countries not see a gain? While every region in the world saw improved food security, quality and safety considerations together with reduced diet diversification and political difficulties pulled down South America and Asia.

As with water, there is no formal international agency with governmental or regulatory authority to directly, fairly, and efficaciously manage global food production and distribution. There is no international scientific authority equivalent to the IPCC, nor have there been a series of international meetings to set forth goals for food security and management, as was the case with the Kyoto Protocol. The FAO, UNICEF and other international groups, and NGOs carry out studies, often support research, and set forth policy regulations, but they have no binding authority. For the purposes of this book, what sets food apart from the other four horsemen—where no significant progress has been made to stop their upward trajectory—is that progress has at least temporarily been achieved. Just how long, however, the present decline in malnutrition and stabilization of food prices will continue is uncertain. But not uncertain is the projection that food production is falling well behind projections for population increase. Only a large increase in food production will meet coming food needs.

While there have been some calls for the WTO to take up enforcement of food agreements, there is no movement under way to make that happen. The WTO's Dispute Settlement Body has, however, dealt with disputes among countries on particular agricultural matters, but that is about as far it has gone.[39] The Codex Alimentarius Commission in Rome has developed standards for food safety and fair trade practices but operates by consensus only.[40] In the case of water, particularly with transnational sources of water, strong treaties and agreements are patently necessary. Handshakes and informal agreements are not strong enough. Robert Parlberg has noted the main reasons the international organizations have limited power: national governments have uncommon power because food and farming systems are local; most of the world's food is grown and consumed within national borders; agricultural systems depend upon immobile assets, such as land and water for irrigation; and in developed countries, as in the United States, agriculture is heavily sustained by government subsidies and research. In poor countries, state intervention is even more directed and interventionist, but often limits the freedom of farmers and harms their livelihood.[41]

Two stories provide some good insights into the global food situation, ethanol production from corn, a technological move, and the fate of an important food project, probably the most important event yet on food security.

## Ethanol

When ethanol derived from corn was initially introduced as an additive to gasoline it was seen as a relatively easy and painless way to lower gasoline use and thus reduce U.S. dependence on oil. It was also believed that the ethanol-gasoline blend would reduce carbon emissions and be an economic boost to farmers. By 2010 the production of ethanol as an additive used 40% of American corn production, not the so-called white corn used for human consumption but the corn used as animal feed, the major global market for most corn. In the United States the government provided considerable subsidies for ethanol in the form of tax credits, protective tariffs, and a consumption mandate imposed by a government Renewable Fuel Standard (RFS). The RFS, established in 2005, called for a rising consumption of biofuels. It seemed to have everything going for it.

Yet little thought was apparently given at first to its potential impact on white corn production, to the reduction of arable land increased corn production would bring about, and to the effects on the price of corn.[42] The price of corn internationally rose from less than $100 million in 2000 to $376 million in 2012, for developed and developing countries alike. The other impacts were—apart from diversion of farmland from other food crops—expanded use of farmland for corn production, increased prices for food crops used as dietary substitutes for high-priced biofuel crops (wheat for corn), declining food inventories and thus 'vulnerability to sudden drops in supply, and a rise of speculative trading in the international commodities market that creates global price volatility.[43] Most damning were other consequences. They included the realization that the 40% of corn devoted to ethanol production brought only a 10% reduction in gasoline consumption; that climate change benefits in reducing emissions may be small at best and possibly negative; that corn crops consume considerable water and through the use of fertilizers contribute to water pollution; and that international "land grabs," in which countries needing more land buy it from other countries, often harm indigenous populations.[44]

The embrace of ethanol, designed only to reduce the use of gasoline, thus has multiple consequences that effect land use, water availability and quality, commodity markets, and purchase of land from one country to benefit another; this is not to mention its illusory benefits. It has not worked out as hoped. In 2013, the EPA lowered the RFS, bringing a complaint from the agricultural interests, but it signaled some second thoughts about ethanol. In 2013 the European Parliament voted to reduce biofuel production from 10% to 6%. According to EU research, if biofuels received no EU support, the price of food would be 50% lower in 2020.[45] Beyond the European Union the evidence is accumulating that biofuels, particularly ethanol, are having a harmful impact on food prices in developing countries. The ethanol bubble may be collapsing.[46]

## The Fate of the IAASTD

In 2003 the International Assessment of Agricultural Knowledge, Science, and Technology for Development (IAASTD) was established as an outcome of the earlier World Summit on Sustainable Development. Its purpose was to "forecast and propose responses to the agricultural challenges facing the world from [then] until 2050."[47] That was an ambitious goal, very similar in its organization and participants to the IPCC. There were some 400 participants, a mix of scientists, government representatives, NGOs, and the industry sector. It was a model that thus balanced scientists; representation of all relevant interest groups, regionally balanced with a majority from developing countries; and various international organizations. The two cochairs were prominent agricultural scientists, one from Switzerland the other from Kenya. It was seen as a fine example of science on the frontiers of science, dealing with new and "wicked" problems, and its procedures were well organized.

It set as the context within which it was working, among others: "(1) current social and economic uncertainties, (2) uncertainties about the ability to sustainably produce and access sufficient food; (3) uncertainties about the future of world food prices . . . and (6) increasing chronic diseases that are partially a consequence of poor nutrition and poor food quality as well as food safety."[48] It took up a variety of topics, including poverty and livelihoods, food security, human health and nutrition, equity, biotechnology, trade and markets and investments, traditional and local knowledge, and community-based innovation.[49] It provided various options for action, all of a reasonable and forward-looking kind. On the face of it the IAASTD's report was a model of its kind, and it concluded with a final plenary session held in Johannesburg in 2008.

Yet the report was a failure, in part politically, in part ideologically, and in part from efforts to reconcile different viewpoints that resulted in weak and vague language. Studying the response to its work some four years later, one astute analyst wrote that the "IAASTD began with strong political support, institutional legitimacy drawn from its eight co-sponsors, and experienced leadership. But almost anything that could go wrong, did. Civil society representatives clashed with agronomists over the value of physical science versus ordinary knowledge. Business delegates fought with civil society representatives over the merits of large-scale agribusiness vs. small-village farming systems. . . . Everyone deplored the fact that hardly any 'real farmers' attended the gathering. . . . There were public walkouts by some business representatives, while other NGOs simply stepped out of the process. . . . In the end, three key agricultural producers—Canada, Australia, and the United States—declined to endorse the synthesis report."[50] Another commentator, writing in 2008, noted that two corporate participants, Monsanto and Syngenta, resigned from the project but gave no reasons for doing so; it was suggested they were put off by its lack of support for genetically modified organisms.[51]

Once again I can note by saying the obvious, that a combination of affected industries and business-oriented market advocates can wreak havoc on gaining agreements that work against their interests—with Canada, Australia, and the United States, as reliable partners in that effort. But most of the other arguments also showed the play of different interests. It may also show the hazards of too capacious a mixture of minimalists and deniers in the name of democratic diversity.

## CHRONIC ILLNESS

When I began writing about chronic illness some years ago as part of my work on health-care costs and reform, it never occurred to me to look directly at it as a problem of global governance. As had many others, I framed it in terms of the practice of medicine, the cost and use of technology, and the organization of health-care systems in my country and others. When I turned to it in the context of this book, where the focus is international, I expected that I would find authoritative global governance and regulations of a binding kind. Yet it turns out—as with food—that for all the organizations working in the field, for all the myriad reports, coalitions, and recommendations, there is only one at the global level with any binding power. And that one is not focused on global illness as an *aggregate* of all of its causes, but on only one cause of chronic illness, tobacco—the UN Framework Convention on Tobacco Control (FCTC).

### Varieties of Governance

Instead of governance in a conventional sense (at least as I had understood it), I found that those working in the noncommunicable disease (NCD) field have come to use a very broad definition of that word, putting aside the tobacco control FCTC treaty model as the gold standard. As Roger Magnusson put it in an influential 2010 article, governance should be understood as "process": "the engine of change for initiatives at the global level will simply be political will, formalized through partnership agreements, goals and targets, global reporting, meeting and other exchanges. . . . Nor should we minimize the aggregate impact, over time, of advocacy by thought leaders . . . media campaigns . . . and knowledge exchange networks."[52] He then anchors that strategy in "core national and public health functions," which include among other things evidence, health promotion, financing, training, and research.

Law and regulation at the national and local level is another matter. Two professors at the Georgetown University Law Center, Bryan Thompson and Lawrence Gostin, wrote a paper with the ambitious title "Tackling the Global NCD Crisis: Innovations in Law and Governance." National governments and their local counterparts "will of necessity be the primary actors . . . as they alone possess the sovereign authority to implement needed legal and regulatory measures."[53] They then survey a number of implemented initiatives, for example, full disclosure of the health effects of consumer goods (food package labeling, health warnings) and regulation of advertising (for food, cigarettes, alcoholic beverages). These have already been put in place in many places, but the authors then move on to what they call "direct regulation," far harder to implement, with limits on trans fat in food one of the few examples of success.

There is, finally, one other category, that of "performance-based regulation," one that puts "great faith" in "command and control" methods of "engineering [a] healthy life style." It has two disadvantages, one of which is that it "may have unforeseen negative consequences" and the other, I would add, is to provoke charges of unacceptable paternalism. While many European countries with strong welfare programs have citizens prepared to accept a strong government role and some paternalism, this is far less true in the United States. The Oxford Health Alliance in the United Kingdom has made a strong case that the economic costs of chronic disease provide a "justification for governments to intervene in the private sphere of the individual," adding that "'non-rational' behavior," a lack of good information, and the reality that "individuals accept instant gratification at the expense of their long-term best interests" all contribute to chronic disease. Those "internal costs" to individuals matter as much as the "external costs" to society."[54]

Those bluntly paternalistic arguments are fighting words in America. Former mayor Bloomberg of New York gained national and controversial attention by, in the words of Lawrence O. Gostin, using his "sweeping mayoral power to socially engineer the city and its inhabitants."[55] That effort led Gostin to say that the mayor was the victim of an "unjustified paternalism" charge—a "public health approach rejects the idea that there is such a thing as unfettered free will."[56] While many of Bloomberg's efforts were successful (diabetes surveillance), that was not the case with his effort first to impose a tax on sugared beverages, which failed, and then a limitation of serving sizes for such drinks. The latter failed as well, not only because of industry opposition but also because of a lack of public support (no doubt pushed along by the industry efforts). Free "choice" by individuals in the care of their bodies was a simple and effective selling point made by opponents, aided by mockery of "nanny government." It gained opponents support by some 60% in public opinion polls. I will return to the paternalism issue in my next section, on obesity.

I will end this section with some comments on chronic disease and prevention, taking off from a statement by Thomas and Gostin in their innovations paper. After pointing out the harm of NCDs, they say it is a mistake to see them as just a danger in old age, noting that more than 50% of NCDs afflict those under 70. They then say that "the moral tragedy lies in the fact that this is largely preventable. The primary risk factors for NCDs are well known, and could be reduced or eliminated, given the political and social will."[57] That contention needs modification. He may be correct, if overoptimistic, that younger people can by prevention be saved from chronic disease, but that still leaves us with the present elderly in developed countries, steadily increasing, and the much larger number of elderly coming along in the years ahead in the developing countries.

They will not be spared bad health, gradual decline, disability, and finally death in old age from NCDs. The present future trajectory and costs of NCDs in developed countries are already unsustainable. As a 2011 *Lancet Oncology* study showed, cancer is now the leading cause of global NCD costs, and those costs are staggering.[58] It is not just the "increase in absolute numbers, but also the rate of increased expenditure." This is a feature of cancer care but also of the other major lethal NCDs. I put the crisis this way: "medicine has learned how to expensively keep elderly people alive and in poor health for a longer and longer time and, in the process, has kept them alive also with more disabilities in old age."[59] As I contended earlier, if this trend will not be sustainable for developed countries, surely the poor countries will not get anywhere if the goal is emulation and equity. A much beloved set of values must be changed to resolve this dilemma.

## OBESITY

Law and regulation are imperative to manage obesity, but industry opposition has been exceptionally effective in the United States and elsewhere (save for Mexico and a few other places). Hardly less obstructive has been the public's unwillingness to accept law and regulation of what they eat and drink, embracing a sovereignty of the body. No progress of any consequence has taken place in the fight against obesity. I want to contend that efforts to cope with it must make use of what the public health community has tried zealously to avoid, the use of social pressure, and that pressure will also be a good pathway for law and regulation to be accepted.

Obesity has a long, interesting, and neglected history. It was a subject addressed in Greek, Egyptian, and Asian medicine in ancient times, and has received fluctuating attention ever since, ordinarily understood to be within the province of physicians.[60] By the eighteenth century, different theories about its causes were more actively and steadily debated, as were various cultural views of what a healthy body should be and look like. Obesity has rarely been thought good for human beings. By the 1970s it was coming to be taken much more seriously—just when the post–World War II affluence had settled in—as a source of rapidly increasingly illness, disability, and death, and particularly as evidence mounted that obesity is a cause of diabetes, cancer, and heart disease. The World Health Organization (WHO) drew attention to it in the 1990s, and by 2013 the American Medical Association had declared it a disease. Globally, obesity has doubled since 1980, to about 500 billion people. The numbers have not stopped rising.

Obesity came to be known as an "epidemic," although it is not caused by an infectious agent as the attribution suggests. What mainly changed was its characterization by the public health community as an "obesogenic" malady, brought on by changing diets and lifestyles. Requiring both cultural and social change, obesity reduction needs a variety of tactics, most at the national and local level. There is no global governance of obesity control, although the WHO plays an important role with its research and policy recommendations. There are three leading NGOs: the Global Obesity Forum; the World Obesity Federation; and the Global Alliance for the Prevention of Obesity and Related Chronic Disease. That last organization is a coalition of other groups devoted to various consequences of obesity, such as heart disease and obesity.

But with most of those efforts an impasse has emerged. The public, with a strong push from the food and beverage industry, says no one should paternalistically tell them what they can put into their bodies. Regulating eating and drinking are out. Nor does the public health community want government or anyone else to tell people how their body should look. Stigma as way of social change is out. Is there a way out of this impasse? Neither of those statements is wholly correct. Exceptions can be found and distinctions made. But in terms of developing effective policy for obesity, they are the main roadblocks.

## Stagnant Public Opinion

There is no good evidence, however, that public opinion is changing. A 2013 study found that a strong majority (88%) gave three major reasons for the rise of obesity: too much time spent in sedentary activities, inexpensive and easily available fast foods, and people's unwillingness to change. Along with other diagnoses it appears the public is aware of the obesity problem and takes it seriously. But it is no less evident that the public is not keen on a strong government role, supported by only 23% of the respondents. In contrast, 88% consider that individuals have a "very large responsibility" for obesity.[61] That is not a good combination, reminiscent of a similar impasse with global warming: yes, it's a problem but don't ask us to do much about it.

As a 2011 paper in an important series by the British medical journal *Lancet* concluded, "Unlike other major causes of preventable death and disability, such as tobacco use, injuries, and infectious diseases, there are no exemplar populations in which the obesity epidemic has been reversed by public health measures."[62] "Not only is obesity increasing," the report concluded, "but no national success stories have been reported in the past 33 years" (1980–2013).[63] I cannot help noting another irony. None of the measures pushed by the obesity expert community, part of public health, has been successful, but the approach most often rejected, that of personal responsibility, is embraced by the public. Does the latter know something the former does not know—or is it an example of the experts knowing more than a misguided populace? I will leave that question hanging, returning to it at the end of this section.

In the meantime, I want to take a closer look at the role, real and proposed, of regulation in controlling obesity, and why it is so troubled. Lawrence Gostin has provided a useful inventory of regulatory possibilities (to which I add one of my own at the end, although it is more an example of governance than regulation):[64]

- Required disclosure of nutritional information on food and beverage products (product labeling and calorie figures on menus)
- Mandatory surveillance
- Tort liability for industry practice or false claims (false benefits of weight-loss products)
- Regulation of food marketing (what can and cannot be sold)
- Taxation (sugared beverages)
- School and workplace practices (school lunch programs)
- Zoning to influence the built environment (parks and bike lanes)
- Food prohibition (trans fat)
- Agricultural food subsidies (my addition)

One can find examples of every one of those possibilities somewhere in the world and the United States, especially at the local level. But not many. With the exception of healthier school lunch programs, and some barriers to TV and other forms of media advertising directed at children (where it has not been eliminated), industry opposition has been most pronounced at the adult level. Most striking was the intuitively unlikely 2013 assistance given to industry working against Mayor Bloomberg's effort to control portion size for sugared beverages. Hispanic and African-American civil rights groups (including the National Association for the Advancement of

Colored People), health advocacy groups, and some business associations joined the industry opposition. I call it "intuitively unlikely," because the minority groups are those most harmed by unhealthy food and beverages; support from Coca-Cola or PepsiCo was far less surprising.[65] It seems the minority groups were effectively persuaded that such regulation would do them harm. A similar coalition was able to stop food stamp legislation in Florida in 2012—the aim of which was to make unhealthy foods ineligible for food stamp purchase—and a national group succeeded as well with the food stamp program at the federal level.[66] Free choice would be denied to the poor, a discriminatory act. Just as "choice" became the successful rhetorical value to defeat Mayor Bloomberg, it was no less successful in the defeat of food stamp limitation legislation in 2012–2013 in Illinois, Oregon, California, Vermont, and Texas.

If there was any bright spot in the dismal failure of most regulatory possibilities, it was the April 2014 farm bill signed by President Obama—one of his few bipartisan legislative successes with a fractious, ideologically divided Congress. It made a number of changes in agricultural subsidies. Organic farmers and fruit and hemp growers gained, while subsidies for more traditional crops were cut by 30% over ten years. Fruit and vegetable subsidies were increased by 50% over the same period. The food stamp program, SNAP, was bolstered with better support of those same two crops, designed to encourage consumption. Lower prices are a very useful nudge. A study of the effectiveness of food subsidies carried out by the Rand Corporation with evidence from seven countries (including the United States) found them to be effective and "to significantly increase the purchase and consumption of promoted products."[67] This compares most favorably with evidence of a lack of impact with a more publicized use of calorie counts on restaurants and food boards.[68] If the public has been unwilling to support tough regulations, the passage of that bill suggests that the national discussion of obesity is seeing some movement.

A perceptive 2008 study examined the roles of policy, laws, and regulation in obesity prevention. "Virtually," Boyd Swinburn wrote, "all the important hard policy options [laws, regulations] are directed to the environment (making the healthy choice the easy choice) and virtually all the policies that directly target the population are softer choices (encouraging people to make the healthy choice) . . . governments have not shied away from requiring certain behaviors of the citizens when the public health threat is high—seatbelts, workplace safety, smoke free areas, and illicit drugs are common. . . . But requiring certain eating and physical activity behaviors is highly unlikely to happen."[69] Nothing has happened since 2008 to change that judgment, save for isolated efforts in the Scandinavian countries, Hungary, and Japan. "There is not," Swinburn notes, "a groundswell of overweight and obese people calling for action—the pressure is predominantly coming from the professional sector." Occasionally, there are claims of improvements with childhood obesity. Two important studies reported in 2013 and 2014 came to conflicting conclusions, one saying yes, there has been improvement, the other saying no. It turned out that they were using different time frames—and that the longer time frame showed a continuing increase in obesity.[70]

## Using Social Pressure

But that change and coordination cannot, I would contend, come about without finding ways to break the obesity impasse noted at the beginning of this section. Despite the failure of conventional public health efforts to induce change by health education alone, that community of professionals has not come up with any potent alternative. Obesity experts are quite sure that social pressure and a focus on personal behavior making use of stigma is both unfair and counterproductive, a position argued most passionately by Rebecca Puhl and Chelsea Heuer at the University of Connecticut.[71] Yet the obesity expert community is strongly supportive of efforts to bring change by law and regulation. The public is persistently resistant to that strategy, here and abroad, siding with industry; and not, I think, just victimized and brainwashed by industry—but also because that stance cleverly draws upon a deep cultural strand of personal freedom and the inviolability of the body.

Moreover, most stigma comes from within the family and from doctors, that is, those who know these obese people best, not just nasty and censorious outside critics of their behavior. Nor do they much like themselves. Could it not be that those closest to the obese see the way they behave in the company of food? They do not see obesogenicity as the real cause—even though they probably know (as public opinion surveys suggest they do) that there are social reasons that matter as well. More-over, the obesity expert story—as Lawrence Gostin put it, "free will does not exist in public health"—is itself compromised by its embrace of the language of "making good choices," reading calorie counts, exercising more. The use of such language assumes, as does the judgmental family member, that there is such a thing as personal responsibility, no doubt hard and even impossible for some but open to most of us. In addition, when it is said that the obese are stigmatized, I take this to mean that the attack dog of censorious public opprobrium is already out of the kennel, not to be put back in. Once stigmatized, always stigmatized. In any case, in obesity studies the emphasis is now heavily on prevention, aiming to stop it from spreading further. But a case can be made that the fear of stigma has stood in the way of persuasive and effective prevention efforts. I will call it a paradoxical case. On the one hand, health scares have not worked to stop obesity, and prevention is not likely to be fully effective without social pressure against obesity also. If stigma is harmful for all of those already obese and weight loss exceedingly difficult (with low odds of success), then prevention is the obvious choice for the future: helping people avoid becoming obese in the first place. On the other hand, when there is some evidence that a fear of stigma is itself a motive for choosing a healthy way of life, that signals the possible utility of using social pres-sure as an explicit motivation to live a healthy life: don't let that happen to you!

There is such evidence, ironically provided by Puhl in another article, of a different way to think of stigma. Citing a study in response to efforts to control stigma by those fearful of it, she and two colleagues note that "108 million Americans attempted to lose weight in 2012" and that some "63% of overweight and obese women reported coping with stigma by dieting . . . even though others responded by eating more and refusing to diet."[72] Moreover, the article they cite to support that judgment showed that those citing the fear of stigma as a motivator were in the majority.[73]

## Choice and Free Will in an Obesogenic Society

When it comes to preventing obesity from taking hold in the first place, shaming and other forms of social pressure are justifiable. It is for a good cause, health and the saving of life. The common use of the phrase "making good health choices" surely implies (and entails) free will. We already use acceptable social pressure and shaming in much of our common life: it is not against the law to fart in a noisy public way, nor to eat with our fingers, nor to use nasty language to describe different racial and ethnic groups; but most of us avoid those behaviors, censoring our activities and language. Although rejected by obesity experts as a valid analogy, stigmatization has been effective in the anti-smoking campaign.[74] Politicians, corporate CEOs, TV news anchors, and other public figures also receive a signal: do not become obese and beware of even being noticeably overweight. In sum, stigma can serve a good end, and the pervasive use of the phrase "making good choices" is a striking contradiction to the belief that "obesogenicity" has us all in its steely grip; not quite.

The psychologist Michael R. Lowe has made an important distinction between a "population food environment"—that is, an obesogenic society beyond a person's control—and a "personal food environment," which is within one's control. Some people will not be able to resist eating too much once seated in a restaurant but will have the choice not to enter the restaurant in the first place. And I might note, if they do, the current push to list calories on a menu or food board will also offer the possibility for healthy choice.[75]

A closer look as the social history of obesity is far richer and more nuanced than the obesogenic story taken alone, with stigma taken to be its contemporary virulent disease. The essence of that social history is that obesity has for centuries usually evoked a negative judgment, but not always a wholly hostile one. It started with the Old Testament: "God found them in the wilderness/a wasteland of howling desert. . . . He nursed them with honey from the crags . . . [and] with butter from cows and milk from the flock/with fat of lambs and rams. . . . So Jacob ate and was satisfied/Jershurun grew fat and kicked/You became fat and gross and gorged/They forsook the God who made them."[76] And Greek sculptors, we might recall, had many statues glorifying the human body, but never the fat body.

After that, the history of obesity has shown interesting fluctuations of appraisal—but rarely marked by indifference. In his rich study *Fat: A Cultural History of Fat*, the psychiatrist Sandor Gilman observes that "the boundary between 'just right' (well proportioned) and too much (obesity) is constantly negotiated with the general culture and within medicine. The litmus test is whether the 'fat' is a sign of actual or potential ill health."[77] The title of Abigail Saguy's provocative book, *What's Wrong with Fat?*, is an illustration of such a negotiation, arguing (with good evidence) that some degree of fat is healthy, while extreme obesity is a threat. Gilman offers a number of examples of a long-standing interest in obesity: Shakespeare's Falstaff, the eighteenth-century philosopher Immanuel Kant, nineteenth-century medical and other efforts to reduce obesity, and Chinese concerns going back then also. Gilman cites an 1819 writer's observation that the "Missouri Indian is symmetrical and absent . . . and [that prevents] the unsightly obesity, so often a concomitant of civilization, indolence, and serenity of mental temperament."[78] The second American president, John Adams, a corpulent man, was sometimes called "his rotundity," while early in the twentieth century President William Howard Taft's weight of more than 300 pounds was noted but apparently not politically held against him. Queen Victoria was famously obese and no less famously beloved. The prominent novelist Edith Wharton, whose work spanned the last decades of the 1800s and the first decades of the 1900s, spoke unflatteringly of women with "fat fingers."

The French social scientist Georges Vigarello, covering much the same obesity territory as Sander Gilman, notes that our present time has seen an intensified preoccupation with our body. Such interest has never been absent historically, but it picked up speed first in the sixteenth century, gained further momentum with the advent of the bathroom scale in the nineteenth century, and is now epitomized as reaching "the point of becoming a central obsession."[79] "More than ever," he writes, "identity comes from the body, and more than ever before we have the anxious feeling that this body can double cross us." I would add, however, that those of us of an advanced age would say the same thing about our deteriorating bodies. The difference I note is that there is a volatility to a discussion of obesity not present with aging. I learned early on with this project to tread carefully with obesity, especially in even having a discussion of stigma and social pressure. The good of the body brings out emotions of a more intimate and personal kind than the good of the planet.

## Social Pressure and Individual Freedom

However else we might like to talk about it, obesity is *bad* for human beings, and it is something they should avoid. But is it acceptable to use social pressure, even at the risk of harmful stigma, to achieve that goal? John Stuart Mill, the great nineteenth-century philosopher, pressed for acceptance of the "harm principle," now widely accepted in free and democratic countries: that "the sole end for which mankind are warranted, individually and collectively, in interfering with the liberty or action of any of their number, is self protection. That the only purpose for which power can be rightfully exercised over any member of a civilized community, against his will, is to prevent harm to others."[80] Yet he had an important qualification of his harm principle: "Human beings owe to each other help to distinguish the better from the worse, and encouragement to choose the former and avoid the latter."[81] Prevention efforts for the nonobese to keep them from obesity belongs in that category. Its prevention is a human good, both because of the personal harm it can bring to a person, but no less for its impact on health-care costs.

I recognize that the stigmatization or shaming of any group can be demeaning if their condition is well beyond their control—but then fear also can motivate many others in a good direction when they have not gone that far. Unfortunately, at least in my own experience, many obese people take *any* public discussion whatever of the harm of obesity—even when intended for prevention purposes for others—as indirectly a condemnation of them. That can be hurtful. When we say obesity is bad, not intending to insult the already obese, they can and do apply it to themselves. I see no way of avoiding that possibility. But bad is bad. Care must be taken to say that the message is not directed at them, but that will not always work. The nervousness of walking that fine line should be exorcised.

The decision by the American Medical Association to call obesity a disease brought this dilemma out in the open in a 2014 *New York Times* op-ed article by two psychologists, both obesity researchers. They found from their research that "on the positive side . . . the obesity-disease message increased body satisfaction among obese individuals, probably removed the shame of obesity as a moral failing [while there was also] a significant negative consequence, suggesting that one's weight is a fixed state [that] made attempts at weight management seem futile. . . . Ideally," concluded the article, "we would have a public health message that leads to a decrease in self-blame and stigma while at the same time promoting adaptive regulation and weight loss. We've yet to find an answer to this dilemma."[82]

That is variant on the dilemma I discerned at the beginning of this section, and here is my resolution. The public should learn that as individuals and citizens they must openly recognize and label the harms of obesity to themselves and their society. At the same time they cannot avoid the risk of harming the already obese by using social pressure and stigma to motivate through prevention those who are not yet obese to not let it happen to them. Is the psychological and social harm of possible stigma greater than the physical and social harm of potential obesity? I say it is not. To fail to use social pressure, a method so often helpful in public policy, is to invite a continuing public health crisis. In her illuminating book *Childhood Obesity in America: Biography of an Epidemic*, Laura Dawes specifies three requirements for success in reducing childhood obesity, but applicable to adult obesity as well: "Any intervention would have to be *effective . . . acceptable . . . and possible*."[83] No present strategy is doing that. Careful use of social pressure is a necessary new ingredient.

Moreover, if law and regulation are to be brought to bear, nearly impossible now, then a willingness to recognize obesity as both an individual and public health hazard must include law and regulation as a way of coping with it. That is already done with many other public health threats, and there is no reason to assume it ought not be used in this context. Just as the public health professionals should rethink the flat-out rule against social pressure, the public needs to understand that law and regulation are necessary to combat obesity.

## CHINA AND THE UNITED STATES

I want to end this chapter by noting a different perspective on governance, one bearing directly on the five horsemen, and I will pose it as a question: What kind of political system is most likely to best manage them? I will use China and the United States as my contrasting examples. I was drawn to this question by uneasiness in recent years whether a democratic country as fractiously divided politically as the United States, and particularly on the power to be ceded to government, could effectively deal with issues requiring massive social change.[84] History shows that the threat of war can mobilize a nation. But what about in peacetime, when a similar level of mobilization is needed to meet a different kind of threat?

I use China as my contrasting example because it has an authoritarian system radically different from that of the democratic United States—and it has threats from the five horsemen as serious, probably more so, as those facing us. Many of them reflect China's turn to the market and its rise in affluence in recent decades (a subject worth exploring, but not here). Each is at or near the top of the global danger list of environmental and health problems. Each has active programs and policies to do something about them. The parallels between the two countries are striking, but no less so than the different resources each brings to the table.

Here are some of the parallels. At the top of the list is that the Chinese now lead the world in carbon emissions, with the United States second. China's agreement to work to set stronger carbon emission goals was matched by the United States. Its one-child family policy, if often evaded and undergoing change, has kept its population growth rate low, but nonetheless it now leads the world with 1.3 billion people in 2019 compared with 3.3 million in the United States, the leader in the developed world.[85] The average life span is 76 years compared with 79 in the United States. Its aging population now has the fastest growth rate in the world, reaching 202 million in 2013 (figure 9.1). The 23 million of those over age eighty make up the largest growth rate of that category in the world. It is only now beginning to develop a pension plan for the majority of its elderly citizens, and the caretaker burden will increase as well, with little more than family care available for the foreseeable future. If not per capita, China at 5.4 million suffering from dementia has the highest figure in the world, with the United States second at 3.9 million.[86] A 2014 Pew Research survey found that the Chinese are high on the global aging anxiety list with 23% of those polled concerned about aging. The comparable U.S. figure of 21.4% is fairly close, despite the fact that the United States is far more prepared for aging than China. China's GDP growth rate, one of the highest in the world in recent years, has come down from 10.4% in 2010 to 7.4% in 2014, and is expected to go lower.

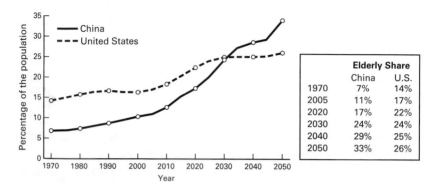

| | Elderly Share | |
|---|---|---|
| | China | U.S. |
| 1970 | 7% | 14% |
| 2005 | 11% | 17% |
| 2020 | 17% | 22% |
| 2030 | 24% | 24% |
| 2040 | 29% | 25% |
| 2050 | 33% | 26% |

**FIGURE 9.1**

Elderly (aged 60 and over) as a percentage of the population in China and the United States, 1970–2050. (United Nations, 2007, cited in Richard Jackson, et al., *China's Long March to Retirement Reform*. Washington, D.C.: Center for Strategic and International Studies (CSIS), 2009: 10.)

With aging societies comes a rise in chronic illness. In China it accounts for an estimated 80% of illness, with cancer and heart disease leading the mortality list. The aging of the population is expected to see a 200% increase by 2040. The Chinese health-care system is moving closer to some form of universal health care but has a long way to go. It will not be able anytime soon to approach the spending levels now common in developing countries for high-tech care for chronic illness. Urbanization has brought with it an increase in smoking and many of the other liabilities of large cities. A cause of chronic disease is, of course, obesity, and China is second in the world to the United States, with some 46 million obese and 300 million overweight adults. Chinese children are not quite as obese as American children but are on the way to catching up.[87]

Both water and food problems are serious in China and closely linked. Its water availability and water pollution are already well into the danger range. Some 50% of rivers that existed in the 1990s have disappeared, and 60% of its groundwater is polluted. Water consumption, most heavily for agriculture (70%) and coal (20%), continues to rise, while its water resources are down 13% since 2000. Yet the average water consumption by people is very low. There is also a serious logistics problem, with some

60% of water in the southern part of the country but both agriculture and coal consumption in the water-scarce northern part.[88] The water troubles also spell trouble for the food supply. China has to feed one-fifth of the world's population with only one-fifteenth of its arable land and to do so with a growing population and with declining land, water, and other production.[89] So far, China's food production has stayed ahead of its food needs, but that is now threatened.

In the context of the five horsemen, the United States is much better situated than China, with an ample food supply, water availability and cleanliness, but no less a need to reduce obesity, to work out a long-term plan to control chronic health-care costs and better welfare support, and to find a way out of its poor global-warming status. China has one great advantage, that of an autocratic society of long historical standing and public support. Though the autocracy did no better in preventing the five horsemen when embracing economic development, it is bringing considerable energy and organization to the challenges it faces. One can both decry its lack of democratic values and at the same time understand its advantages for a strong government role. It is already on a close to war footing with the five horsemen and, in that sense, ahead of the United States in speed of action.

Can a polarized democracy like the United States, beset by politics and a population with too little fear of global warming, manage its "wicked" demands as effectively as a strong authoritarian government?[90] President Franklin D. Roosevelt in his second inaugural address responded to an identical worry, that the dictatorship of Germany had enormous industrial and military power, well beyond that of the United States at that time. Roosevelt was doing all he could to get industry, heavily conservative and anti-interventionist, on the side of supporting assistance to the United Kingdom. "There are some," he said, "who say that democracy cannot cope with the new technique of government developed in recent years by a few countries which deny the freedoms which we maintain are essential to our democratic way of life. This I reject."[91] He knew business had to be brought around to be an active partner. He made that happen; so can we. All things considered, it would be wonderful if authoritarian China and democratic America decided to limit their competition to mastery of the five horsemen, not to military power.

# III

# TOWARD THE FUTURE: PROGRESS, HOPE, AND FEAR

——

# 10

## PROGRESS AND ITS ERRANT CHILDREN

---

### More Is Never Enough

I have come to the end of the road. The first stage, part I of the book, aimed at planting my feet solidly on the ground by attempting to describe each of the five horsemen in a fair and balanced way, putting aside my own judgments. Even though I knew beforehand how hard that would be to do, it was drummed into me again and again in my research how fuzzy the lines are between facts and values, between declared truth and tacit convictions, between hope and reality. That is doubly so when dealing with "wicked" problems. Part II tried to analyze (and often to autopsy) the arguments of the myriad actors who make up the cast of characters in the five global dramas. In part III, chapters 10 and 11 are meant to look back at where we have come from and where we need to go.

The general evidence on global welfare supports the oft-proclaimed belief that human life in general is better now than in the past, that *significant* economic and health progress has been made. But it is hard to find anyone who would say that about any one of the five horsemen, despite the word "promising" being everywhere deployed with any hint of forward movement (with food perhaps a borderline case).

My intuition from the outset in trying to understanding the five horsemen was that the ultimate problem lies at the cultural level, that of the values, ideologies, habits, and sociocultural patterns that make up the modern world in which we live, what I call "life as usual." They started with the developed countries but have now been imported to even the poorest countries, sometimes for the better, sometimes for the worse; typically some of both in ways difficult to disentangle. How did we get in this situation? Part of the reason is a familiar story: the eighteenth-century

Enlightenment with its interest in science begat the nineteenth-century Industrial and Scientific Revolutions, and those begat the twentieth-century intensification of technological innovation in all areas of life and commerce, including rapid medical advances. In the last half of the twentieth century, life expectancies began to rapidly increase and world population growth sped up (even if family size among the more affluent began to decline). Worries began to arise about environmental harms, but not much attention was given to global warming. Rachel Carson's 1962 book *Silent Spring* focused mainly on harmful pesticides and polluted water, not the rise of $CO_2$ in the atmosphere.

My aim in this chapter is to put forth what, for want of a less pretentious word, I will call a theory about how and why the five horsemen came into existence: how their roots were put down, why they have been so hard to pull up, and why the deepest root of all, the idea of progress, cannot be dislodged.

## THE IMPACT OF THE 1970s

For many, the 1960s stand out as the great decade of cultural change in the post–World War II era, overthrowing the old order and obstreperously introducing the new. A better case can be made that it was the 1970s that brought the seeds introduced in that earlier decade to full germination. George Packer, citing a number of studies and books on the 1970s, nicely summarized that scene:

> The seventies turned out to be the decade when the country began its transformation from steady economic growth to spasms of contraction, from industry to information and finance, from institutional authorities to individual freedoms, from center-left to right. Global competition happened in the seventies, and so did populist politics, special-interest money, the personal computer, and the cult of the self. The obsessed-over sixties seem increasingly remote and sui generis, while the trademarks of the seventies are strangely persistent.[1]

It was the 1970s that brought the five horsemen into sight, not so much to the public eye, but to the attention of the World Health Organization

(WHO) and a small cadre of climate and agricultural researchers and health surveillance agencies.

By the 1980s the horsemen were capturing public, government, and media attention. Yet they were not of much interest to the writers who focus on the 1970s, looking mainly at the politics and cultural shifts in other domains. Nor does the burgeoning place of medicine during that time get the attention it merits. Medicine went through a remarkable period in the late 1960s and intensified in the 1970s. The contraceptive pill of the 1960s and the emergence of family planning efforts marked a sharper downturn of population growth (already under way earlier) in developed countries, and gradually moved downward in low- and middle-income countries.

Then there was the dramatic flowering of postwar medical technologies: organ transplantation; kidney dialysis; more effective treatments for cancer and heart disease (with a decline in death rates but not in the incidence and prevalence of each); a wider range of antibiotic and antiviral vaccines, powerfully effective with global infectious disease; the advent of ICUs and advanced technologies in every area of medicine (notably PET and MRI scanning devices); and a burst of new drugs and the vigorous and notably profitable pharmaceutical and medical device industries. Rising life expectancies as a result of that progress and greater affluence, and a corresponding increase in the aged population, intensified the incidence and prevalence of chronic disease. It also stimulated technology research for its (expensive) treatment. Nixon's 1971 declaration of a "war on cancer" is a good emblem of the times, but an even better emblem is the skyrocketing annual budget of the National Institutes of Health (NIH), a congressional bipartisan favorite, which has only flattened in the past few years.

Why were the 1970s a turning point for the five horsemen? I suggest two likely explanations, globalization and the market, and affluence. First, that same era saw the rapid growth and economic force of globalization and intensified market forces and ideology. Leo Strauss, Ludwig von Mises, Friedrich Hayek, and Milton Friedman (helped along by the popular novelist Ayn Rand) became prominent international figures in laying an economic foundation for a reinvigoration of market theory and advocacy. It was then taken up by nascent neoliberal leaders in the United Kingdom (Sir Andrew Bem, Oliver Smedley) and the United States (William F. Buckley Jr., Irving Kristol). Those ideas were pursued politically and most successfully by Margaret Thatcher and Ronald Reagan—and not too long after by the leaders of China and India. In the United States, corporations adopted a sharper emphasis on the bottom line and shareholder returns. The earlier inclusion of social obligations and responsibilities was either rejected or minimized. Globalization became a cause no less than an effect of increased international trade.

At the base of that movement was a reinvigorated belief in the free market as superior to government programs and policies. International trade had a long history going back centuries (the Silk Road from Asia to Europe was a striking sign). But the difference during the 1970 period was the emergence of massive global industries, affecting all of the horsemen, improving the international marketing of products, bringing more easily available capital for industry research and financial growth, faster, more reliable transportation systems, and the exploitation of the natural resources and cheaper labor of developing countries.[2] The United States took the lead in that development.

## THE ONSET OF AFFLUENCE

The second explanation for the shift during the 1970s was the compounded effects of rising affluence in developed countries, the result of full employment and high GDP growth. There was money to spend and people spent it, moving to sprawling new suburbs, buying houses, having more children during the baby boom (1946–1964), but declining to do so thereafter—but beginning to raise their children more expensively in the 1970s. There was more eating out at restaurants, and going from one-car to two-car families, and more enhancement programs for children. More women came into the workforce, giving couples yet more disposable income. More easily gained bank loans and credit cards were quickly taken up. Fast food-restaurants (McDonald's) and massive stores selling low-cost goods (Walmart), usually produced in poor countries, proliferated. Consumerism was now in the saddle, even embraced in practice by its critics who did not find it easy to give up automobiles, air conditioning, international travel, and good educations for their children. I speak from experience here, bemoaning the ubiquity of impersonal shopping malls one day and buying something from one of them the next day because the prices are lower (thanks in part to foreign products) and more choices are available.

As a child born in the 1930s, a time of depression for poor and middle-income workers and of constant scrimping for all but the rich, I was raised in an atmosphere of scarcity: socks were darned, trousers were patched, and we even took in boarders for a time. That was all the more true of Europeans, who had also gone through the prewar depression, wartime rationing, and slow recovery during the immediate postwar era. By the 1970s, much had changed, and it came fast and furiously. Affluence became the new normal in developed countries, rapidly planting the kind of deep roots that nurtured an emergent culture of high expectations, now and for the future.

Those changes had multiple major consequences. The combination of rapid population and GPD growth gestated and nurtured all of them in the postwar years, later adding the impact of aging and increased longevity. The intensified industrial growth brought climate change that simultaneously affects the atmosphere, the food supply, and the availability of water. Early-onset chronic illness in the developing countries is caused by changing diets, especially more meat, which in turn is a cause of obesity. Those diets require more farmland and water than earlier and healthier ones. Affluence raises the likelihood of obesity, and global economic production the desire for more of everything: better food, housing, health, endless technological innovation, improved tablets and iPhones . . . this list could go for pages; that is, for everything we can think of to spend money on—our imagination expanded by affluence, which knows only the limits to the money in our pocket. During this time, new houses featured enlarged closet sizes to accommodate more stuff, and restaurant plates were expanded to allow more food. The traditional 6-ounce Coca-Cola bottle gave way to 12 ounces, which gave way to 20 ounces, now a common standard serving and bottle size. The huge increase in use of high-fructose corn syrup beginning in the 1970s and its disastrous effect—obesity—show the harm of sugared beverages.[3] If the synergies of the five horsemen do not add up to a perfect storm (no one blames obesity on the rising $CO_2$ level), they come close enough for acute discomfort.

## WHEN THE SWEET MILK TURNS SOUR

There is another feature of the horsemen worth noting. Mae West was quoted as once saying "too much of a good thing is wonderful." She did not say what good things she had in mind (though we can imagine), but it is probably not true. The hydrogen bomb made many a zealous believer in scientific progress pause for a second thought. Is the invention of bombs that can kill millions of us really a desirable "progress"? The five horsemen are splendid models for further pauses. Every one is an example of what I liken to sweet milk turning sour, that of something initially good that had unforeseen, often irreversible, harmful consequences.

Let me count some ways. Global warming is the result of much desired progress and human benefit, extracting coal and oil for heat in cold weather, air conditioning in hot weather, and electricity for both of them and much else, including light and communication. That same progress has given us longer lives and a higher quality of life (though we can surely argue about the right definition of "quality"). More than a few hundred millions of us like automobiles, fast air travel by those planes that pollute the air, and air conditioning. They long ago ceased being seen as luxuries.

We just have too much of those good things and cannot decide how to find a line not to cross, a place to stop and say enough is enough. Obesity is driven by cheaper and more readily available food, the latter hardly an evil in itself, but too much of it is. It is good to have inexpensive ways to extract river and aquifer water, but not if it invites excess use of aquifer water that cannot be replaced. That water is being used for agricultural irrigation, again not an evil thing, but hardly beneficial when the water runs out. Save for early-onset cases, chronic diseases grow in intensity with age, and the elderly now live long enough due to medical and social progress to inevitably contract them, surely a mixed blessing. Heart attacks were once common and a quick way of dying in middle age. Now they have been greatly reduced in incidence, allowing those with chronic heart disease to live longer, permitting them to endure long enough to contract one or more additional chronic illnesses, which will eventually do them in. Antibiotics have saved millions of lives but are now often overused for humans and animals, and even if they were more carefully used they would still over time result in more antibiotic resistance. The pharmaceutical companies have not helped, slowing research on antibiotics, not a profitable product.

As if all that is not enough, a vexing feature of the milk going sour is that of progress-driven trade-offs, where an initial benefit turns out to have a deleterious and often, if not always, unforeseeable downside; or, perhaps worse, a foreseeable downside that must be endured. A large amount of water was needed for the Green Revolution, and the food needed in the future will likely require even more. Inexpensive water pumps in water-stricken regions will often deplete irreplaceable fossil aquifer water. In the absence of cheap renewable energy, coal is required for continue GDP growth. Other examples are easily found.

For all their similarities, the five horsemen have different impacts. Global warming has the capacity—if not slowed down—of affecting the entire global community because of its direct and indirect impact. With aging societies everywhere and earlier onset of chronic diseases in developing countries, no nation will be free of chronic disease; it will become more of what it has already become, the world's leading cause of death. Food shortages may cause starvation in some regions, malnutrition in others, and sharply higher food costs in developed countries. Any country can be affected by obesity, and I know of few countries (France, Japan) not already experiencing it to some degree. Water shortages will be regional in many but by no means all places and most dangerous for countries and regions already experiencing water shortages and often dirty water (India).

I will add to this overall picture a significant feature of some of the horsemen, a division between those for which there no regaining the status quo ante, and later correction is unlikely, and those for which correction is possible. The first include global warming, where the $CO_2$ already in the atmosphere is there to stay for many centuries; obesity, with its small 5% to 10% of successful treatment rate; depleted fossil aquifers, groundwater gone for good; and the chronic diseases of aging societies, inevitably destined to rise in both total and per capita terms. Childhood obesity may be more preventable in the future (which is where most of the effort now goes), but no likely successful scenario has so far presented itself for adult obesity.

## A VARIABLE RANGE OF HEAT AND EMOTIONS

Another feature of the five horsemen that I have found particularly intriguing is the varying degrees of emotional temperature and extent of discussion and debate about them. That temperature ranges from hot and heavy with global warming, among both experts and laypeople; to a little more temperate with others, though with a number of hot spots in each: food (buying up of land in poor countries for agriculture to benefit richer countries), water (building dams), and scorn for any signs of stigma (obesity). In comparison, chronic illness when couched in aggregate terms is quiet and reserved, save for indignation among its advocates that it is ignored by well-known wealthy philanthropists. Surprisingly, there has yet to emerge a strong public opinion movement or advocates to make readily available already successful inexpensive prevention strategies for poor countries.

Four ingredients affect the heat and extent of most of the debates. There is the extent and depth of arguments among the acknowledged experts in a particular field, the volatility of political interests, the range of public opinion, and media interest and depth of coverage. Global warming is the obvious winner in this derby, and by more than a nose. There is a wide range of conflicting judgments on the different policy options for global warming, assorted technological solutions, and different national responses. Political ideologies raise their heads in an obvious way, playing out the larger global struggles between market and government predilections. It is hard to find any of the horsemen that have advocacy leaders as vociferous, outstanding, and well-known among the interested general public as Al Gore, James Hansen, and Bill McKibben; or subject to as much (generally muted) criticism by some of their colleagues in the field (Hansen for his pessimism and McKibben for his flamboyance and frequent lack of scientific nuance).

The media, with some different ideological strands, has followed the environmental debates with far more systematic, sustained attention than is given to the other four horsemen, no doubt educating the public more about this issue than the other four. At the other end of the volatility and contention scale are chronic illness and obesity. They both have grievances about outside obstacles, and both receive only sporadic media attention. But they are relatively calm inwardly, with small professional squabbles of a technical kind, well out of the public and media eyes. Neither tends to look to the media to be of help, possibly because they have little to report in the language of "promising breakthroughs" or "exciting new research findings"; that is, of the kind regularly found in reporting and writing on individual chronic diseases. Scientific arguments about the science of obesity reaching the public are rare (the body mass index debate), and in any case present research does not portend promise of great gains or media attention in the future.

## CULTURAL CHANGE

There are many references in the literature of the five horsemen, uttered in a serious tone, on the necessity of massive social changes needed to affect reduction of risk and to advance serious mitigation and adaptation efforts. They imply the need for deep cultural change. Yet the idea of cultural change, as I have defined it—our underlying values, traditions, ways of life—is now and then fleetingly recognized then usually put aside in favor of governance, law and regulation change, or a faith in technology, allowing an end run on cultural change and a seemingly easier route to take. Deeply penetrating and acceptable governance and law as a way of managing and changing behavior require a deep and corresponding change in values. But law and values can and often do have a synergistic relationship. Accepted laws can change behavior, and changes in behavior can change values. In fact it is hard to imagine any long-term serious change without that relationship.

### History and Cultural Change

Nothing less than a significant cultural change will be needed to cope with the five horsemen. While I was an assistant in graduate school to a historian of cultural change (Christopher Dawson) he was most interested in much earlier history, but he hooked me on the subject. I returned to thinking about cultural change much later, using examples from my own lifetime and beginning with the rise of feminism, civil rights, and environmentalism in the 1960s. All required cultural change. They seemed to burst out suddenly then, with great publicity for early leaders (notably Betty Friedan, Martin Luther King Jr., and Rachel Carson). But each had a history, with feminism going back to Seneca Falls, New York, in 1840 and then later the struggle for women's voting rights in the United Kingdom and United States. The struggle for civil rights for blacks began in the 1780s in England with the movement to ban the slave trade and in the New England abolitionism movement in the pre–Civil War years. The work of John Muir and others in the nineteenth century to glorify nature and protect wilderness areas from encroachments by migrants laid a foundation for environmentalism. Each at its core aimed to change culture and values, and was accompanied by proposed legislative reforms, most of which came later.

But why did the 1960s and 1970s see such an acceleration of what had been slowly developing? Part of what held back the pioneers of change was the centuries of neglect and bigotry that had been life as usual for preceding generations. It took a Civil War to uproot slavery and a civil rights act in the 1960s to cope with lawless racism, and it took World War II to speed up the others: service by blacks in that war, women factory workers during the war and a fast rise in female education afterward, and environmentalism pushed in response to inescapably visible water pollution and land degradation in the 1950s and 1960s. More broadly, rapid change was in the air, itself becoming a part of life as usual, and it has stayed that way. The sexual revolution came on rapidly, and even faster have been the gay rights movement and that of single-sex marriage. The legislative changes came simultaneously or shortly thereafter.

Yet some qualifications are necessary in citing the latter as examples of cultural change pertinent to the five horsemen. At least in the cases of civil rights, feminism, gay rights, and single-sex marriage, they had the behind them the grievance of large and badly treated groups. They could invoke the potent language of human rights, even stronger after the war with the UN Declaration of Human Rights. My second example is a feature now of a present life as usual, that of a fast-growing polarization of political and cultural ideologies. The struggle over the role of the market in national welfare, never fully absent since at least the eighteenth century and initiated by Adam Smith and his book *The Wealth of Nations*, has been greatly intensified by the gains of twentieth-century conservatives. For the most part, that influence has been most sharply felt in the opposition by various industries to regulations and taxes to deal with the five horsemen, mainly driven by a concern for profits and other interests.

"Interests" in this context, however, can be understood in two ways, direct and indirect. A threat to profits and jobs is a direct threat, as are threats to successful competition. An indirect threat, but equally powerful with many, is that of government intervention in market freedom, usually overlapping direct threats. In chapter 8 I defined a variant on the latter as a fear of a threat to a cherished way of life. Taken together, these threats are potent, particularly when blessed with money, lawyers, and segments of the media to defend against them. And all the more reason to find ways to turn industry into a ally, the missing link in the imperative troika of social movements, governance, and the business community.

## Modernization and Cultural Transformation

Important work over the years by the social scientist Ronald Inglehart and colleagues on cultural change offers some illumination. Inglehart pointed out that some years ago a debate had broken out on what was called modernization theory, looking into the future of industrial society.[4] One side contended that economic development brings pervasive cultural change, while the other side held that cultural changes are an enduring and autonomous influence on society shaped over centuries by their traditions, their politics, their religions (or lack of them), and their economics, among other things. Neither side would deny massive changes brought about by modernization in the way we live our lives, as individuals, families, and communities. These trends include a movement to smaller families, greater gender equality, service economies rather than manufacturing, greater emphasis on personal fulfillment, a move to rationalistic and away from institutionalized religious ways of thinking, and heightened social individualism.

Yet Inglehart and colleagues have found evidence that most societies graft modernism onto traditional values. They do not just throw out the old and start over again. Traces of earlier Protestantism in northern European countries can easily be found, even where surveys can discern a huge decline in church membership and more secular ways of thinking.

But they are now accompanied by a sustained interest in what is often now called spirituality, a kind amorphous religious outlook severed from church or sectarian connection. The United States, Inglehart has found, is unusual among other modernized countries. "In fact," he writes, "the United States is a deviant case, having a much more traditional value system than any other advanced industrial society . . . with levels of religiosity and national pride comparable to those found in developing countries."[5] The combination of religiosity and political conservatism has in recent years been pronounced in the southern and southwestern United States—the same areas most opposed to global-warming reform and the least concerned with obesity.

Alexis de Tocqueville in his 1835 book *Democracy in America* noted the same combination of religion and individualism (drawing on Protestantism) in the opening years of the new country. The love of science and technical progress, championed by Benjamin Franklin and Thomas Jefferson, added to the mix, with some of the founding fathers also determined to limit the power of government and taxation. It does not require much imagination to see traces of that mix in the American global-warming debate. The famous struggle between Thomas Jefferson and Alexander Hamilton over the size and power of the federal government continues to this day, supporting Inglehart's view of the persistence of old values over time despite many changes. Over the past 100 years China and Russia have undergone a number of revolutions, but despite them hold on to autocratic governance as tenaciously as in their past, and with considerable public support. No confident prediction has been made more steadily for at least fifty years than how acceptance of market practices and theory would force a change in those autocratic societies, thus fulfilling a much embraced conservative theory about the inherent and necessary connection between democracy and a free market. The wait for that change to take place in China and Russia goes on and on. It turns out to be possible to have the one without the other, at least so far.

## Can Cultures Be Changed?

The answer to the question about whether cultures can be changed is easy: of course it is possible, and global history is full of examples of it. Jared

Diamond and Joseph Tainter have astutely explored disasters of a kind that can simply entirely wipe out or badly damage—sometimes physically, sometimes culturally—whole societies.[6] The disasters often came about by a careless misuse of natural resources, or plagues, or harsh conquest by another society with a different religion or values. But cultures are also changed by internal political or religious upheavals: the Reformation of the sixteenth century, beginning in Germany but eventually spreading through Europe; the coming of communism to Russia in 1917; the American Revolution creating a new form of democracy; and the rise of Islam in the seventh century in the Arab countries of the Middle East. Sometimes great changes are reversed. The French Revolution aimed to overthrow regimes of monarchies that went back many centuries. It succeeded for a time, only to be undone by the terror and slaughter it employed. Before long, however, Napoleon Bonaparte came to power, supportive of many of the gains of the earlier revolution but also some of its repressions—and he called himself Emperor Napoleon I.

An important question for our modern life as usual is whether what gains a public consensus on the high value of progress—which I take to be our situation—can be turned back or significantly changed? Not easily. The transformations of our modern life and culture in the pursuit of progress that came on fast and strong in the 1970s had earlier historical roots but spread them much more broadly and deeply thereafter. As de Tocqueville said in *Democracy in America*, "The idea of progress comes naturally into each man's mind; the desire to rise swells into every heart at once, and all men want to quit their former social conditions."[7] Following in the footsteps of Jefferson and Franklin, he saw progress as the great driving force of American life.

The roots of health and economic progress, the core of contemporary life as usual, have been well and deeply planted and will be nearly impossible to uproot. Some people eschew medical care and some leave modern life for communes or a rural life. But not many. That half of the world population now living in cities will not easily go back to a bucolic rural life (even if they want to). A few painful health emergencies usually turn a medical denier into, if not a true believer, at least someone glad to get some treatment. Thoreau, we may recall, eventually left his beloved Walden Pond to go back to his comfortable and affluent family.

The role of industry and the fertilization it has supplied to the roots of economic affluence and improved health can hardly be minimized. It has given people what they want in the name of progress and economic and social growth, but it has also sold them on things they did not hanker for or need, persuading them to purchase things they had never contemplated. And it has too often been a hostile obstacle to reform efforts with each of the five horsemen. Industry's ambition has spread globalization, built the factories to make cars, the Internet, jet airplanes, drugs and medical devices, air conditioners, and just about everything else now available for our material comfort and economic advancement. It also provides for its millions of employees jobs and an income. This book has not been sparing in pointing to all the downsides and unforeseen harmful consequences of that affluence and scientific progress. But that should not be the end of industry's story.

## CAN PROGRESS BE FORSAKEN?

Even so, modern people do not readily give up what they perceive to be progress, warts, pain, and all. I once asked my mother, when she was in her eighties (she was born in 1895), whether she thought life was better now than when she was growing up. She responded by saying in about one-tenth of a second: now! The public opinion polls I have cited, and the low priority they give to action on the five horsemen, tend to support that judgment. The disasters of the present are preferable for many people to those that can be foreseen in the future, even if judged likely. As for the future—our duties to our progeny and future generations—they are likely to be superseded by any significant present losses. An employee responsible to feed and raise his children is likely to fight to save his job in the coal industry if threatened with its loss in the name of future environmental benefits. As the biblical saying puts it, "Sufficient to the day is the evil thereof." It is perfectly possible for people to understand scientific findings and yet reject their implications when they impinge on their lives.

Before he was elected president, Abraham Lincoln said in a campaign speech against slavery in the 1850s something I found uncommonly perceptive in our context of looking for the benefits and burdens of progress.

He said about Southerners that "they are just where we would be in their situation. If slavery did not now exist among them, they would not introduce it. If it did exist among us, we should not instantly give it up. . . . When it is said that the institution exists, and that it is very difficult to give it up, in any satisfactory way, I can understand and appreciate the saying. I surely will not blame them for not doing what I should not know myself."[8] The income and economic survival of the South was dependent upon the slavery necessary to support its cotton industry. Lincoln understood precisely why they could not give it up. It was not just racism, but it took a deadly civil war to resolve the issue, even though an equally important reason for the war was that the Southern states wanted to secede from the union to protect themselves and the North wanted to protect the union. As it turned out, when the cotton industry eventually declined in later years, as much from international competition as from the loss of slaves, the South found other ways to support itself. Should the country be sorry that the "New South" has provided good jobs and wealth by luring foreign automakers to build thousands of $CO_2$-producing automobiles (one incentive for which has been that region's success in keeping unions out of factories)?

It also took a long time, with much travail along the way and some other historical shifts as well, to change the culture on women, the rights of blacks, and the environment. But progress itself as a value was not as directly at stake as it is with the five horsemen, the kind of progress motivating and undergirding modern life itself and particularly developed countries. More than just wanting progress to keep going, every bit of available evidence shows that people do not want to go backward or even to stand still for too long. The Great Depression in the 1930s forced millions to change or simplify their lives. They did not like it and could not wait for it to end. The unhappiness brought along by the 2008 recession and the continuing difficulty of bringing millions out of their reduced state has spread continuing unhappiness in the United States and Europe. The loss of ground on income for the middle class since 2008 and the slow recovery from the recession has caused a backlash against national political leaders. People do not like to regress economically in bad times or see their income losing out to inflation. It undoes the expectation of continuing progress and hurts all the more when younger generations start adulthood economically disadvantaged in comparison with their parents at the same age.[9]

## The Idea and Value of Progress

Built into the core of the idea of progress is nothing less than a struggle against human finitude. We do not like war, domestic strife, poverty, a body that falls apart, threats to our financial security, or a natural world that can be "red in tooth and claw." The value of progress and that struggle against finitude is the base for the troubles of the five horsemen. Progress can be likened to a vehicle in which human life has been riding for centuries, gradually picking up speed in recent times, but of late moving too fast, going down the track at a dangerous pace; and with its current engineers—us—struggling to get the brakes to work well enough to avoid a crash. The late sociologist Robert Nisbet in his illuminating study of progress noted that late in the sixth century B.C.E. Xenophanes said that "the gods did not reveal to men all things in the beginning, but through their own search [they] find in the course of time that which is better."[10] Nisbet goes on argue that "the secular forms in which we find the idea of progress from the late seventeenth [century] on are inconceivable in the historical sense apart from their Christian roots."[11] That view had been rejected earlier by the distinguished Cambridge historian, J. B. Bury in his 1920 book *The Idea of Progress*. But Nisbet believed that in both Greek thought and Judaism one could find evidence supporting the idea of progress (sometimes put in the language of growth). Nisbet knew, however, that he had taken arms against many earlier historians, who rejected much before the seventeenth century, some holding that it was impossible for earlier generations to understand the concept at all. On the contrary, he responded, it "was necessary for the rationalists of the modern era to secularize, to remove God or relegate God to a distance and the modern idea of progress would have been achieved."[12]

I will not attempt to adjudicate the struggle among the historians. Of late, the interest in the idea of progress has focused on its secular roots, even though often controversial—just what counts as progress?[13] In any case, by the early twentieth century the belief in technological progress was, as Nisbet put it, "at high tide, just as Condorcet had predicted in the eighteenth century."[14] The turn to science easily led to a belief that science was the key to progress. The concept of and faith in progress, once the idea was established and deep roots planted, helped us to take arms against our finitude. We can have a better body, find ways to make use of nature for our chosen ends, and raise human aspirations, not settling for what is, but striving for what might be and accepting no stopping point. A desire to overcome our natural shortcomings is not an act of hubris at all but a kind of redemption of faith and hope in secular garb. Yet progress is also what I think of as a weasel concept, that is, one that can be turned to different ends if we choose to pursue them. The creation of nuclear weapons was considered a masterpiece of science and innovation helping to end a war. In the aftermath, as those weapons began to be pursued by many nations, progress came to be understood as arms control, stopping that spread.

If technological innovation in industry, medicine, and agriculture eventually got us in trouble, innovations could be designed to combat the problems. If carbon emissions grew because of clever uses of oil and coal to create energy, they could be controlled by clever use of carbon sequestration to limit those emissions. One of the seductive charms of the idea of progress is that it is self-renewing. By that I mean if it goes wrong it can be used to undo the harm and find another road to progress. In that sense, progress is an indestructible concept. Its errors can always be corrected, There is always more to be had if we use our scientific knowledge and imagination.

My problem with the protean nature of progress is that, once started down roads that looked safe and attractive—economic growth, better diets, life extension—there seems to be no way to set limits, to stop it or to turn it in a new, safer direction once it has become the settled and usual way of life. It is not so easy in practice to undo its excess once we have become habituated to it. We thus are now in the process, already too late and too comfortable, to easily change the direction of progress. Worse than that, a part of the difficulty of changing course is that some could be the losers if they are expected to pay too high a price. That is exactly the terrible dilemma of the poor countries, asked to reduce their emissions and thereby risk losing the chance to bring their poor out of poverty. They are willing to take the eventual risk, even likelihood, of global-warming disasters to fight the present harsh reality of poverty. The rich countries have a different dilemma, that of insuring they can work against global warming but not in a way that risks a significant loss of the benefits that progress has brought them. Hardly anyone, the public opinion polls persistently tell us, wants to give up much of what he or she has, the benefits of the affluence progress has given us. Obesity is harmful, but losing our unfettered autonomy over our bodies, or even protecting us with regulations, is an assault on our rights to live our lives as we see fit.

Changing the trajectories of earlier visions of progress that turned sour can be done, but not easily, and maybe not at all in some cases. It is like trying to rapidly change the course of a 900-foot cargo vessel in rough seas or to stop in a hurry a long and heavy freight train running down the tracks at 60 miles an hour. Moreover, just as it is a reasonable aspiration for poor countries to want the better, longer, and healthier lives enjoyed by the rich countries, it is unthinkable that those who already enjoy those benefits want to change places with the poor. While some Americans with family roots in India do return there, Americans without that heritage do not migrate there to improve their income or quality of life; nor do they migrate to sub-Saharan Africa for that reason either.

## Progress and Capitalism

Technological progress was easily grafted on to the free market vision of constant movement and change and the spread of capitalism. If business is capitalism in practice, the base it rests upon is the value of progress. As a Dutch economist Bob Goudzward has documented, "the pursuit of, and faith in, progress is a crucial component of the social structure of western society—capitalism." Earlier, a famous Austrian American economist, Joseph Schumpeter, had described capitalism as a "form or method of economic change."[15] Silicon Valley, making and profitably selling constant technological innovation, could well be called "the Vatican of Capitalism." Can anyone imagine Apple several years from now advertising its iPhone6 as "still the same wonderful phone as ever, never changed"? General Motors is in the midst of a campaign to sell more Cadillacs, which they have openly said has meant selling the idea to young people that they are no longer the dowdy car your grandparents cherished, but sales are not going well.

My answer to the question of whether cultures can be changed might be differently stated. Can long-standing cultures at their core be changed, as with progress, and can capitalism as a set of values rooted in the idea of progress be changed? Yes, some cultures can be changed but only very slowly if the change is to be peaceful, or rapidly and peacefully once in a while, as with single-sex marriage. Otherwise, history at least suggests it takes cataclysms, natural or man-made, of one kind or another to make it happen. Where does capitalism stand? Can it be eliminated or changed? It cannot be eliminated if it continues on the base value of progress; and is hard to imagine how it can be separated from that base. It requires progress to fuel it and, in turn, feeds progress to continue by moving it along. In the end progress as a value cannot be uprooted. But it can be refined as a value by a constant exploration of what counts as beneficial progress and what does not, and with better and worse ways of adapting to it.

The libertarian CATO Institute in Washington, D.C., recently created a program and website to promote the idea of progress, which it believes is being weakened by attacks on the market and capitalism. Its premise is that the "world is getting better," a truth obscured by skepticism about the state of the world and messages of fear about the future. Why is there skepticism? A likely reason, they suggest, is that we receive an excess of biased information. "Our species has evolved to prioritize bad news," helped along by a media that knows that "pessimism sells."

That may be true, but I want to add that the idea of progress as a limitless benefit plays its role as well. It has tutored us that, whatever the state of the world at any given time, it both can and ought to be changed. We should settle for nothing less than always moving forward. The idea that there ought to be limits to progress is the ultimate heresy. So far, not one of my horsemen has been mentioned on the CATO website. They would not be helpful to the progress cheerleaders. And cheerleaders there are, clustered in Silicon Valley and with nighttime reading from such books as Robert Bryce's book *Smaller Faster Lighter Denser Cheaper: How Innovation Keeps Proving the Catastrophists Wrong*; Peter Diamandis and Steven Kotler's *Abundance: The Future Is Better Than You Think*; and Diane Ackerman's paean to the future, *The Human Age: The World Shaped by Us*.[16]

## GLOBALIZATION

One further ingredient is necessary to make full sense of the mixture of progress, capitalism, and the troubles of the five horsemen, that of globalization. "Globalization," the political economist Dani Rodrik has written, "is the world-wide extension of capitalism. Indeed, so intertwined has it become with globalization that it is impossible to talk about the one without discussing the future of the other."[17] There has for centuries been trade between countries and regions, the most important probably the Silk Road, initiated by the Chinese Han Dynasty (220 B.C.E.–200 C.E.). That trade was an early form of the market in action. Our local farmer's market is nothing more or less than a small-time instance of capitalism.

Globalization can be defined as the linking up of the world through trade, investment, and production networks. That is not new, but what is new is the economic intensification and complexity of those networks and the way they influence almost every feature of modern states and cultures. The technologies of information and communication, and transportation, are the chief means of using the networks. As Jeffrey Sachs points out, "The leading protagonist of the new globalization is the multinational company (MNC), with operations straddling more than one country, and sometimes a hundred or more."[18] In America, these companies include General Electric, Exxon-Mobil, Chevron, Ford Motor Company, ConocoPhillips, Procter & Gamble, Walmart Stores, IBM, and Pfizer. Many of those companies have half of their workforce in other countries. Globalization makes it easily possible to promote smoking in poor countries to make up for a decline in U.S. shares. Globalization makes possible the sale of unhealthy processed foods and fast food in those countries. Globalization allows coal companies to export coal, encouraged by a diminishing use in the United States, thus nullifying some of the $CO_2$ emissions reduced here by allowing it to be increased elsewhere. Globalization has allowed American pharmaceutical companies to patent new drugs in other countries, depriving the United States of tax revenues on new profits from innovations, a practice called inversion.

Foreign companies with international clout include Nestlé, Royal Dutch Schell, Samsung Electronics, Sandoz, HSBC Holdings, Agricultural

Bank of China, and Bayer. By far the top global firms by market capital- ization in 2015 were headquartered in the United States ($9.32 trillion), China ($1.97 trillion), and the United Kingdom ($941 billion).[19] One or more of these companies sells products that directly or indirectly affect the five horsemen. Every horseman has felt the brunt of their power and influence along with those of other multinationals: global warming in the international trading of coal and oil and massive deforestation for palm oil harvesting; chronic illness a result of well-advertised changes in diets and cheaper food and the sale of tobacco in poor countries; water short- ages from rising agricultural needs; obesity as a result of changes in eat- ing habits globalization made possible; and international sales campaigns by global food and beverage companies. In the name of their corporate progress, profit and competitiveness, these companies constantly improve old products and create new ones. Companies that fail to do so, or do so less well than competitors, see a decline in sales and shareholder profits. And failure is not well tolerated in business. It is a powerful force working against all the horsemen.

Jeffrey Sachs notes three effects of globalization that he calls "globally transformative."[20] The *convergence effect* that allows emerging economies to leapfrog technologies and to narrow the gap between rich and poor countries, notably the U.S. companies (cell phones, now ubiquitous in poor countries). Then there is the *mobility effect*, the difference between mobile capital and immobile labor. "In the ensuing competition among governments there is a 'race' to the bottom," he writes, " in which govern- ments engage in a downward spiral of taxation and regulation in order to keep one step ahead of other countries." The *labor effect* has been that of bringing millions of low-skilled workers in developing countries into the labor market, affecting the entire global workforce and "pushing down the wages of low-skilled workers around the world."

I will not get into the debate about how much so-called free trade agreements actually help or hurt those workers but there is no doubt that it has cost a few million Americans their jobs. This has often been accom- plished, on occasion ironically and painfully enough, with the help of American companies in competition with one another. The 2014 book *Factory Man* by Beth Macy describes in great detail the way the Chinese were able to wrest jobs from the American furniture industry. They did so by copying American furniture without permission and then selling their own Chinese knockoffs more cheaply, often half or more less expensive than the originals. A third-generation member of a family furniture company, John Bassett III, one the America leaders in that industry, fought with the Chinese about stealing his company's furniture designs and even persuaded the federal government to extract payments from the Chinese to compensate companies like his (though insufficiently to make up their loss). But it was the American retailers who were most resistant and even harder to work with than the Chinese. They could make a much larger profit from the cheaper Chinese products, in the process showing utter indifference to the loss of American jobs and vilifying Bassett as the leader of the anti-Chinese effort.[21]

## Reforming Globalization

For the Nobel laureate in economics Joseph E. Stiglitz, it is not globalization that is bad. It is the way it has been managed. "Globalization," he contends, "has brought better health, as well as an active global civil society, fighting for more democracy and greater social justice. The problem is not with globalization, but how it is managed. Part of the problem is with the international economic institutions," that is, with the International Monetary Fund (IMF), World Bank, and World Trade Organization (WTO), which help set the rules of the game. They have done so in ways that all too often have served the interests of the more advanced industrialized countries—and competing interests within those countries—rather than those of the developing world.[22] Stiglitz writes as one who worked as chief economist for the World Bank beginning in 1997. He could see and experience the infighting and political forces that had hampered that agency but, even more, the IMF. The latter undertook in the late 1990s to introduce "structural adjustment," by which it meant the flourishing of the free market and the provision of loans to countries that pursued such policy. A colleague and I examined its effect on health care and drug availability in our book *Medicine and the Market: Equity v. Choice* and found, along with Stiglitz and others, not only bad health policy, but failure in many other areas of economics and welfare as well, doing considerable harm and not much good.[23]

As James Gustave Speth succinctly put it, "Globalization is indeed occurring but it is the globalization of market failure."[24]

Stiglitz lays out in his book a number of steps necessary to reform the international financial system: acceptance of the dangers of capital market liberalization, bankruptcy reforms, less reliance on bailouts, improved banking regulations and improved risk management, and improved safety nets and response to crises.[25] While none of those reforms bears directly on the five horsemen, they can affect the fiscal stability of poor countries. Countries mired in poverty, heavily dependent on outside money to help them, and beset with poor, often corrupt governments, need the help of the World Bank and the International Monetary Fund to do so.

George Soros also complains of bias and imbalance, but chooses the WTO as his target. The mission of the WTO is to facilitate the international exchange of goods and services among willing partners by means of binding rules and to develop effective enforcement mechanisms by granting or withdrawing market access. Yet it is biased toward rich countries and multinational corporations. "The bias," Soros writes, "is not due to the mechanism of the WTO . . . but to the way it has been used, and to the absence of similarly effective structures for the pursuit of other social goals such as the protection of the environment, labor rights, and human rights."[26] No less important, it is not suitable for the provision of public goods because member states will not or cannot provide them. Per Pinstrup-Anderson and Derrill Watson observe that "with food trade, effective and fair globalization has three obstacles: loss of value of local and indigenous cultures from mass production and cultural homogenization; . . . large-scale [and volatile] movement . . . causing social disruption; and chaotic swings in labor demand and wages."[27]

Yet however sharp the criticism of globalization and its close kin, capitalism, few say they are inherently bad or should somehow be altogether abolished—and in favor of what plausible alternative?—even if possible. Globalization, along with progress and the market, is here to stay, its own deep roots intertwined with the others. After all, the farmer's market movement is a market-driven development, aiming to improve diets, to strengthen local farming for better food, and to make the latter economically viable. As for capitalism, I like a phrase in a nineteenth-century poem by the poet Henry Wadsworth Longfellow, about a mischievous "little girl who had a little curl" who "when she was good, she was very good indeed, but when she was bad she was horrid." Capitalism lacks only the curls.

## SOCIAL MOVEMENTS

I have tried along the way in this book to take notice of social movements. I will not even try to present a summary of others, a near impossible task since many are wholly informal, are unnoticed by the media, and attract no wider public notice; and there is a whole branch of the social sciences devoted to the topic. I will instead do two things. First, to note the work of Robert Keohane and David Victor on what they call a "regime complex" for climate change. Second, to summarize the work of the late Nobel laureate in economics, the political scientist Elinor Ostrom, one of the few women (and one of the very few people from outside of the field of economics) ever to win that prize. Her leading and pioneering contribution was to explore the idea of social movements.

In a 2011 article, Keohane and Victor observe that "for two decades, governments have struggled to craft a strong, integrated, and comprehensive regulatory system for managing climate change. Instead their efforts have led to a varied array of narrowly-focused regulatory regimes."[28] There is an alternative route. They call it a "regime complex," meant to concentrate on more "focused and decentralized activities." That last phrase seems to contain a self-contradiction, but it is in line with more recent ideas to move away from a treaty model for global governance to a more informal model allowing room for sovereign nations to accommodate themselves to their own needs while taking account of pressure to reach an agreement. "Flexibility and adaptability," Keohane and Victor write, are the "advantages" they offer to nations not ready to agree to a more rigid model. The UN Framework Convention on Climate Change (UNFCCC) would continue to play an important role, much as it did with the Kyoto Protocol and the organization of its coming 2015 Paris successor. They see the possibility of, but no certainty that, the "UNFCCC could evolve into an integrated and comprehensive political regime." In any event they conclude that "there is little room for optimism" that, in their present state, the emergent complex regime change could stop "global warming at two degrees above pre-industrial levels."

Elinor Ostrom's aim in her most influential book, *Governing the Commons: The Evolution of Institutions for Collective Action*, is to take on the challenge of "how best to limit the uses of natural resources so as to ensure their long-term viability."[29] That issue is, if anything, more insistent and timely now than when it was published in 1990. Even more pertinent is her effort to offer an alternative to the three conventional solutions: central regulation by government, privatization (the market), and the regulation of those who endanger a rational use of scarce resources. Based on my survey of the solutions offered to the five horsemen, those are the dominant and most debated options, coming down to a choice between a dominant leviathan state—or global governance by treaty versus a free market to maximize choice and competition among the users of resources. Her point of departure was Garrett Hardin's famous 1968 article in *Science* on the "Tragedy of the Commons";[30] that is, those situations in which those who raise cattle and herd them on land open to all others who do the same. "Therein," Hardin wrote, "is the tragedy," wherein each herder seeking his own unlimited gain in competition with others eventually ruins the commons. Hardin's argument, I might note, was also much cited in the population control debate and by the Zero Population Growth movement of the 1960s. The proliferation of people is itself the main danger to human survival in a world of limited resources.

Ostrom's approach was to examine a diverse range of situations in which those affected by a coming or actual threat to needed natural resources (particularly water and fish) organized themselves to take collective action; that is, they put together a social movement. Their success in doing so not only developed cooperation among those otherwise competing with one another—with lobstermen in the state of Maine once shooting at each other—but also brought about or influenced local regulations. Ostrom's work was a sophisticated way of formalizing in the language of economics and political science what is now widely identified as "social movements." Those movements have been defined as "conscious, concerted, and sustained efforts by ordinary people to change some aspect of society by using extra-institutional means. They are more conscious

and organized than fads and fashions."[31] The affinities of these relationships seem to me a corollary of the kind of regime complex described by
Keohane and Victor but at a different level. Social movements come in
many forms, some are looser than others and some are hybrid.[32] Social
movements abound in global-warming activities and are common at a
local level with food and water, but are rare in chronic illness and obesity. While they do not emphasize it as such, the books *A Perfect Moral
Storm* by the philosopher Stephen M. Gardiner, which focuses on global
warming as essentially an ethical issue, and *Eco-Republic* by the political
scientist Melissa Lane, which casts it in the language of virtue and ethics,
speak in a way that resonates well with the social movements of global
warming—where passions and language reach well beyond the cool and
impersonal rationality of cost-benefit calculations.[33]

One might ask, finally, whether social movements are important even
if their direct impact is not great; that is, can they have a symbolic importance even if they bring little change with industry? The movement over
the last few years for university divestiture of investments in coal and
oil companies is a case in point. It has gained some university support,
mainly at smaller colleges, not large elite universities. The most common
argument against divestiture is that, for a university with a large and broad
portfolio of investments, dropping a few industries will have a negligible
impact on the companies dropped. That may well be true, but if the divestiture does no more than stir up faculty, student, and alumni discussion
and interest in global warming that in itself could represent an important
educational gain. During World War II, even organizations with little if
any defense activity felt it important for the sake of solidarity to find ways
to show they were doing their bit, even if it was no more than hanging
patriotic posters and publicizing how many of their former employees
were then serving in the military. I have a personal feeling for gestures of
that kind. My father was hired by the Treasury Department during World
War II to manage publicity for the war bond campaign, and a key effort
was to get posters advertising them in every possible place. It was a great
success, not only for raising money but no less to inspire patriotism. The
iconic poster, captioned "Uncle Sam Needs You," showed Uncle Sam with
his finger pointing at the reader.

## MAKING USE OF WHAT CANNOT BE CHANGED

In looking back over the thesis developed in this chapter, I seem to have come to an odd kind of position. At the bottom of the troubles with the five horsemen is the idea and value of progress. But then I argued that it cannot be uprooted as a value, nor should it be, and neither should its often errant offspring, capitalism. I tried to acknowledge all of capitalism's failings, the way progress can go too far and turn sour. Even so, it has brought more benefits than harms, leaving to the side those frequent occasions when the capitalist version of progress gets excessively infatuated with the market as the savior of human freedom and future prosperity. Then, as with the "structural adjustment" pushed by the WTO in the 1990s, it falls on its face and leaves a trail of victims. The power of the market romance has of late been intensified in American politics. Combined with its fraternal twin, a dislike of "big government," it has meant a blessing of the inequalities of the 1% rich and the cutting of important social, welfare, and infrastructure programs—and the money needed to support work on renewable energy and to help poor countries gain what they need to struggle against global warming and other ills. Even so, we are better off with progress and capitalism than without them.

Naomi Klein's belief that it is a matter of "Capitalism vs. the Climate," as a way to sum up the struggle over global warming, goes too far (other factors are at work), and her embrace of social movements as a decisive was to cope with it is insufficient.[34] Social movements can be very helpful at local levels, but far less so in the international arena. But even if she is only partially right in her emphasis on the power of social movements, that is an important contribution, as is her book. Strong, insistent, and persistent social movements have time and again moved mountains. But it is only partially correct to think they are the decisive key to change. They have to influence voters and legislators, where power ultimately lies in democratic societies; and that does not necessarily or reliably happen.

# 11

## THE NECESSARY COALITION

---

### Social Movements, Legislatures, and Business

n September, October, and November of 2014, just as I was beginning to write my final draft for this book, four events occurred that rattled me and no doubt also those working to reduce global warming. The events during that period commanded considerable media attention and could not be ignored, more than for all the other horsemen taken together. The events marked high points and low points, and each of them calls for a closer look to see the portents for the future. Meanwhile, with the exception of a temporary relief of food shortages, no movement of any importance in 2014, or 2013 for that matter, could be discerned with global water and food problems, chronic illness, and obesity.

## THE SWINGING PENDULUM

*The New York March.* That march, called the People's Climate March by its organizers, attracted 310,000 participants, far beyond the expected 100,000. It had called upon some 1,400 organizations drawn from religious, environmental, and business groups and state representatives to help bring it off. Bill McKibben celebrated the power of social movements, noting along with the New York march the successful efforts of thousands in public protests in China to get the government to act against air pollution.

*The US-China Agreement and a $3 Billion Pledge to the UN Green Fund.* Although immediately dismissed as meaningless by the newly elected Republicans, the agreement reached during the president's trip to Beijing was hailed by most others as a great success, not only the agreement itself but its significance for global warming. It promised to be a helpful and needed stimulus to the 2015 meeting in Paris to replace the Kyoto Protocol. While the follow-up stories noted that China was already on track for the substance of the agreement, two new elements had been added.[1] China agreed to pursue policies to reach its carbon emissions peak for coal in 2030 and to reach 20% use of renewables as a source of energy by then. Obama made one new pledge as well. He had earlier said the United States would aim to reduce its carbon emissions by 17% by 2020, and he changed that to 26% to 28% by 2025. At present, China has 26% of world carbon emissions; the United States, 16%; Europe, 13%; and India, 6%. Emissions in the United States and Europe have been declining while those in China are on the rise. India's decision, not just to make no effort whatsoever to reduce coal use but actually to increase its use of coal, has introduced a highly negative development for overall global reductions.

Evident as well was that the president was not going to let the loss of the midterm elections deter his strategy on global warming. He intended to fight for it, even if he was more conciliatory on other issues. After reaching the agreement with China he almost immediately pledged $3 billion to the lagging UN Green Fund to assist global-warming efforts, but did not say how he would come up with that money. In any case, one of the aims of the U.S. contributions to the fund was to stimulate other countries to contribute toward its annual goal of $100 billion.[2] A November conference in Berlin gained nearly $10 billion in pledges from a number of European countries. Obviously there is still a long way to go to reach the final goal. Unfortunately, other earlier studies have said that $200 billion will be needed. Mixed news had come in September from the International Renewable Energy Agency. It reported that, while there had been a drop from an earlier high point in 2011 of investments in renewable energy of $270 billion to $214 billion in 2013, there had been a shift upward since 2001 from 19% of total power generation to 58% by 2013. It then added, however, that $500 billion a year would be needed to restrict the rise of temperature to 2°C. That was a jolt.

*The U.S. Elections.* The November 2014 midterm elections were not only badly damaging to the Democrats, they were probably even worse for global-warming efforts. If not spending as much as their wealthy

opponents, there were large contributions from well-heeled liberal supporters of Democrats committed to climate control efforts, including the
billionaire Thomas Steyer. That support did not help most of them, and
there seemed to be some reluctant agreement in the aftermath that supporting global-warming efforts was not a winning position. Republican
opponents had an even greater advantage of an old-fashioned kind, voters.
The broad plank of the Republicans was that of lower taxes, the downsiz-
ing of government, and job creation. Apart from pathologies of hope,
a serious effort to reduce global warming cannot promise any of those
benefits. Moreover, recall those persistent public opinion polls over the
years and right up to the end of 2014 that showed global warming has very
low priority. No less persistently, job creation has been at the top of the
priority list. Mitch McConnell of Kentucky, the new Republican leader of
the Senate from a state heavily dependent on the coal industry, said after
the election that he would work against global-warming initiatives and
legislation based on their threats to jobs.

*The 2014 Intergovernmental Panel on Climate Change Report.* It is perfectly possible that the very harsh message at the core of the report—that
despite much progress with the science the efforts to reduce global warming
are nowhere sufficient at present to change the continuing upward trend of
the emissions—will finally make some skeptics fearful enough to change
sides. So far there is no sign of that happening in the United States. The
Republican dominance in Congress will only strengthen the resistance.
Public concern may be rising, but it is not changing the persistent low
priority given to global warming and the unwillingness to pay higher taxes
and tolerate more regulation, the influential and decisive part of the polls.

Information coming out of India in November 2014 added a distressing coda—but one softened in September 2015. India, already the world's
third highest emitter of $CO_2$, had made no pledges whatever in previous
years about lowering emissions, and a 2014 statement by India's power
minister, Piyush Goyal, left no room for hope that it would. India's development imperative," he said, "cannot be sacrificed at the altar of potential climate changes many years in the future."[3] He has also promised to
double India's use of coal from 565 million tons in 2013 to 1 billion tons by
2019. Yet India is also planning to greatly expand its solar panel building

program—putting back, it might be said, with one hand, what had been taken away with the other. Yet on balance those moves and statement seemed a great loss for global-warming reduction and a major setback for 2015 Paris hopes.

But by September 2015 a shift came about, no doubt in responses to pressure.[4] India waited until close to the last possible minute and was the last country to submit its plan to be included in planning for the Paris conference. Its plan is neither robust nor venturesome but did represent a change. It committed to reducing carbon emissions by 33% to 35% from its 2005 level to 2030, but still leaving its emissions triple what they had been in 2005—not notably helpful from a global perspective. Yet it requested no assistance from other countries or the UN Green Fund with the cost of renewable technologies, which they will continue to pursue.

Where does that leave us? At the moment, to use a World War I phrase, in no-man's-land; that is, caught between the battle lines, unable at the moment to go forward with any confidence or to go back to safety. Will some new event make it easier to go ahead? My reverie of hope is to hope so, looking for some important change and betting on the business community to speed up the momentum, as I will explain in the last section of this chapter. But there is no plan on any table that could introduce rapid change. All will take time and money, a much more aroused citizenry, and more galvanized leadership and legislators.

How might we best and most prudently assess the progress to date of efforts to undo, or limit, the harms of the five horsemen? I use the word "progress" with some caution in the context of answering that question. On the left, where I stand, progress means finding a way to stop the harms of the horsemen that result from the endless insatiably of industrial progress, profligate and thoughtless waste of water, poor diets brought about by cheap processed food, and the health risks in poor countries of unregulated but well-advertised tobacco products—and most generally returning to simpler modes and scales of living in the modern world. In my terms, the progress desired is that of finding ways to limit progress and knowing how to live within those limits. For the right, progress means enhancing the freedom of individuals, the flourishing of a free market, locally and globally, and a smaller role for government. It longs also for a simpler kind of life, like that before the growth of governments since World War II and the profusion of laws and regulations they generate. They fear the loss of market freedom more than they fear the ravages of global warming, or sugared beverages, or obesity, or water shortages—and often even when they concede their harms.

## SMOKING REDUCTION: THE ONE AND ONLY SUCCESS STORY—ALMOST

Among all the maladies that cause the five horsemen, smoking stands out. Smoking is a potent cause of cancer, chronic obstructive pulmonary disease, and heart disease, the leaders in chronic illness. While smoking is on the rise in developing countries, it has declined in the rich countries. In the United States the percentage of the adult smokers has declined from 42.4% in 1966 to 17.8% in 2013.[5] For high school smokers, it has gone from 27.5 in 1991 to 18.1% in 2011. That still leaves a large number of adult smokers, some 42 million, hardly a trivial number. But considering the obstacles it had to overcome in the 1960s when the antismoking campaign gained speed, it is a remarkable victory, even if a still imperfect one. Yet it is from a public health perspective a partial and still incomplete victory. The millions[6] who still smoke, the increase in established causes of death, and its disparate impact on the poor, leaves much to be done. That is why I used the word "almost" in my subheading.

But in the company of the five horsemen, where there is a severe shortage of even partial successes, it does stand out. The antismoking campaign had to prove to the public that smoking is a health menace to be feared, to insist that the rights of individuals had to give way to the burden on public health, and to beat back a tobacco industry that fought it with advertising and legal pressures that were long effective. It had to have regulatory wins and strong government support, and it had to change the cultural favor that smoking enjoyed. It took years to bring that about. Not only did celebrities smoke but so did doctors (often used in advertising). For a large proportion of the population, smoking is now considered intrinsically bad—what would have been called a sin in earlier times—not to be tolerated under any circumstances.

I was once chastised by a former leading public health foundation leader for saying I thought terminally ill elderly patients should be allowed to smoke in nursing homes. I might as well have said they should be allowed to have automatic rifles. Stigma had its place in the tobacco war as well. I can testify to that as a longtime smoker, and no less as a stigmatized one, who suffered from responses ranging from a look of well-telegraphed facial disgust by some, to peremptory requests by others in frosty, stern tones to "just go outside, please!" It took me years to finally quit. I offer smoking as the paradigm case of what it takes to deal with the causes of the five horsemen: many things must come together, and especially an emotional and personal punch.

With that model in mind, singular and flawed though it is, it is time now to see the path I have taken from the beginning of this book to the end. I will begin with a summary of where I have come from in my earlier chapters, noting the bread crumbs I left on that trail, and trying to assess the present status of the five horsemen. Where do we stand at the moment in the efforts, globally and locally, to do something about them? What are the remaining steps to be taken, if not to decisively solve the problems, at least to move them along in a plausibly strong way? What are the most likely routes for the future and the most imposing obstacles along the way?

## CRITERIA FOR ASSESSMENT

This is, however, an appropriate point to interject a consideration that has gnawed at me ever since I began writing this book. We need to find a plausible balance between hope and pessimism in the evaluation of evidence and the assessment of policy proposals and ideas. There are plenty of good ideas being advanced (particularly with global warming), but too few of those who put them together pass judgment on their plausibility and the realistic likelihood of their implementation. The author or committee just puts them out there, it seems, and hopes for the best. The language of hope and references to "promising" policy ideas or lines of research abound in the literature of my five horsemen. It is the common emotional fare in circumstances of desperation, and no doubt a part of human nature in the face of dangers that cannot easily be dodged. The foremost rule of ancient Hippocratic medicine was for the doctor to inspire hope in the critically ill. It was taken to be a virtue even when cure was of low probability or even not possible. A similar perspective is not hard to discern in efforts to overcome the five horsemen. But what counts as reasonable hope in response to difficult and dangerous problems?

Here are my criteria for judgment in answer to that question: (1) *governance, law, and regulation*: technical and policy plausibility; (2) *political feasibility*: the support of legislators, the public, and industry; and (3) *the probability of success*: taking plausibility and feasibility jointly into account. In a landscape filled with many good and often competing plans and ideas, distinguishing among them is important in determining success, though often not done.

As a way of keeping score, I want to borrow from cancer diagnosis the characterization of different stages of cancer spread. They serve as well as stand-in markers for prognosis with the five horsemen. In stage 0 the cancer cells are malignant, needing only local treatment and careful watching in the future. Stage 1 is when the cancer cells are small, the risk low, treatable by surgery. Stages 2 and 3 show increasing local and regional growth without evidence of movement to other parts of the body, although risk of later spread is high. Additional treatment beyond surgery, such as radiation or chemotherapy, is often needed. Stage 4 is the most deadly, as the cancer has spread to other parts of the body. While not always fatal, it usually is. A complete remission is the disappearance of all evidence of the cancer. Whether a remission signals a complete cure or low probability of return is often uncertain, and oncologists, aware of that uncertainty, are unlikely to promise a definitive cure. A score of 3 to 4 signals a borderline case, which could go either way, better or worse.

In moving now to assess the status of each of the five horsemen, my standards for assessment will be a combination, then, of the balance of hope and pessimism and the criteria just listed. I will not articulate in detail how each of them applies but trust that the previous chapters will suffice to support my judgments.

## POPULATION GROWTH, AGING,
## AND GROSS DOMESTIC PRODUCT

Global population growth is, despite some seriously lagging regions, on the decline and could reach zero growth rate or close to it shortly after 2100. Even so, the projected population will likely have reached 11 billion or slightly more by then, and at 7 billion it is already a large and growing burden on resources. In the meantime there will have been a massive explosion in the number and proportion of the aged, and with it a corollary increase in chronic disease and social dependency. There are few scenarios, even of a speculative kind, about how they will be economically and socially managed. Just how fast future GDP growth will be in both developed and developing countries is uncertain, but it will most likely be lower than the past few decades. Its pace will make a difference, particularly in the capacity of developed countries to provide aid to poorer countries for family planning programs, and for parental aid and for aging relief.

Save for the continuing need for financial aid from developed countries, mitigation of the high birthrates in some regions is possible and necessary (but only with external financing). The total fertility rates are already low in a majority of countries. Aging trends can go nowhere but up, and GDP growth will continue. **Stage 3**.

## GLOBAL WARMING

If the Intergovernmental Panel on Climate Change (IPCC) reports are solid, as I believe they are, then global warming will continue. None of the optimistically reported developments of renewables or declining emissions rates in many countries come close to being sufficient at present to stopping it globally. I give global warming an uncertain score as a way of balancing hopeful and pessimistic projections. The hopeful ones assume success with a combination of technological advances and a strong increase in the so-far wavering public seriousness about the problem—including a willingness to tolerate tax increases and more freedom-limiting laws. The pessimistic perspective notes an ambivalent public ready to acknowledge it as a problem but with a persistent unwillingness to take painful steps to alleviate it.

A strong minority are resisters because they understand that a cherished way of life is at stake, threatening their livelihood and increasing the menace of big government. A great uncertainty is whether and to what extent India and the United States will moderate their opposition and whether China will honor its recent conciliatory agreements. The 2015 Paris conference to devise a replacement of the Kyoto Protocol will be a key indicator of which way the global wind is blowing. In early 2015, as I write this, there are rising hopes, with some evidence, that the Chinese will take strong steps to reduce the use of coal-generated power. The failure to halt global warming is already leading to adaptation efforts in many of the countries most threatened. But they will need to be intensified. The rapid introduction of renewables, notably wind and solar power, is a hopeful sign, although still short of needed financing. I call global warming a borderline case because there is good evidence of some successful reform at the grassroots and national levels, and because the dangers will be of different intensity at different places The hard question to answer is whether that can happen with sufficient force at the global level, guaranteeing **stage 4** if it does not. As I write this, in the fall of 2015—and despite the rise of renewables—the long-term outlook is still for a continued rise of $CO_2$ to dangerous levels. Harm will be done. While the importance of adaptation is widely recognized it has so far only erratically been pursued. **Stages 3–4.**

## FOOD

Even if there are only a few food crises at the moment, the projected gap between agricultural productivity and continuing population growth remains a serious challenge. Add to this the threats posed to some regions from global warming, a decline in agricultural land, and the absence of any promise of technological breakthroughs of significant importance. But since the most serious food problems are likely to be regional, mainly affecting poor countries, I put it as a **stage 3** issue globally, understanding that for the most affected regions it could be **stage 4**. The greatest need will be for strengthening food security, restoring recently weakened safety nets for afflicted countries, getting the leaders of poor countries to take food and agricultural research more seriously—and recognizing an ongoing need for financial help from wealthy countries to do all that. **Stages 3–4.**

## WATER

Like food, the future of available and clean water is likely to be a regional need and thus **stage 3** globally, but surely **stage 4** for the most seriously afflicted areas. In those areas likely to face drought, the most dangerous threat will be to aquifers from which water is being drawn faster than it can be replenished by enough rain or other aboveground sources such as mountain runoff, leading to a steady decline. The danger is particularly great for agriculture, which requires proportionately more water, but hardly good for humans in need of drinking water and a number of industries. There is a wide range of ways to use water more efficiently, a number of technological means to lower water usage and to recycle it, and a number of possibilities for reducing water pollution. People in threatened local areas have shown themselves to be amenable to voluntarily reducing their own use of water and willing to support regulations. That may well be because the threats are the most obviously visible and require action. Even so, and depending upon globa-warming patterns, there are limits to what can effectively be done, especially in poor countries but in some developed regions as well (the western United States). **Stages 3–4.**

## CHRONIC ILLNESS

It took medical progress to allow more people to avoid deadly infectious diseases across the life span and thus live into old age, in which they will mainly die of a chronic disease. A new historical era had appeared in the post–World War II era, gestating slowly, with the first movement of this new child palpable in the 1970s. There was about it no historical necessity or inevitability of a kind that could easily have been seen coming decades earlier. Yet it was not seen as a problem earlier. Death is inevitable, and if one lives into old age then one or more of the chronic diseases will most likely be its cause. But as life expectancy increased rapidly in the post-war era (close to ten to twelve years), more people made it into old age, with those over eighty the fastest-growing segment. Alzheimer's disease, rapidly gaining ground, is an inevitable part of aging societies, and with no treatment at present it may in many respects turn out to be the worst chronic disease.

In short, chronic illness is a biological reality with death its inevitable end point, but by our response we have also medically turned it into another troubled feature of our life as usual. Hence, for developed countries I put it into the **stages 3–4** category. For developing countries, it is **stage 4**, not simply because it starts earlier in life and will be considerably harder to deal with, but because the present health-care systems lack, and will continue to lack, adequate economic and other resources to manage it.

## OBESITY

Obesity is easily the disaster case of the five horsemen. There are plenty of long lists for reducing obesity, most sensible enough, but none that have shown any progress of consequence. There is scientific research of interest on possible genetic causes or exacerbaters of obesity, yet no major breakthroughs for clinical application (though we know it can run in families). Nor are there any technological salvations on the horizon, save for a few drugs whose value in the long run is not yet known and therefore uncertain; and what is known does not suggest success with large numbers. Intensive clinical and counseling programs are preventive in aim and useful for those who are overweight; thes programs have had some success with the already obese. But they are expensive and time-consuming, hence, they are of limited use. The best that can be said for the many commercial weight-loss programs and for over-the-counter nostrums is that they can help many make progress but with no certainty of success. The data showing that the uses of weight loss programs can encourage multiple attempts, mostly not successful, is not encouraging. There is some minor if conflicting evidence to support optimism about reducing child obesity. While I believe a careful use of social pressure in prevention efforts is necessary, it would not remove the obesogenic pressures at work, thus making its effective only in part. **Stage 4**.

## IS THERE A WAY FORWARD?

Looking at the five horsemen as a group the most optimism I can generate is a **stage 3–4**. I began this project wondering why these five, almost uniquely, have proved so intractable, so resistant to solutions, even with an abundance of good ideas with which to tackle them. I think I have discovered why, going back to the modern and well-rooted love of progress, unlikely to be abandoned but perhaps possible to move in a more helpful direction. But another common feature gradually caught my eye, which I will call the "intensity level." By that I mean a combination of emotional push, research and policy drive, and an agitated public, eager for a solution and prepared to accept some high economic and personal costs, understanding their necessity. Even global warming, beginning to have that kind of intensity, does not come close now to having a good and equal mix of all three. At a less lofty level, the emotions helped generate movement: willingness at local levels to tolerate higher costs and rationing of water; in a country hardly noted for citizen marches and protests, the Chinese public's agitation against air pollution; and countries in low-lying areas near oceans galvanized to work hard at adaptation to manage the threat of flooding.

In the remainder of this chapter, I want to concentrate on two issues, pertinent to each of the horsemen, and each a potentially important way to achieve the necessary intensity level: the embrace of fear at the emotional level and the business community as powerful participant with influence and considerable energy.

## FEAR AS A STRATEGY

In earlier chapters of the book, I noted the disagreements over the use of fear to motivate change and could understand the resistance to using it as a strategy, possibly harming rather than helping. I wavered back and forth about that for some time, but gradually came to change my opinion. Could it not be fear as such, but the intensity of the fear, with *weak* fears doing harm? Could it be that the level of fear must be so high and unavoidable that it overcomes resistance, denial, and assorted other reasons not to take a problem seriously? Wars do that, natural disasters happening before one's eyes do it, and a knowledge of the harm of not doing something now for the sake of future harms can on occasion do it. Recall the mammography debate I discussed in chapter 8. Women turned out ready to run and experience the present risks of serious harm from screening to avoid the even greater harm of future breast cancer. That is just the kind of reasoning lacking with the five horsemen, where there is concern but at an insufficient level of threat and emotional heat to take risks.

But are the dangers posed by the five horsemen sufficient to justify raising the heat? The IPCC reports answer that question for global warming and decisively so. I would not want to invoke the war model for the other horsemen, but only a slightly lower level of heat and intensity. Food shortages are already appearing in various places with far more projected for the future. Water shortages and pollution are present in many places and the death rate from pollution is now in the millions. Chronic illness is now the leading cause of global deaths, and even allowing for its inevitability with the old, is already claiming the lives of people not yet old, becoming an even bigger threat with the rise of childhood chronic diseases. Obesity is already cutting lives short all over the world as a major cause of diabetes and heart disease.

I should add an important qualification about each of my examples, making it different from the mammography example. Every woman will probably know firsthand of a family member or friend who is now sick with or died of breast cancer. But most of us in developed countries have no experience of large numbers dying from early onset chronic illness in poor societies. I live in a small but relatively affluent New York City suburb where there are few obese adults or children: the few obese stand out. Many of those who live in poor areas see obesity in such profusion that it seems altogether natural. It does not stand out. Hurricane Sandy gave us a good feel for the damage storms and bad weather can bring (even though experts were unwilling to call it clear evidence of global warming). But we experience drought rarely, and our water supply is clean and abundant. Our supermarkets are so overstuffed with food that they have a 30% wastage rate. In other words, it takes a serious attention and an empathic imagination for affluent Americans who live in safe place and lead generally healthy lives to grasp the harm of the five horsemen. Those of us who are now old have a good feel for chronic illness (not nice!), taught that message by our failing bodies; but it is harder to grasp that when one is young.

It will take the memory of some history and some present empathetic reading and imagination to give most of us insight into the five horsemen, providing imaginative occasions where fear itself is a rational and powerful motivator. Ironically, there is a willingness to label something as dangerous—which the looming devastation that the harm done by the horsemen can bring, especially global warming. But then to be anxious about using fear as an incentive, simply diluting danger itself into something less threatening, not so dangerous after all, seems the worst possible compromise. What is dangerous is to be feared; that's all. But something more is also required, finding ways to get the business community, generally respected and powerful as a social force, to join forces with those already working on the five horsemen.

## A DUAL STRATEGY: FEAR AND BUSINESS

My strategy proposal rests on two points. The first is to recollect the decade before World War II, touched on in chapter 8. It was marked in the United States and some European countries by many clear warnings about the rise of Nazism and German rearmament. The other is the necessity of gaining business support. Recall that the warnings about Germany were ignored or downplayed, fed by an isolationist ideology; fear could not be generated sufficiently to begin intensive military preparation. It took the first German invasions to awaken Europe and, later, for Pearl Harbor to do the same for the United States. Fear took hold at last. It is also worth looking carefully, moreover, at the U.S. World War II effort to observe the considerable extent to which it had a profound influence on the economic, industrial, and technological strength of the country despite the high price of death and suffering it took to win the war. While not using the language of war, an international group of economists, led by Sir Nicholas Stern, developed a report that, in effect, proposes a number of policies that would not only stem global warming but would also be a powerful economic stimulus reminiscent of a kind produced by World War II. It would do so without the horrors of that war.[7]

An economic historian, Christopher Tassava, has summed up the impact of the war:

> For the United States, World War II and the Great Depression constituted the most important economic event of the 20th century. . . . The war effectively ended the depression. . . . The federal government emerged from the war as a potent economic actor. . . . The organized labor movement . . . became a major counterbalance to both the government and private industry. . . . The war's rapid scientific and technological changes created a permanent expectation of continued innovation on the part of many scientists, engineers, government officials and citizens.[8]

Is something like that what it will take to curb global warming in a decisive way? And was not fear the principal stimulant, something that could not be denied or evaded? The decision to go to war was not the result of a social movement in its favor. Instead it meant the only way to end what had actually been the well-organized social movement of nonintervention, which framed the war as simply being a European problem.

The IPCC reports and other scientific findings of the dangers of global warming, and the exasperatingly sluggish response of the public, show the need to put ourselves on a war footing, with a perfectly reasonable fear of a future disaster. Just as the business community once financed a campaign to stop global-warming reform efforts, fortunately failing to do so, that community could finance a massive social media campaign in the opposite direction. Its aim would specifically be to scare people, to lay out the disastrous consequences of a failure to act rapidly and decisively, and to motivate and allow legislators to pass bills and regulate industry behavior in a way not now possible. Progress would be sought, but progress to make good use of technological innovation for global-warming efforts, currently not well financed. The advent of a strong and growing business presence opens the way for movement in a way sorely missing until now, particularly if industry sees a good business model in doing so. The potential is there.

I would also look for a spillover effect on the other horsemen. Not one of them generates the energy, color, and passion the global-warming debate has elicited. Not one of them has brought business in to help rather than hinder them. Not one of them commands public concern of a kind that could organize a 310,000-person march. Earlier in this chapter I gave my own estimates of the present state of success with each of the horsemen. All have a low probability of success.

All of the horsemen will, however, have to surmount three obstacles. The one I put first is persuading the public that the struggle against the five horsemen must emotionally be put at or at least close to a wartime level. Still another is the continuing resistance, with few signs of letting up in early 2015, to major change on the part of the United States and India, two of the three world leaders in $CO_2$ emissions, with some lingering uncertainty about Chinese follow-through. The Paris meeting late this year will be an epochal event in determining whether that resistance will undermine whatever else is accomplished. The third obstacle will be that of finding sufficient start-up funds from governments to intensify technological research and implementation, now sorely lacking. That need applies to each horseman but one, chronic illness. Those diseases have, taken individually, been blessed with good research funding. But medical care has been too much dominated by industry power, which has pushed up costs in caring for the chronically ill and promoted economically unsustainable high-cost technologies for rich and poor countries—and too little for less expensive prevention technologies, such as vaccines.

## THE GROWING ROLE OF BUSINESS

I now turn to the business community. It is the last of my candidate three pieces that together can make a huge difference in the global-warming struggle, and to varying degrees with each of the horsemen: the combination of (1) grassroots work from the bottom up; (2) national and regional groups in the middle; and (3) global agencies and nations working from the top down. For that to succeed, however, it imperative to gain the support of business, which cuts across those three levels and can push things along with money and influence—energetically if it chooses to—in ways not yet possible. Is it plausible to even entertain that idea? The previous chapters of this book have detailed the many, and successful, ways some industries stand in the way of progress with each of the horsemen. Industries have done so with political influence, money and lawyers, working to thwart necessary laws and regulations, often corrupting professional organizations, and using the media to serve their obstructive goals.

The gap between business and the research, academic, and policy communities (I group all three together) is large and often venomous. In most of the literature from those communities I have cited in this book, the hostility toward business is pervasive and polarizing. Three items catch the flavor. Kelly Brownell, former director of the Center for Food Policy and Obesity at Yale, published an article in 2012 in which he said, "When the history of the world's attempt to address obesity is written, the greatest failure may be collaboration with and appeasement of the food industry."[9] In 2006, William Wiist wrote about the urgent need in public health (including obesity and chronic disease) to strengthen what he called "the anticorporate movement," which he characterized as a "social movement."[10] There is little evidence in that literature of useful and helpful contributions from that camp on the part of industry. I cannot fault the judgment, which I share, that industry opposition and obstruction have been poisonous. Marion Nestle—whose criticism of the food and beverage in industry is unsparing and unparalleled in its details—is nonetheless someone who stands out for noting that there are moral dilemmas inherent in trying to work with industry for reform; can't live with them, can't live without them. Noting the intensified "ferocity" of industry resistance to reform in the mid-2000s, she finds some slight gains: "the ferocity must also be seen as a tribute to the effectiveness of the current food movement . . . may it flourish."[11]

Only when I went outside of that organized opposition did I find the makings of a possibly more positive and friendly relationship between business and social change, much of it unknown to the general public and to most of its most vehement critics, and rarely touched on in the media.

## From an Adversarial to a Diplomatic Model

The first step is to persuade the research, academic, and policy communities to accept what I will call the diplomatic model of relationships, typically now seen between and among nations, and to open a serious dialogue with the business community. The 2008 *Pocket Oxford American Dictionary* definition of diplomacy is this: "that of the conduct by government officials of negotiations and other relations between nations; the art or science of conducting such negotiations; and the skill in managing negotiations so that there is no ill will."[12] I would like to see that diplomatic model adopted and lines of communication established, and in that model a mixture of public and private discussion and bargaining. To use another word from diplomacy, the time has come for a détente, that is, the easing of hostilities or strained relations.

Business is changing and increasingly appears open to a détente. In recent years there has been a strong upsurge of business interest in what has most broadly been called the social corporate responsibility movement. Prior to the 1970s industries were expected to take a philanthropic approach in supporting their local communities, and the larger national companies were major contributors to nonprofit organizations on a wide range of educational and cultural fronts. This was especially the case with a whole generation of corporate leaders immediately after World War II.[13] But pressure from the financial sector in the 1970s stripped the freedom of management to allocate profits at will.[14] The advent of predatory corporate raiders pushed public corporations to change their management practices, forcing them often to change executives, downsize their operations, or even liquidate their companies. The term "leveraged buyout" comes from that period. The corporate raiders, such as Carl Icahn, T. Boone Pickens, and Victor Posner, purchased equity stakes in businesses to influence their boards and managements. Eventually, corporations devised ways to protect themselves, but much philanthropic activity fell by the wayside. Many of the earlier corporate supporters of my own organization made clear that they would no longer provide general support, entertaining proposals only for projects that would directly benefit them. We went from twenty or so corporate contributors to less than five in the 1980s.

## The Beginning of the Business Shift

Yet in the late 1980s and during the 1990s some change was under way, most notably with global warming, though less so with the other horsemen.

It began in an odd fashion, with the failure of an impressive group of leading American and global corporations, the Global Climate Coalition, to hold on to its supporters. In 1987 that coalition began with a big bang, launching an intense television advertising campaign aiming to prevent any endorsement by the United States of the reduction of global carbon emissions.[15] Its strategy was to frighten the public, in part by holding out the specter of a huge increase in taxes on gasoline. For a time the coalition's efforts seemed to be working, but internal dissent began to occur, stimulated in part by the Kyoto Protocol being taken more seriously by the greater public. The chairman of the prominent international petroleum company, BP, was one of the first to leave, not because global warming had been proven but that it was coming to be seen as a possibility that could no longer be ignored. By early 2000 DaimlerChrysler, Texaco, General Motors, and Ford had left. The organization was on its way to failure, and by 2002 it had reached such a low point that it ceased to exist, surviving only as a trade association.

Almost simultaneously with the start of the Global Climate Coalition in the late 1980s and early 1990s, a number of other major corporations began taking global warming seriously. In 1998 the Pew Charitable Trusts established the nonpartisan Pew Center on Global Climate, aiming for a closer relationship of business leaders and policy makers at the domestic and international levels. It gained the support of a number of the corporations that had changed their mind and abandoned the Global Climate Coalition. In 2011 it changed its leadership and severed its earlier ties in a friendly way with the Pew Trusts, becoming an explicitly corporate-managed organization with Royal Dutch Shell, Hewlett-Packard, and Entergy Corporation as its strategic sponsors and with Alcoa, Bank of America, Duke Energy, the Energy Foundation, the Rockefeller Brothers Fund, and the William and Flora Hewlett Foundation its major contributors. Its name was changed to Center for Climate and Energy Solutions, which spawned a markets and business strategy program to directly engage business (Business Environment Leadership Council [BELC]).

On a parallel track was the corporate social responsibility movement (CSR). It has a broad agenda of interests to improve relationships between industry and society, with global warming a high priority.

In 1989 the Coalition for Environmentally Responsible Economies was founded, with the goal of gaining business support for collecting information about company sustainability practices, and compiling that information in annual sustainability reports based on a series of environmental sustainability standards (called the Ceres Principles). It spun off a nonprofit named the Global Reporting Initiative (GRI), which continues to develop information on economic, environmental, social, and governance performance. Some 4,000 organizations around the world use the GRI guidelines, including 82% of the world's 250 largest companies. While "greenwashing" has been a concern—using the guidelines simply to look good—the continued push for transparency has worked against that practice.

Investors have also tracked corporate performance in this area—referred to as the "triple bottom line": people, planet, profits—through the "ESG" rankings, reflecting the data reported by corporations in their annual environmental, social, and governance reports. Socially minded investors are not just concerned citizens. They seek—and find—higher premiums from corporations successfully addressing these ESG issues, the more so as the risks inherent in the issues of global warming, environmental degradation, resource depletion, and human impact become widely recognized.

The corporate sustainability officer, a novelty in the 1990s, has also become a standard player in the corporate landscape, responsible for a broad portfolio of environmental and social programs. Reflecting the embrace of these practices as simply "good for business," a majority of sustainability officers are executives now working closely with CEOs. As Christine Bader writes in her 2014 memoir, *The Evolution of a Corporate Idealist*, sustainability has become so well integrated into standard practice that it is no longer a stand-alone division in many companies.[16] Bader built her sustainability office at BP in the 2000s under a CEO who embraced the philosophy of the triple bottom line. But she did note the vulnerability to management changes. After her departure in 2008, new leadership stepped in and gave a very different face to the BP of April of 2010 when its offshore drilling rig, Deepwater Horizon, exploded with the loss of lives and vast environmental damage.

## The Scope and Power of Business

George Kell, executive director of the UN Global Compact program, has made the strongest possible case for the importance of business taking a leading role: "With business already responsible for the bulk of climate finance flows, and producing far and away the greater part of global GDP and jobs, its active support will be the key to sealing any deal that governments make, or to move ahead if negotiations falter. Moreover, corporate engagement and activity can be pivotal factors in disarming climate skeptical outlooks."[17]

There are, in brief, many positive changes afoot in industry, far too many for all of them to be listed and described here. Business representatives played an important role in the 2014 New York climate march and a number of satellite meetings connected with it. I give here a brief list of a number of groups working toward climate control, with a thumbnail sketch for each:

- *GreenBiz Group*: large-membership group with annual detailed reports on global industry and environment efforts
- *Risky Business Project*: a potent group of U.S. business leaders working to get business to prepare for global warming
- *The New Climate Economy*: an international group of economists aiming to achieve lasting economic growth while also attacking the risks of climate change
- *UN Global Compact*: UN-business partnerships with voluntary corporate responsibility
- *The International Business Leader's Forum* (UK): founded by the prince of Wales to focus on the role of business in society, embracing social responsibility as "core business"
- *World Economic Forum*: producing studies and reports and holding an annual meeting in Davos combining economic growth and risk of global warming
- *The Climate Group and the CDP (formerly known as the Carbon Disclosure Project)*: formed *RE100* with the aim of getting 100 companies to pledge to switch to 100% renewables

- *New York Declaration on Forests*: thirty-four companies pledged to cut deforestation
- *Business for Innovative Climate & Energy Policy (BICEP)*: twenty-nine consumer businesses that have the goal of getting widespread U.S. bi-partisan energy and climate legislation; its *2013 Climate Declaration* gained 800 companies and had 1,000 signatories by September 8, 2014, just before New York climate week and march
- *World Business Council for Sustainable Development (WBCSD)*: a CEO-led organization of companies working to create a "sustainable" future for business
- *The Global Commission on the Economy and Climate*: supported by seven countries around the world and aiming to combine economic growth and the risks of climate change
- *United States Climate Action Partnership (USCAP)*: business and environmental organizations to get national legislation to require significant emission reductions
- *The B-Team*: not-for-profit organized "to catalyze a better way of doing business for the well-being of people and the planet"
- *The Divest-Invest Campaign*: led by the Wallace Global Fund to pledge divestment from fossil fuel stocks and move money into clean energy investments; announced gaining 800 global investors with combined assets of $50 billion by 2014, and aiming for $150 billion by 2015
- *We Mean Business*: a coalition of organizations working with the world's influential businesses and investors to accelerate the transition to a low carbon economy

Here are a few examples pertinent to the other four horsemen. There are not as many organizations doing for four of the horsemen what business is doing for global warming:

- *Anheuser-Busch*: SmartBarley benchmarking to improve crop yield through better water efficiency
- *Sustainable Agriculture Guiding Principles*: Coca-Cola, requiring its supply-chain members to practice these principles to maintain farm-lands and communities

- *Coca-Cola*: with Nature Conservancy and World Wildlife Foundation, collecting water fees from companies operating in the developing world to restore natural ecosystems
- *Better Cotton Initiative*: founded by partnership between World Wildlife Fund and IKEA to address water-gobbling cotton production
- *WATERisLIFE*: the Drinkable Book holding twenty filter pages, each capable of filtering up to 100 liters of water at a cost of about 10 cents per page
- *Global Alliance for Clean Cookstoves*: partnering 1,000 groups with 45 national governments to built a sustainable market for clean cooking solutions
- *Kraft Foods*: pledging to cut portion sizes, fat levels, and marketing to children
- *Walmart*: committing to remove all trans fats and reduce salt by 25% and added sugars by 10% by 2015, and to make healthier food more affordable
- *Darden (Olive Garden, Red Lobster)*: committing to reduce calories and salt in menus and include vegetables, fruit, and milk as default dishes for kids' meals
- *Healthy Weight Commitment Foundation*: a CEO-led partnership to reduce obesity, comprising more than 250 active corporations and nonprofits

## Assessing Motives and Seriousness

It is easy to question the motives and seriousness of industry efforts. One can wonder whether the motives are simply trying to look good ("greenwashing"); feeling forced by public pressure to take social obligations seriously; getting internal pressure from its workforce (especially from younger employees); recognizing that there may be economic advantages to doing so, out of a cautionary expectation that the problem is getting worse and preparations for possible dangers ahead is simply prudent; or just a CEO who is interested.[18] The diplomatic model I propose understands that different and often conflicting motives are at work. No CEO is going to ignore the need for profits and satisfaction of shareholders. But that leaves much room for negotiation.

An obsession to pry loose from industry their "real motives" is not help-ful. Like most things in life involving human beings, they may be mixed. Good diplomacy lives with and expects the other side to sometimes lie, conceal, mislead, engage in posturing, play to its own constituents, and engage in backstairs private talks often at odds with their public utterances. But those likely realities are rarely sufficient to stop negotiations and break off diplomatic relations altogether. While in Beijing, President Obama was trying to get China to cool its territorial push without much success, while affirming its agreement for mutual reductions of carbon emissions.

But it is obvious also that those industries with a direct interest in pro-tecting their jobs and profits, and with a conservative political ideology as well, are likely to remain oppositional, and some will not be open at all to diplomacy. The Koch brothers are not noted for a conciliatory style. The coal and fracking industries have not joined the CSR; their absence stands out among the wide range of business represented in the groups and companies joining that movement.

In the contemporary U.S. context, there are likely to be two different social movements in industry, those who believe they will gain (whatever their motives) or at least not lose from taking global warming seriously, and those who believe they will lose. There are a few in the middle, like Coca-Cola and Kraft, who do stand to lose if the opposition to the sugar, fat, and salt they purvey wins in the short run, but they will try to find ways to mollify their critics, take account of public opinion, and yet con-tinue in business at a profit. Coca-Cola and PepsiCo have pledged to all but eliminate their present products, selling instead beverages with few or low calories only.[19] But that is a pledge about how they will act over a ten-year span, shrewdly allowing time enough, I presume, for the company to develop a new line of beverages. And there are those who, like Robert Stern's group, believe that industry should see the present situation as an opportunity to innovate and come out as winners, not losers.

Joel Makower, the chairman and executive editor of GreenBiz, and an important leader in stimulating business to engage the global-warming problem, wrote an interesting and revealing article in the aftermath of the New York march. There were many people from industry and representa-tives of a number of the businesses in the march, and many satellite meet-ings took place during that week. Makower undertook, in effect, to take the pulse and temper of those participants, summarized well in the title of the article, "The Audacious Optimism of Climate Week."[20] The results surprised him: "unlike the government leaders and citizen marchers . . . the corporate crowd seemed unrelentingly positive about the pathway to solving the climate crisis."

How could this be, he wondered, given the unrelenting bad news about efforts to stem ever-rising emissions and the fact that 2013 saw the level of greenhouse gases reached a record high and 2014 was on track to be the hottest year ever? Earlier in this book I said I wished I could see in global-warming circles the kind of zeal and enthusiasm that has been the hallmark of Silicon Valley. Makower found just that spirit in the meetings he attended. One of the corporate participants, himself a CEO, said something that might have been taken from the Silicon Valley playbook: "A lot of business leaders are optimistic because they see and understand the fact that climate change presents an opportunity for innovation, and they are making changes to their business that are helping them to capture the future." Put another way, they want to use the spirit of capitalism— progress, innovation, and profit—to do for global-warming efforts what has been done by the technological entrepreneurs, youthful visionaries, and billionaires of Silicon Valley.

I am impressed by that kind of enthusiasm, seeing in it a way to better manage the painful solutions that seem to me inevitable and necessary. It is all to the good to gain business as an ally. To what extent it can succeed in reducing the sway of the corporate deniers and minimizers, and be a source of influence on the public and legislators is uncertain, but I hope they will step into the ring and throw some punches at the deniers. Or if that option is not attractive, have enough influence to overwhelm the resisters.

## WINNING OR LOSING? A COMPLICATED QUESTION

I opened this chapter by recalling four important developments in the fall of 2014, all bearing on global warming. I have been waiting expectantly to see if there will be more on the way to the 2015 Paris meeting. I also uneasily felt when reviewing them how unbalanced toward global warming my book has been. I did not expect that when I began my research nearly four years ago. None of the other four horsemen have caught fire at all since then in a comparable way. That was a disappointment. My early ambition was that the comparison of strategies to cope with them would be suggestive of moves among each other they could emulate. I do believe they are interesting and worth knowing—and that it is important to understand their shared cultural roots—but offer few ideas for those hoping to find useful gems in what others are doing. Global warming offers ideas for the other horsemen, but they have little to offer to it or to each other. It is the volatility and sense of real urgency that is most lacking in the others.

Despite all the drama of global-warming developments, it is far from finding some definitive solution. The 2013–2014 IPCC reports dampen that hope. Global warming does offer a model of how to capture public, media, and legislative attention, which none of the others come near. My succinct advice for the other horsemen, having no more enlightenment to add, is this: watch what the forces working on global warming do, and do thee likewise; and I have offered suggestions and parallels along the way in this respect, some of which are being embraced without the global-warming model in mind.

The greatest need with each of the horsemen is to raise the emotional heat, at the highest level for global warming, using justifiable fear as the accelerant. Cultivate business interest and draw upon their resources. Carefully use social pressure, whether in influencing other countries, as is done with global warming, or with individuals, whether to reduce smoking or to divert people away from obesity. Strengthen social movements and, even more, create some where they do not now exist, as with early onset chronic illness and obesity. Pressure the pharmaceutical industry to increase research on vaccines and reduce the cost of drugs in poor countries. Cultivate the fervor and enthusiasm of Silicon Valley, seeking through research and implementation the technological innovation and skill in raising money for it. Persuade government to finance start-up efforts, to subsidize them when necessary, and remind legislators how government support has historically been critical in nurturing new technologies.

As it turned out, there were some further developments in addition to those I noted at the opening of this final chapter. I noted four significant events in the fall of 2014, observing that two of them offered good portents for the future of global-warming reductions, and two of them made it look much harder. I finished that section wondering whether any other event would take place before I finished this book that would have equal consequences, good or bad. As it happened, three events of note did happen.

The most important was a March 2015 report by the UN International Energy Agency (IEA) that global carbon emissions did not increase in 2014.[21] There had been such years before, in the early 1980s, 1992, and 2009, but all had been associated with an economic downturn. In 2014, however, there had been 3% global income growth. That flattening of emissions was greeted as a hopeful sign that perhaps the mitigation efforts are finally paying off. Moreover, there had been earlier news on the positive side that China was coming along even faster than many had expected, and that the prospect of a successful Paris meeting in December 2015 to set new emission targets was greatly enhanced. For its part, China set forth plans to eventually close down its coal-fueled power plants. Both the 2015 IEA report and the Chinese developments, but particularly the former, did elicit some wariness and cautionary responses; what counts the most are long-term trends not year-to-year swings. Yet, counterbalancing the IEA report are reports on data for March 2015 of the National Oceanic and Atmospheric Administration (NOAA) that (1) the average temperature over global land and ocean surfaces was the highest for that month since record keeping began in 1880;[22] and (2) the atmospheric $CO_2$ level for April reached 400.3 ppm, surpassing for the first time 400 ppm for a month since it began record keeping in the late 1950s.[23]

Oddly enough, adding to that mixed news was a survey of 1,300 CEOs in seventy-seven countries, undertaken in preparation for the annual World Economic Forum in Davos in 2015, which found that only 6% believe that mitigating global warming should be a global priority.[24] That result came as a surprise, since the forum and its annual meetings in Davos have given considerable attention to global warming. Commentators on that result did note, however, that the priorities revealed in the polls centered on immediate business needs, such as taxation and regulation reduction, not long-range global problems. That was not a great consolation.

While not quite as disturbing, Joel Makower's 2015 *State of Green Business* report was troublesome enough.[25] The core finding of the report was that "companies' progress in greenhouse gas and emissions, air pollution, water use and solid-waste production are all leveling off or declining." In his opinion, that trend "is a tad distressing, if not depressing." That judgment, of course, comes from the same Joel Makower I just cited for his enthusiastic, wholly upbeat reading of industry's emerging role. He then went on, in the face of his more sobering report, to say, "And still. It's a good time to be in the sustainability business." He makes a good case for that, citing many examples of positive initiatives. Add to that the world business community's adoption of a set of "Sustainable Development Goals for 2030" during the September 2015 UN session.[26] Developed jointly by the World Business Council for Sustainable Development, the Global Reporting Initiative and the UN Global Compact, "the SDGs explicitly call on all businesses to apply their creativity and innovation to solving sustainable development challenges [presenting] an opportunity for business-led solutions and technologies to be developed and implemented to address the world's biggest sustainable development challenges . . . both by minimizing negative impacts and maximizing positive impacts on people and the planet."

As I was writing that last paragraph I naively assumed that no more surprises were coming, with only the Paris United Nations meeting on the horizon. I should have known better. That paragraph was written in 2015. In June there were two large breaks in the clouds. The first was the news that six large European oil and gas firms called for a coordinated price on $CO_2$ emissions. They are the BG Group, BP, ENI, Royal Dutch Shell, Statoil, and Total.[27] In effect, they are supporting William Nordhaus's long-pursued idea of a carbon tax and also adding his more recent proposal to use tariffs on imported goods from nonconforming countries. To be sure, it is a proposal only and will have to run the gauntlet of national legislatures, cool in the past toward carbon taxes. The absence of American companies from that group was conspicuous.

The second piece of good news was the June 2015 encyclical of Pope Francis on the environment, "Laudato Si: On the Care of Our Common Home." It was a powerful statement of the human responsibility for our planet and its resources, and was also a call for action on global warming. Its critique of the market and embrace of vigorous government action was important, drawing, as expected, conservative criticism. But it also surprisingly included cap and trade efforts among its criticisms of the market, not a helpful note.

Among my five horsemen, global warming stands out. Efforts to stop or slow it show great progress (renewables) but no fundamental change in its overall dangerous trajectory. The other horsemen are doing no better, and probably even worse, lacking the kind of serious agitation that climate change has inspired. Their future must be considered uncertain, but not bright. If I am correct in my diagnosis of the deepest underlying obstacle in coping with each of them—that of the embedded modern embrace of progress, still cherished—radical change will not be easy to achieve. We will have to do with something less, waiting for an era that learns how to live within limits, a new Enlightenment to supplant the Enlightenment we now live in, which goes back to the seventeenth and eighteenth centuries.

## PARIS: DECEMBER 2015

Books are meant to have a shelf life longer than a few days, and authors should keep that in mind. As the writing of this book moved along, however, I found myself disproportionately mesmerized by the global-warming debate. Unlike my other horsemen, it kept gathering momentum and, as the last few pages have laid out, pouring forth a constant stream of good and bad news. My abiding interest in the idea of progress and cultural change as part of the whole five horsemen story was almost daily whetted by that stream. All the while, I was waiting for the early December UN meeting in Paris, bringing together representatives of 195 countries who then worked to draw up agreement to develop a global-warming strategy for the future that would replace the Kyoto Protocol.

Widely hailed as a historic moment, the Paris agreement is a masterful diplomatic example of achieving consensus in the face of divergent national values and exceedingly different economic and political needs. I will not summarize the agreement here but will instead relate some of its key features to the major themes of this book. The agreement has at its core the goal of "holding the increase in the global average temperature to well below 2°C above its pre-industrial level and to pursue efforts to limit the temperature increase to 1.5°C." In line with chapter topics in part 2 of this book, here are some important parts of the agreement:

*Law and governance.* The most important point is that it is an agreement, not a binding treaty. The latter would surely have made it more difficult for some countries, notably the United States, to get it through their legislatures. Each country sets its own emission reduction goals, which are then visible to other countries. But it wisely put in place a variety of procedures to ensure that that each country will periodically reveal its success in meeting those goals and also, in concert with other countries, will be prepared to raise its goals in 2020 and 2030. The agreement recognizes that even achieving the present goals will leave the globe far from radically lower harmful emissions, much less zero emissions.

*Technology.* If mitigation of global warming is a central aim of the agreement, it lays a heavy emphasis on technological research for both mitigation and adaptation. Coordinating that effort will be a technology mechanism, coordinating public and private technology contributions and efforts. It specifically recognizes that developing countries will need funding for their technological adaptation and mitigation efforts. Technological assistance is assumed as part of the larger requirement that "developed countries should continue to take the lead in mobilizing climate financing from a wide variety of sources" (Article 9).

*Business.* I had earlier noted in this chapter some evidence of a cooling trend in the business community toward global-warming efforts, reported by Joel Makower of GreenBiz in his annual survey. Just days before the Paris meeting, some other news stories noted a decline in the stock value of most of the companies that have taken the lead in developing technology to address climate change; and for some, their debt is piling up. Yet about the same time as that news was coming out, Bill Gates announced that a number of prominent private investors and twenty countries had organized the Mission Innovation to double the investment in clean energy projects, particularly wind and solar. Gates said the coalition expected to fund 100 new companies over the next ten years. A parallel Breakthrough Energy Coalition was organized by the private investors to provide support for companies to bring innovative new ideas to the marketplace. Mark Zuckerman, Jeff Bezos, and George Soros, together with a number of multi-billionaires from other countries joined together as part of that effort.

Michael R. Bloomberg—obviously aware of the hazards for industry in efforts to ramp up the role of business in global-warming efforts—is the leader of the international Financial Stability Board's task-force on the disclosure of climate-related financial risks. As Bloomberg has emphasized, the transition to a low-carbon future will carry both business risks and gains, and not of a trivial kind. Apart from companies focused on technology research and dissemination, it is now possible to add to the list of business groups listed earlier in this chapter. The coming of the Paris agreement conference was a further stimulus to expand that group. President Obama announced that 154 American companies have joined the American Business Act on Climate Pledge to demonstrate their support on climate change and to support the Paris agreement. Taken together, these companies have operations in all fifty states and some $4.2 trillion in annual revenue. As part of a program called Divest for Paris, 500 organizations (some outside of business) agreed to purge their fossil fuel investments.

## Social Movements, Public Opinion, and GDP Growth

While it is hard to quantify the impact that social movements had in the large number of countries taking part in the Paris agreement, it is reasonable to assume they played a role. Bill McKibben's "350.org" group has helped to organize citizen groups around the world. Public protests against air pollution surely influenced important policy shifts in China and India, and a number of global public opinion surveys showed a gradual rise of concern about global warming, high enough to catch the eye of legislators. As pointed out in earlier chapters, a rising concern about climate change has not necessarily meant a willingness to accept tough laws or higher taxes to bring it about. Future public opinion, possibly stimulated by the Paris agreement, will show whether a meaningful shift can take place.

But the ideological gap between Democrats and Republicans is huge. Every one of the Republican presidential aspirants has stated opposition to climate change efforts. However the 2016 elections turn out, it will be a continuing struggle for the United States to play a global leadership role and to follow through on pledges of financial support for developing countries—much less to meet its own stated carbon reduction goals. It may well be that the success of the Paris agreement and the rising support of industry will bring about a change. Maybe.

One ideological struggle discussed earlier in this book, that of an effort to reduce GDP growth, apparently received little attention in Paris. On the contrary, it was implicitly agreed that the future welfare and prosperity of those countries depends on continued high GDP growth. Moreover, far from rejecting capitalism, the emphasis placed on technological innovation and strong industry support for global-warming reduction gives capitalism a stronger role in the future.

## Fear and Emotional Temperature

Why did increased interest in global warming and many new initiatives emerge over the past few years (give or take a few negative fluctuations)? I offer a few guesses. The steady stream of IPCC scientific reports gave greater and greater credibility to their judgment that carbon emissions are increasing, as are ocean and air temperatures, all of them surpassing earlier preindustrial levels. If not all decisively proving global warming, a number of events of record high local temperatures, emptying aquifers, excessively heavy rainfalls, persistent droughts, and less and less arctic ice, gave added credibility to the IPCC reports. People everywhere can see some evidence, however different in kind. My small town on the Hudson River suffered unprecedented harm from Hurricane Sandy. Thirty years ago our town pond had ice skating for three to four weeks every winter. Once a winter is lucky now, as uncommon as the once reliable freezing over of the river. All that is trivial compared with what is happening in Alaska, northern China, and low-lying island nations.

Almost certainly, China was persuaded by fear of global warming, with unmistakable signs of growing danger. Fear works once the hazards of inattention give way to screams for attention.

■ ■ ■

The feature that first drew me to an interest in the five horsemen was their obdurate resistance to significant change. Why were all of them steadily worsening despite efforts to bring them under control? My answer to that question is that all of them reflect the deeply rooted value of progress, which has seen its earlier benefits too often turning sour but, at the same time, resistant to complete reversal. In the intricate tapestry of modern life, how can what remains valuable be preserved while trying to eliminate what is harmful, particularly since they are tightly interwoven? Not one of the research efforts to combat the five horsemen has revealed a clear pathway for doing that. And by "clear" I mean not just plausible theoretically—not all that hard—but able to cut through the complex mix of politics, cultural variation, vested interests, and disagreements about tactics even with shared goals, which is also a mark of our modern life (even if an old human story also). If my other four horsemen can learn anything from the success of the Paris agreement for global warming, it may be that success is possible with the following approach: work diplomatically to gain the cooperationof industry, regardless of how obstructionist it has been in the past; intensify efforts to find technological pathways that work while also lowering prices; develop citizens' groups at the local level and broader grassroots social movements to break through the barrier that too often succeeds in raising interest and concern but fails togenerate action and legislative attention; if there are real dangers and hazards, do not hesitate to evoke them (but do not exaggerate); and, most of all, don't give up. The problems of the five horsemen are most likely chronic, to be lived with and combatted simultaneously. That can be done.

# NOTES

## PREFACE

1. Gregory Easterbrook, *The Progress Paradox: How Life Gets Better While People Feel Worse* (New York: Random House, 2004).
2. United Nations. Millennium Development Goals. http://www.un.org/millenniumgoals/
3. "Mapping the Progress of Millennium Development Goals—VII," Guizetti & Associates. http://www.guizzetti.org/article_view2.asp?id=132

## 1. OUR OVERHEATING, FRAYING PLANET

1. Anthony Giddens, *The Politics of Climate Change* (Cambridge: Polity), 2009.
2. John Theodore Houghton, *Global Warming: The Complete Briefing*, 4th ed. (Cambridge: Cambridge University, 2009); Spencer R. Weart, *The Discovery of Global Warming*, rev. and expanded ed. (Cambridge, Mass.: Harvard University Press, 2008); Dale Jamieson, *Reason in a Dark Time: Why the Struggle Against Climate Change Failed—and What It Means for Our Future* (New York: Oxford University Press, 2014).
3. Houghton, *Global Warming*, 17.
4. Jeremy D. Shakun et al., "Global Warming Preceded by Increasing Carbon Dioxide Concentrations During the Last Deglaciation," *Nature* 484, no. 7392 (2012): 49.
5. Tim Jackson, *Prosperity Without Growth: Economics for a Finite Planet* (London: Earthscan, 2011), 12.
6. Houghton, *Global Warming*, 20.
7. Ibid., 230; Nicholas Herbert Stern and Her Majesty's Treasury, *Stern Review: The Economics of Climate Change*, vol. 30 (London: HM Treasury, 2006).
8. Houghton, *Global Warming*, 227.
9. Mike Hulme, *Why We Disagree About Climate Change: Understanding Controversy, Inaction and Opportunity* (Cambridge: Cambridge University Press, 2009), 100–106.
10. Ibid., 104.

11. Daniel Sarewitz, "How Science Makes Environmental Controversies Worse," *Environmental Science & Policy* 7, no. 5 (2004): 396.

12. Ibid., 400.

13. Gwyn Prins et al., "The Hartwell Paper: A New Direction for Climate Policy After the Crash of 2009" (London School of Economics and Political Science, London, 2010).

14. William D. Nordhaus, "Why the Global Warming Skeptics Are Wrong," *New York Review of Books* 59, no. 5 (2012), 32–34.

15. Houghton, *Global Warming*, 316.

16. U.S. Environmental Protection Agency, "Draft Inventory of U.S. Greenhouse Gas Emissions and Sinks: 1990–2013," February 2015, www.epa.gov/climatechange /ghgemissions /usinventoryreport.html.

17. Carbon Disclosure Project, *Sector Insights: What Is Driving Climate Change Action in the World's Largest Companies? Global 500 Climate Change Report 2013;* 9/12/13. https://www .cdp.net/cdpresults/cdp-global-500-climate-change-report-2013.pdf.

18. Carbon Disclosure Project, *Sector Insights*, 8.

19. Union of Concerned Scientists, *A Climate of Corporate Control: How Corporations Have Influenced the U.S. Dialogue on Climate Science and Policy* (Cambridge, Mass., 2012).

20. Kiley Kroh, "Conservative Donors Pump $1 Billion a Year into Climate Denying Groups, Study Finds," *ThinkProgress*, December 22, 2013, http://thinkprogress.org.

21. Naomi Oreskes and Erik M. Conway, *The Collapse of Western Civilization: A View from the Future* (New York: Columbia University Press, 2014); Naomi Oreskes and Erik M. Conway, *Merchants of Doubt: How a Handful of Scientists Obscured the Truth on Issues from Tobacco Smoke to Global Warming* (New York: Bloomsbury, 2010).

22. Pew Research Center, "Modest Rise in Number Saying There Is 'Solid Evidence' of Global Warming," December 1, 2011, www.people-press.org/2011/12/01/modest-rise-in-number -saying-there-is-solid-evidence-of-global-warming.

23. Anthony Leiserowitz, "Climate Change Risk Perception and Policy Preferences: The Role of Affect, Imagery, and Values," *Climatic Change* 77, nos. 1–2 (2006): 45–72.

24. Andrew Dugan, "Americans Most Likely to Say Global Warming is Exaggerated," Gallup, March 17, 2014, http://www.gallup.com/poll/167960/americans-likely-say-global -warming-exaggerated.aspx.

25. Jan C. Semenza et al., "Public Perception of Climate Change: Voluntary Mitigation and Barriers to Behavior Change," *American Journal of Preventive Medicine* 35, no. 5 (2008): 479.

26. Pew Research Center, "Global Warming Seen as a Major Problem Around the World Less Concern in the U.S., China and Russia," December 2, 2009, www.pewglobal .org/2009/12/02/global-warming-seen-as-a-major-problem-around-the-world-less -concern-in-the-us-china-and-russia.

27. Houghton, *Global Warming*, 320.

28. Ibid., 298.

29. Ozone Secretariat, United National Environmental Programme, "Montreal Protocol on Substances That Deplete the Ozone Layer, 2010–2011," http://ozone.unep.org/new_site /en/Treaties/treaties_decisions-hb.php?sec_id=5.

30. Hulme, *Why We Disagree About Climate Change*, cited on 295–296.

31. Mike Zajko, "The Shifting Politics of Climate Science," *Society* 48, no. 6 (2011): 457.

32. Houghton, *Global Warming*, 353.

33. Roger A. Pielke Jr., *The Climate Fix: What Scientists and Politicians Won't Tell You About Global Warming* (New York: Basic, 2010), 57–58; Houghton, *Global Warming*; Jeffrey Ball, "Tough Love for Renewable Energy: Making Wind and Solar Power Affordable," *Foreign Affairs* 91 (2012): 122; Michael Specter, "The Climate Fixers—Is There a Technological Solution to Global Warming?" *New Yorker*, May 14, 2102, 96–106; Peter J. Irvine, Ryan L. Sriver, and Klaus Keller, "Tension Between Reducing Sea-Level Rise and Global Warming Through Solar-Radiation Management," *Nature Climate Change* 2, no. 2 (2012): 97–100; Ernest J. Moniz, "Stimulating Energy Technology Innovation," *Daedalus* 141, no. 2 (2012): 81–93.

34. Daniel Sarewitz and Richard Nelson, "Three Rules for Technological Fixes," *Nature* 456, no. 7224 (2008): 871–872.

35. Ibid., 872.

36. David C. Mowery, Richard R. Nelson, and Ben R. Martin, "Technology Policy and Global Warming: Why New Policy Models Are Needed (or Why Putting New Wine in Old Bottles Won't Work)," *Research Policy* 39, no. 8 (2010): 1011–1023.

37. Prins et al., "The Hartwell Paper."

38. Ibid., 12.

39. Ibid., 34.

40. Pielke, *Climate Fix*, 71.

41. Michael Schellenberger and Ted Nordhaus, "The Death of Environmentalism—Global Warming Politics in a Post-Environmental World," *Geopolitics, History, and International Relations* no. 1 (2009): 121.

42. Donella H. Meadows, Jørgen Randers, and Dennis L. Meadows, *The Limits to Growth: The 30-Year Update* (White River Junction, Vt.: Chelsea Green, 2004).

43. Bjørn Lomborg, "Environmental Alarmism, Then and Now: The Club of Rome's Problem—and Ours," *Foreign Affairs* 91 (2012): 24.

44. Ibid., 39.

45. Naomi Klein, "Capitalism vs. the Climate," *Nation*, November 28, 2011.

46. Ibid., 14.

47. Hulme, *Why We Disagree About Climate Change*, 340–358.

## 2. FEEDING A GROWING POPULATION

1. Raymond F. Hopkins, "Responding to the 2008 'Food Crisis': Lessons from the Evolution of the Food Aid Regime," in *The Global Food Crisis: Governance Challenges and Opportunities*, ed. Jennifer Clapp and Marc J. Cohen, Studies in International Governance (Waterloo, Ont.: Wilfrid Laurier University Press, 2009), 79.

2. H. C. Godfray et al., "Food Security: The Challenge of Feeding 9 Billion People." *Science* 327, no. 5967 (2010): 812.

3. Julian Cribb, *The Coming Famine: The Global Food Crisis and What We Can Do to Avoid It* (Berkeley: University of California Press, 2010), 103.

4. Per Pinstrup-Andersen and Derrill D. Watson. *Food Policy for Developing Countries: The Role of Government in Global, National, and Local Food Systems* (Ithaca, N.Y.: Cornell University Press, 2011), 90.

5. Food and Agriculture Organization of the United Nations, *The State of Food Security in the World 2014* (Rome, 2014).

6. Pinstrup-Anderson and Watson, *Food Policy for Developing Countries*, 60–61.

7. Intergovernmental Panel on Climate Change (IPCC) Working Group I and Thomas F. Stocker. *Fifth Assessment Report: Climate Change 2013: The Physical Science Basis— Summary for Policymakers* (Geneva, 2013), 485ff; Justin Gillis, "Climate Change Seen Posing Risk to Food Supplies." *New York Times,* November 2, 2013.

8. Cribb, *Coming Famine*; Food and Agriculture Organization of the United Nations, World Food Programme, and International Fund for Agricultural Development, *The State of Food Insecurity in the World 2012. Economic Growth Is Necessary but Not Sufficient to Accelerate Reduction of Hunger and Malnutrition* (Rome, 2012), 51.

9. Cristina Tirado et al. "The Impact of Climate Change on Nutrition," in Clapp and Cohen, *Global Food Crisis*, 129; Anuradha Mittal, "The Blame Game: Understanding Structural Causes of the Food Crisis," in Clapp and Cohen, *Global Food Crisis*, 13.

10. Rudy Ruitenberg, "Global Grain Stocks Seen at 15-Year High on Corn to Wheat; 6/26/14," *Bloomberg Business*, June 26, 2014, www.bloomberg.com/news/articles/2014–06–26 /global-grain-stocks-seen-at-15-year-high-on-corn-to-wheat.

11. Cribb, *Coming Famine*, 10ff.

12. Tirado et. al., "Impact of Climate Change on Nutrition," 136ff.

13. Cribb, *Coming Famine*, 190.

14. Pinstrup-Andersen and Watson, *Food Policy for Developing Countries*, 72.

15. Kimberly Ann Elliott, "U.S. Biofuels Policy and Global Food Price Crisis: A Survey of the Issues," in Clapp and Cohen, *Global Food Crisis*, 59–78.

16. Noah Zerbe, "Setting the Food Dinner Table," in Clapp and Cohen, *Global Food Crisis*, 190; Pinstrup-Andersen and Watson, *Food Policy for Developing Countries*.

17. Cribb, *Coming Famine*; Tony Weis, "Fossil Energy and the Biophysical Roots of the Food Crisis," in Clapp and Cohen, *Global Food Crisis*, 151.

18. Jennifer Clapp, *Food* (Cambridge: Polity, 2012).

19. M. Dinesh Kumar and Om Prakash Singh, "Virtual Water in Global Food and Water Policy Making: Is There a Need for Rethinking?" *Water Resources Managemewnt* 19, no. 6 (2005): 759–789.

20. Cribb, *Coming Famine*, 176–177.

21. Clapp, *Food*, 19.

22. Zerbe, "Setting the Food Dinner Table," 161.

23. Carlos A. Monteiro and Geoffrey Cannon. "The Impact of Transnational 'Big Food' Companies on the South: A View from Brazil." *PLoS Medicine* 9, no. 7 (2012): e1001252.

24. Zerbe, "Setting the Food Dinner Table," 165.

25. Cribb, *Coming Famine*, 103.

26. Pinstrup-Andersen and Watson, *Food Policy for Developing Countries*, 293.

27. Robert L. Paarlberg, *Food Politics: What Everyone Needs to Know* (New York: Oxford University Press, 2010), 182.

28. "The Politics of Food: Hungry for Votes," *Economist*, January 27, 2011.

29. Pinstrup-Andersen and Watson, *Food Policy for Developing Countries*, 32–33.

30. Ibid., 42–43.

31. Raymond F. Hopkins, "Responding to the 2008 'Food Crisis,' " 92.

32. Anthony Giddens, *The Politics of Climate Change* (Cambridge: Polity, 2009), 4.

33. Alex McCalla, "The Governance Challenge of Improving Global Food Security," in Clapp and Cohen, *Global Food Crisis*, 237–249.

34. Ibid, 249.

35. Clapp, *Food*, chap. 6; Clapp and Cohen, *Global Food Crisis*.

36. Clapp, *Food*, 158.

37. Ibid., 160–163.

38. L'Aquila Food Security Initiative (AFSI), "L'Aquila Joint Statement on Global Food Security," July 10, 2009. http://www.g8italia2009.it/static/G8_Allegato/LAquila_Joint _Statement_on_Global_Food_Security%5B1%5D,0.pdf.

39. Clapp, *Food.*, 302.

40. Cribb, *Coming Famine*, 9.

41. Nikos Alexandratos and Jelle Bruinsma, "World Agriculture Towards 2030/2050: The 2012 Revision," ESA Working Paper No. 12–03 (Rome: Food and Agriculture Organization of the United Nations, June 2012); Food and Agriculture Organization of the United Nations, International Fund for Agricultural Development, and World Food Programme, *The State of Food Insecurity in the World 2015. Meeting the 2015 International Hunger Targets: Taking Stock of Uneven Progress* (Rome, 2015).

42. Daniel Callahan and Angela A. Wasunna, *Medicine and the Market: Equity v. Choice* (Baltimore: Johns Hopkins University Press, 2006), 117–162.

43. Frances Moore Lappé, "The Food Movement: Its Power and Possibilities," *Nation*, October 3, 2011, 11–19.

44. Lappé, "Food Movement," 11.

45. Ibid, 14.

46. Ibid, 15.

47. Raj Patel, "Why Hunger Is Still with Us," *Nation*, October 3, 2011, 17; See also Raj Patel, *Stuffed and Starved: The Hidden Battle for the World Food System*, 2d ed. (Brooklyn, N.Y.: Melville House, 2012), which develops that argument in greater detail.

48. Michael Pollan, "How Change Is Going to Come in the Food System," *Nation*, October 3, 2011, 18.

49. Eric Schlosser, "It's Not Just About Food," *Nation,* October 3, 2011, 14.

50. Pollan, "How Change Is Going to Come in the Food System19.

51.  Paarlberg, *Food Politics*, 6.
52.  Ibid., 149.
53.  Patel, *Stuffed and Starved*, 309ff.

# 3. WATER

1.  Robert J. Wyman, "The Effects of Population on the Depletion of Fresh Water," *Population and Development Review* 39, no. 4 (2013): 687–704.
2.  Justin Sheffield and Eric F. Wood, *Drought: Past Problems and Future Scenarios* (London: Earthscan, 2011), 4–5.
3.  Arjen Y. Hoekstra and Ashok K. Chapagain, "Water Footprints of Nations: Water Use by People as a Function of Their Consumption Pattern," *Water Resources Management* 21, no. 1 (2007): 35–48.
4.  Tony Allan, *Virtual Water: Tackling the Threat to Our Planet's Most Precious Resource* (London: Tauris, 2011), 3.
5.  Sheffield and Wood, *Drought*, 3. World Health Organization (WHO), *UN-Water Global Analysis and Assessment of Sanitation and Drinking-Water the Challenge of Extending and Sustaining Services. GLASS 2012 Report* (Geneva, 2012).
6.  Fred Pearce, *When the Rivers Run Dry: Water, the Defining Crisis of the Twenty-First Century* (Boston: Beacon, 2006), x.
7.  Ibid., 109–110.
8.  Steven Solomon, *Water: The Epic Struggle for Wealth, Power, and Civilization* (New York: Harper, 2010), 421–425.
9.  Ibid., 423.
10.  Global Water Partnership, "Groundwater Resources and Irrigated Agriculture" (Perspectives Paper, Stockholm, 2012).
11.  Charles Fishman, *The Big Thirst: The Secret Life and Turbulent Future of Water* (New York: Free Press, 2011), chap. 8.
12.  Krishna M. Singh, et al. "A Review of Indian Water Policy," ICAR-RCER, Patha, Feb. 14, 2013, https://mpra.ub.uni-muenchen.de/45230/1/MPRA_paper_45230.pdf.
13.  Ibid, 9–11.
14.  Ibid., 14.
15.  Hemant K. Pullabhotla, Chandan Kumar, and Shilp Verma. "Micro-Irrigation Subsidies in Gujarat and Andhra Pradesh Implications for Market Dynamics and Growth" (Water Policy Research Highlight 43, Gujarat, India: International Water Management Institute–Tata, 2012), 43.
16.  Ram Mashru, "India's Worsening Water Crisis," *Diplomat*, April 19, 2014, http://thediplomat.com /2014 04/indias-worsening-water-crisis.
17.  Solomon, *Water*, 323.
18.  David Molden, ed., *Water for Food, Water for Life: A Comprehensive Assessment of Water Management in Agriculture* (London: Earthscan, 2007).

19. Julian Cribb, *The Coming Famine: The Global Food Crisis and What We Can Do to Avoid It* (Berkeley: University of California Press, 2010), 35ff.

20. World Health Organization (WHO), "Facts and Figures on Water Quality and Health, 2014," http://www.who.int/water_sanitation_health/facts_figures/en.

21. Maggie Black, Jannet King, and Robin Clarke, *The Atlas of Water: Mapping the World's Most Critical Resource* (London: Earthscan, 2009), 76.

22. Ibid., 80.

23. Philip Z. Kirpich, "Water Management: The Key Role of the International Agencies," *Water International* 29 (2009): 242–247.

24. Lindsay C. Stringer et al., "Adaptations to Climate Change, Drought and Desertification: Local Insights to Enhance Policy in Southern Africa," *Environmental Science & Policy* 12, no. 7 (2009): 748–765.

25. Peter Rogers and Susan Leal, *Running Out of Water: The Looming Crisis and Solutions to Conserve Our Most Precious Resource* (New York: Palgrave Macmillan, 2010), 186.

26. Solomon, *Water*, 410.

27. Adam James, "The U.S. Wastes 7 Billion Gallons of Drinking Water a Day: Can Innovation Help Solve the Problem?; 11/3/11," *ThinkProgress*. November 3, 2011, http://thinkprogress.org /climate/2011/11/03 /360437/the-us-wastes-7-billion-gallons-drinking-water-a-day-innovation.

28. Burkhard Bilger, "The Great Oasis—Can a Wall of Trees Stop the Sahara from Spreading?" *New Yorker*, December 19, 2011.

29. Cheryl Katz, "New Desalination Technologies Spur Growth in Recycling Water," *Environment 360*, June 3, 2014, http://e360.yale.edu/feature/new_desalination _technologies_spur_growth_in_ recycling_water/2770.

30. Solomon, *Water*, 480.

31. Garrett Hardin, "The Tragedy of the Commons," *Science* 162, no. 3859 (1968): 1243.

32. Yoshihide Wada, "Non-sustainable Groundwater Sustaining Irrigation," Global Water Forum, February 13, 2012, http://www.globalwaterforum.org /2012/02/13/non -sustainable-groundwater-sustaining-irrigation.

33. Pearce, *When the Rivers Run Dry*, 24.

34. S. Joyce, "Is It Worth a Dam?" *Environmental Health Perspectives* 105, no. 10 (1997): 1050–1055.

35. Jacey Fortin, "Dam Rising in Ethiopia Stirs Hope and Tension," *New York Times*, October 12, 2014.

36. Pearce, *When the Rivers Run Dry*, 131.

37. R. E. Grumbine and M. K. Pandit, "Ecology. Threats From India's Himalaya Dams," *Science* 339, no. 6115 (2013): 36–37.

38. Benjamin K. Sovacool and Kelly E. Sovacool. "Identifying Future Electricity–Water Tradeoffs in the United States," *Energy Policy* 37, no. 7 (2009): 2763–2773.

39. Alex Prud'homme, *The Ripple Effect: The Fate of Freshwater in the Twenty-First Century* (New York: Scribner, 2011), 294.

40. Fishman, *Big Thirst*, 135.

41. Prud'homme, *Ripple Effect*, 292.

42. Fishman, *Big Thirst*, 135.

43. Ibid., 293.

44. Prud'homme, *Ripple Effect*, 266–280; Solomon, *Water*, 380.

45. Gregory Pierce, "The Cautious Expansion of Water Privatization in Low and Middle Income Countries," Global Water Forum, January 23, 2013, www.globalwaterforum .org/2013/01/23/the-cautious-expansion-of-water-privatization-in-low-and-middle -income-countries.

46. Alternative World Water Forum, March 2012, www.fame2012.org/en/about/mission/.

47. Solomon, *Water*, 255–257.

48. Gus Lubin, "Citi's Top Economist Says the Water Market Will Soon Eclipse Oil," *Business Insider*, July 21, 2011, www.businessinsider.com/willem-buiter-water-2011-7.

49. Willem Buiter, "Essay: Water as Seen by an Economist," in "Global Themes Strategy: Thirsty Cities—Urbanization to Drive Water Demand," *Global Thematic Investing*, July 20, 2011, 22; http://www.capitalsynthesis.com/wp-content/uploads/2011/08/Water -Thirsty-Cities.pdf.

50. Wenonah Hauter, "Why We Must Fight the Vision of a Global Water Market," December 14, 2012, Alternative World Water Forum, http://www.fame2012.org/en/2012/12/14 /global-water-market/.

51. Solomon, *Water*, 485–486.

52. Prud'homme, *Ripple Effect*, 348.

53. Prud'homme, *Ripple Effect*, 352–354.

54. Ibid., 480.

55. Holger Hoff, "Global Water Resources and Their Management," *Current Opinion in Environmental Sustainability* 1, no. 2 (2009): 141–147.

56. Victor Corral-Verdugo, Blanca Fraijo-Sing, and José Q. Pinheiro, "Sustainable Behavior and Time Perspective: Present, Past, and Future Orientations and Their Relationship with Water Conservation Behavior," *Interamerican Journal of Psychology* 40, no. 2 (2006): 139–147.

57. Rogers and Leal, *Running Out of Water*, 87ff.

58. Cribb, *Coming Famine*, 176–178.

59. UN Educational, Scientific, and Cultural Organization, *World Water Development Report 2012 (WWDR4): Managing Water Under Uncertainty and Rick* (Paris, 2012), http://www .unesco.org/new/en/natural-sciences/environment/water/wwap/wwdr/wwdr4-2012.

## 4. CHRONIC ILLNESS

1. United Nations, "2011 High Level Meeting on Prevention and Control of Non-Communicable Diseases," New York 2011, http://www.un.org/en/ga/ncdmeeting2011.

2. George Weisz, *Chronic Disease in the 20th Century: A History* (Baltimore: Johns Hopkins University Press, 2014), 3.

3. Ibid.

4.  Ibid., see pages 105 ff.

5.  David Stuckler and Karen Siegel, eds., *Sick Societies: Responding to the Global Challenge of Chronic Disease* (Oxford: Oxford University Press, 2011).

6.  Bjørn Lomborg, *Global Crises, Global Solutions*, 2d ed. (Cambridge: Cambridge University Press, 2009).

7.  World Health Organization. Noncommunicable Diseases—Fact Sheet. Updated January 2015. http://www.who.int/mediacentre/factsheets/fs355/en/.

8.  UN Fund for Population Activities, *State of World Population 2014—The Power of 1.8 Billion* (New York, 2014).

9.  U.S. Centers for Disease Control and Prevention, National Center for Health Statistics, accessed March 5, 2015, www.cdc.gov/nchs.

10. Daniel Callahan, "Medical Progress and Global Chronic Disease: The Need for a New Model," *Brown Journal of World Affairs* 20 (2013): 35.

11. Thomas J. Bollyky, "Developing Symptoms: Noncommunicable Diseases Go Global," *Foreign Affairs* 91 (2012): 136.

12. Stuckler and Siegel, *Sick Societies*, 12.

13. Ibid., 111, citing WHO statistics.

14. Ibid., 15.

15. Mohammed K. Ali, et al., "Noncommunicable Diseases: Three Decades of Global Data Show A Mixture of Increases and Decreases in Mortality Rates," *Health Affairs* 34, no. 9 (2015): 1449–1455.

16. Rijo John and Hana Ross, *The Global Economic Cost of Cancer* (Atlanta: American Cancer Society, 2011).

17. George Johnson, *The Cancer Chronicle* (New York: Knopf, 2013).

18. Dana P. Goldman et al., "Substantial Health and Economic Returns from Delayed Aging May Warrant a New Focus for Medical Research," *Health Affairs* 32, no. 10 (2013): 1698–1705.

19. Richard Sullivan et al, "Delivering Affordable Cancer Care in High-Income Countries," *Lancet Oncology* 12, no. 10 (2011): 933.

20. Daniel Callahan, "The Ethics of Rationing: Necessity, Politics and Fairness," in *The Routledge Companion to Bioethics*, ed. John D. Arras, Elizabeth Fenton, and Rebecca Kukla (New York: Oxford University Press, 2014): 33–43.

21. Bollyky, "Developing Symptoms," 135.

22. M. Prince et al,. "The Global Prevalence of Dementia: A Systematic Review and Metaanalysis," *Alzheimer's & Dementia* 9, no. 1 (2013): 63–75.e2.

23. Cleusa P. Ferri et al., "Global Prevalence of Dementia: A Delphi Consensus Study," *Lancet* 366, no. 9503 (2005): 2112–2117.

24. Alzheimer's Association, "2015 Alzheimer's Disease Facts and Figures: Quick Facts Charts," www.alz.org/alzheimers_disease_facts_and_figures.asp.

25. UN Development Programme, *Human Development Report 2011—Sustainability and Equity: A Better Future for All* (New York, 2011), 2.

26. Ibid., 5.

27. Bollyky, "Developing Symptoms," 137.

28. David Stuckler et al., "Politics of Chronic Disease," in Stuckler and Siegel, *Sick Societies*: 135–185.

29. Ibid., 141.

30. Ibid., 157.

31. U.S. Centers for Disease Control and Prevention, National Center for Health Statistics, accessed March 5, 2015, www.cdc.gov/nchs; U.S. Dept of Health and Human Services, *Multiple Chronic Conditions: A Strategic Framework—Optimum Health and Quality of Life for Individuals with Multiple Chronic Conditions, December 2010* (Washington, D.C., 2010); Institute of Medicine, *Living Well with ChronicIllness: A Call for Public Health Action* (Washington, D.C.: National Academies Press, 2012).

32. George Weisz and Etienne Vignola-Gagne, " The World Health Organization and the Globalization of Chronic Noncommunicable Disease," *Population and Development Review* 41, no. 3 (2015): 507–532.

33. Daniel Callahan, *Taming the Beloved Beast: How Medical Technology Costs Are Destroying Our Health Care System* (Princeton: Princeton University Press, 2009).

34. Bollyky, "Developing Symptoms," 139.

35. Tara Acharya et al., "The Current and Future Role of the Food Industry in the Prevention and Control of Chronic Diseases: The Case of Pepsico," in Stuckler and Siegel, *Sick Societies*, 191.

36. William H. Wiist, "The Corporate Play Book, Health and Democracy: The Snack Food and Beverage Industry's Tactics in Context," in Stuckler and Siegel, *Sick Societies*, 204–216.

37. E. M. Crimmins and H. Beltran-Sanchez, "Mortality and Morbidity Trends: Is There Compression of Morbidity?" *Journals of Gerontology: Series B, Psychological Sciences and Social Sciences* 66, no. 1 (2011): 75–86.

38. S. Jay Olshansky et al., "A Potential Decline in Life Expectancy in the United States in the 21st Century," *New England Journal of Medicine* 352, no. 11 (2005): 1138–1145.

39. S. J. Olshansky, B. A. Carnes, and C. Cassel, "In Search of Methuselah: Estimating the Upper Limits to Human Longevity," *Science* 250, no. 4981 (1990): 634.

40. National Institute for Health Care Management, "The Concentration of Health Care Spending" (NIHCM Foundation Data Brief, Washington, D.C., July 2012), www.nihcm.org/pdf/DataBrief3%20Final.pdf.

41. U.S. Department of Health and Human Services, Administration on Aging, "Aging into the 21st Century: Statistics," May 31, 1996, www.aoa.acl.gov/Aging_Statistics/future_growth/aging21/aging_21.aspx.

42. U.S. Centers for Disease Control and Prevention, National Center for Health Statistics, accessed March 5, 2015, www.cdc.gov/nchs.

43. World Health Organization, *Global Action Plan for the Prevention and Control of Noncommunicable Diseases: 2013–2020* (Geneva, 2013).

44. Ibid., 4.

45. World Health Organization. *Assessing National Capacity for the Prevention and Control of Noncommunicable Diseases: Report of the 2010 Global Survey*. Geneva, 2012.

46. Johan Hansen, et al., "Living In A Country With A Strong Primary Care System Is Beneficial To People With Chronic Conditions," *Health Affairs*,34, no. 9 (2015): 1531–1537.

47. Thomas Gaziano, et al., "Cardiovascular Disease Screening By Community Health Workers Can Be Cost-Effective In Low Resource Countries," *Health Affairs*, 34, no. 9 (2015): 1538–1545.

48. Simon Szreter, "The Importance of Social Intervention in Britain's Mortality Decline c. 1850–1914: A Re-interpretation of the Role of Public Health," *Social History of Medicine* 1, no. 1 (1988): 1.

## 5. OBESITY

1. New Zealand Ministry of Health, *Annual Update of Key Results 2013/14: New Zealand Health Survey* (Wellington, N.Z., 2014).

2. U.S. Centers for National Disease Control and Prevention, National Center for Health Statistics, www.cdc.gov/nchs; World Health Organization, "Obesity and Overweight Fact Sheet #311," World Health Organization, Media Center, January 2015, www.who.int /mediacentre/factsheets/fs311/en; Solveig A. Cunningham, Michael R. Kramer, and K. M. Venkat Narayan, "Incidence of Childhood Obesity in the United States," *New England Journal of Medicine* 370, no. 5 (2014): 403–411.

3. Franco Sassi, *Obesity and the Economics of Prevention: Fit Not Fat* (Paris: Organisation for Economic Co-operation and Development, 2010).

4. Tanika Kelly et al., "Global Burden of Obesity in 2005 and Projections to 2030," *International Journal of Obesity* 32, no. 9 (2008): 1431.

5. Steven B. Heymsfield and William T. Cefalu, "Does Body Mass Index Adequately Convey a Patient's Mortality Risk?" *Journal of the American Medical Association* 309, no. 1 (2013): 87–88.

6. World Health Organization, *Obesity: Preventing and Managing the Global Epidemic* (Geneva, 2000); Sharon R. Akabas, Sally A. Lederman, and Barbara J. Moore, eds., *Textbook of Obesity: Biological, Psychological, and Cultural Influences* (Chichester, U.K.: Wiley-Blackwell, 2012), 6.

7. Katherine M. Flegal et al., "Prevalence of Obesity and Trends in the Distribution of Body Mass Index among US Adults, 1999–2010," *JAMA* 307, no. 5 (2012): 491.

8. J. B. Albu et al., "Impact of Obesity During Adulthood on Chronic Disease: Diabetes, Hypertension, Metabolic Syndrome, Cardiovascular Disease, and Cancer," in Akabas et al., *Textbook of Obesity*, 209–220.

9. Barbara J. Moore, I. J. Frame, and Ninia Baehr, "Preventing Childhood Obesity: It Takes a Nation," in Akabas et al., *Textbook of Obesity*, 425.

10. David S. Ludwig and Harold A. Pollack, "Obesity and the Economy: From Crisis to Opportunity," *Journal of the American Medical Association* 301, no. 5 (2009): 533.

11. B. M. Popkin, "The Nutrition Transition and Obesity in the Developing World," *The Journal of Nutrition* 131, no. 3 (2001): 871S.

12. G. A. Bray, S. J. Nielsen, and B. M. Popkin, "Consumption of High-Fructose Corn Syrup in Beverages May Play a Role in the Epidemic of Obesity," *American Journal of Clinical Nutrition* 79, no. 4 (2004): 537–543.

13. Chin Jou, "The Biology and Genetics of Obesity—A Century of Inquiries," *New England Journal of Medicine* 370, no. 20 (2014): 1874–1877.

14. Akabas et al., *Textbook of Obesity*, 99–100.

15. M. D. Klok, S. Jakobsdottir, and M. L. Drent. "The Role of Leptin and Ghrelin in the Regulation of Food Intake and Body Weight in Humans: A Review," *Obesity Reviews* 8, no. 1 (2007): 21–34.

16. Helene Choquet and David Meyre, "Genetics of Obesity: What Have We Learned?" *Current Genomics* 12, no. 3 (2011): 169.

17. Tara Parker-Pope, "The Fat Trap," *New York Times*, December 28, 2011.

18. Albert J. Stunkard, Terryl T. Foch, and Zdenek Hrubec, "A Twin Study of Human Obesity," *Journal of the American Medical Association* 256, no. 1 (1986):51–54.

19. Carl Zimmer, "Gene Linked to Obesity Hasn't Always Been a Problem, Study Finds ," *New York Times*, December 21, 2014.

20. Blanca M. Herrera, Sarah Keildson, and Cecilia M. Lindgren. "Genetics and Epigenetics of Obesity," *Maturitas* 69, no. 1 (2011): 47.

21. Michael L. Power and Jay Schulkin, *The Evolution of Obesity* (Baltimore: Johns Hopkins University Press, 2009).

22. Mai A. Elobeid and David B. Allison, "Putative Environmental-Endocrine Disruptors and Obesity: A Review," *Current Opinion in Endocrinology, Diabetes, and Obesity* 15, no. 5 (2008): 403.

23. Gina Kolata, "Obesity Spreads to Friends, Study Concludes," *New York Times*, July 25, 2007.

24. Kelly et al., "Global Burden of Obesity in 2005 and Projections to 2030."

25. Ibid., 1437.

26. Serena Low, Mien Chew Chin, and Mabel Deurenberg-Yap, "Review on Epidemic of Obesity," *Annals Academy of Medicine Singapore* 38, no. 1 (2009).

27. A. Berghofer et al., "Obesity Prevalence from a European Perspective: A Systematic Review," *BMC Public Health* 8 (2008): 200.

28. Jenny Bua, Lina W. Olsen, and Thorkild I. A. Sørensen, "Secular Trends in Childhood Obesity in Denmark During 50 Years in Relation to Economic Growth," *Obesity* 15, no. 4 (2007): 977.

29. James Hill and Rena Wing, "The National Weight Control Registry," *Permanente Journal* 7, no. 3 (2003): 34–37.

30. Boyd A. Swinburn et al., "The Global Obesity Pandemic: Shaped by Global Drivers and Local Environments," *Lancet* 378, no. 9793 (2011): 806.

31. Robert Wood Johnson Foundation, *F as in Fat: How Obesity Threatens America's Future* (Washington, D.C.: Trust for America's Health, 2010), 55.

32. Allison Aubrey, "Turning to Big Business to Solve the Obesity Epidemic," NPR, December 2, 2011, www.npr.org/blogs/thesalt/2011/12/02/143052025/turning-to-big-business-to -solve-the-obesity-epidemic.

33. J. R. Horwitz, B. D. Kelly and J. E. DiNardo, "Wellness Incentives in the Workplace: Cost Savings through Cost Shifting to Unhealthy Workers," *Health Affairs (Project Hope)* 32, no. 3 (2013): 468–476.

34. James A. Colbert, and Jonathan N. Adler, "Sugar-sweetened Beverages—Polling Results," *New England Journal of Medicine* 368, no. 3 (2013): 1464–1466.

35. L. I. Gonzalez-Zapata et al., "The Potential Role of Taxes and Subsidies on Food in the Prevention of Obesity in Europe," *Journal of Epidemiology and Community Health* 64, no. 8 (2010): 696.

36. Tomas Philipson and Richard Posner, "Is the Obesity Epidemic a Public Health Problem? A Decade of Research on the Economics of Obesity" (NBER Working Paper No. 14010, Cambridge, Mass.: National Bureau of Economic Research, May 2008).

37. Noah Smith, "Big Government, Small Bellies: What Japan Can Teach Us About Fighting Fat," *Atlantic*, September 6, 2012.

38. Catherine Cheney, "Battling the Couch Potatoes: Hungary Introduces 'Fat Tax,' " *Spiegel Online* 6 (2011): 2012.

39. Kelly D. Brownell, "Food Industry Pursues the Strategy of Big Tobacco," *Environment 360*, April 8, 2009, http://e360.yale.edu/feature/food_industry_pursues_the_strategy _of_big_tobacco/2136.

40. Michael Moss, *Salt Sugar Fat* (New York: Random House, 2013), xiiiff.

41. Abigail Cope Saguy, *What's Wrong with Fat?* (New York: Oxford University Press, 2013), 24.

42. Ibid., 163.

43. Ibid., 165.

44. Marion Nestle, *Food Politics: How the Food Industry Influences Nutrition and Health*, 3d ed. (Berkeley: University of California Press, 2013), 120.

45. Margot Sanger-Katz, "Hard Times for Soft Drinks," *New York Times*, October 4, 2015.

46. Bridget Huber, "Michelle's Moves: Has the First Lady's Anti-obesity Campaign Been Too Accommodating Toward the Food Industry?" *The Nation*, October 29, 2012, 11.

47. Ibid., 12.

48. Nestle, *Food Politics*, 383ff.

49. M. Bessler et al., "Surgical Treatment of Severe Obesity: Patient Selection and Screening, Surgical Options, and Nutritional Management," in Akabas et al., *Textbook of Obesity*.

50. Joanna Picot et al., "The Clinical Effectiveness and Cost-Effectiveness of Bariatric (Weight Loss) Surgery for Obesity: A Systematic Review and Economic Evaluation," *Health Technology Assessment* 13, no. 41 (2009): 1–190, 215–357, iii–iv.

51. Lawrence J. Appel et al., "Comparative Effectiveness of Weight-Loss Interventions in Clinical Practice," *New England Journal of Medicine* 365, no. 21 (2011): 1959.

52. Thomas A. Wadden et al., "A Two-Year Randomized Trial of Obesity Treatment in Primary Care Practice," *New England Journal of Medicine* 365, no. 21 (2011): 1969–1979.

53. Rodney Lyn, Barbara J. Moore, and Michael Eriksen, "The Application of Public Health Lessons to Stemming the Obesity Epidemic," in Akabas et al., *Textbook of Obesity*, 58ff.

54. Steven L. Gortmaker et al., "Changing the Future of Obesity: Science, Policy, and Action," *Lancet* 378, no. 9793 (2011): 838–847.

55. Ibid., 844.

56. Colleen L. Barry et al., "Obesity Metaphors: How Beliefs About the Causes of Obesity Affect Support for Public Policy," *Milbank Quarterly* 87, no. 1 (2009): 7–47.

57. Ibid., 9.

58. Stephanie A. Chambers and W. Bruce Traill, "What the UK Public Believes Causes Obesity, and What They Want to Do About It: A Cross-sectional Study," *Journal of Public Health Policy* 32, no. 4 (2011): 430–444.

59. A. Hilbert, W. Rief, and E. Braehler, "What Determines Public Support of Obesity Prevention?" *Journal of Epidemiology and Community Health* 61, no. 7 (2007): 585–590.

60. Chambers and Traill, "What the UK Public Believes Causes Obesity," 441

61. Ibid., 443.

## 6. ALWAYS MORE PEOPLE AND EVER MORE ELDERLY

1. Worldometers, www.worldometers.info (accessed Sept. 2014).

2. Jeffrey Sachs, *Common Wealth: Economics for a Crowded Planet* (New York: Penguin, 2008).

3. John C. Caldwell, "Mass Education as a Determinant of the Timing of Fertility Decline," *Population and Development Review* 6, no. 2 (2008): 225–255.

4. John C. Caldwell, "Toward a Restatement of Demographic Transition Theory," *Population and Development Review* 2, nos. 3/4 (1976): 321.

5. Ibid.

6. Ibid.

7. Michael S. Teitelbaum and Jay M. Winter, *The Fear of Population Decline* (Orlando, Fla.: Academic, 1985), 18.

8. Ibid., passim.

9. Tony Judt, *When the Facts Change: Essays, 1995–2010* (New York: Penguin, 2015).

10. Ibid., 332.

11. Teitelbaum and Winter, *Fear of Population Decline*, 20–21.

12. Gary S. Becker, "An Economic Analysis of Fertility," In *Demographic and Economic Change in Developed Countries* (New York: Columbia University Press, 1960), 209.

13. Caldwell, "Toward a Restatement of Demographic Transition Theory."

14. Daniel Callahan, *The American Population Debate* (New York: Doubleday, 1971). Paul Ehrlich, *The Population Bomb* (New York: Ballantine, 1968).

15. Anke C. Zimmermann and Richard A. Easterlin, "Happily Ever After? Cohabitation, Marriage, Divorce, and Happiness in Germany," *Population and Development Review* 32, no. 3 (2006): 511–528.

16. Paul Demeny and Geoffrey McNicoll, "World Population 1950–2000: Perception and Response," *Population and Development Review* 32, no. S1 (2006): 34–35.

17. Demeny and McNicoll, "World Population 1950–2000," 12.

18. Jonathan Grant et al., "What the Literature Shows—Relationships in the Conceptual Model and Effects of Policy," in *Low Fertility and Population Ageing: Causes, Consequences, and Policy Options* (Santa Monica, Calif.: Rand, 2004), 66.

19. John Bongaarts and Tomáš Sobotka, "A Demographic Explanation for the Recent Rise in European Fertility," *Population and Development Review* 38, no. 1 (2012): 83.

20. United Nations, "The United Nations 2012 Population Projections," *Population and Development Review* 39, no. 3 (2013): 551–555.

21. Paul Sabin, *The Bet: Paul Ehrlich, Julian Simon, and Our Gamble Over Earth's Future* (New Haven, Conn.: Yale University Press, 2013), 226.

22. Quamrul H. Ashraf, David N. Weil, and Joshua Wilde, "The Effect of Fertility Reduction on Economic Growth," *Population and Development Review* 39, no. 1 (2013): 124.

23. Bongaarts and Sobotka, "A Demographic Explanation for the Recent Rise in European Fertility," 83–120.

24. Ronald Lee et al., "Is Low Fertility Really a Problem? Population Aging, Dependency, and Consumption," *Science* 346, no. 6206 (2014): 229–234.

25. Timothy M. Smeeding, "Adjusting to the Fertility Bust," *Science* 346, no. 6206 (2014): 163–164.

26. Wolfgang Lutz, Rudolf Richter, and Chris Wilson, *The New Generations of Europeans: Demography and Families in the Enlarged European Union* (New York: Routledge, 2010).

27. Paul R. Ehrlich and John P. Holdren, "Impact of Population Growth," *Science* 171 (1971): 1212–1217.

28. James Gustave Speth, *The Bridge at the Edge of the World: Capitalism, the Environment, and Crossing from Crisis to Sustainability* (New Haven, Conn.: Yale University Press, 2008), 77–78.

29. Bill McKibben, *Maybe One: A Case for Smaller Families* (New York: Simon & Schuster, 1998).

30. Sachs, *Common Wealth*.

31. William N. Ryerson, "Population: The Multiplier of Everything Else," In *The Post Carbon Reader: Managing the 21st Century's Sustainability Crises*, ed. Richard Heinberg and Daniel Lerch (Santa Rosa, Calif.: Post Carbon Institute, 2010), 153 ff.

32. Paula J. Dobriansky et al., *Why Population Aging Matters: A Global Perspective* (Bethesda, Md.: National Institute on Aging; National Institutes of Health, 2007).

33. Jean-Marc Burniaux, Romaine Duval, and Florence Jaumotte, "Coping with Ageing: A Dynamic Approach to Quantify the Impact of Alternative Policy Options on Future Labor Supply in OECD Countries" (Economics Department Working Papers, No. 371, Paris: Organisation for Economic Co-operation and Development, June 21, 2004), 8.

34. Gary Burtless, "Social Security, Unanticipated Benefit Increases, and the Timing of Retirement," *Review of Economic Studies* 53, no. 5 (1986): 781.

35. John Bongaarts, "How Long Will We Live?" *Population and Development Review* 32, no. 4 (2006): 605–628; Dobriansky et al., *Why Population Aging Matters*.

36. Richard Jackson, "Balancing Adequacy and Sustainability: Lessons from the Global Aging Preparedness Index," *Over 65* (blog), May 3, 2013, www.over65.thehastingscenter .org/balancing-adequacy-and-sustainability-lessons-from-the-global-aging -preparedness-index.

37. Richard Jackson et al., *China's Long March to Retirement Reform* (Washington, D.C.: Center for Strategic and International Studies, 2009).

38. Ibid., 9.

39. Ibid., 2.

40. Dobriansky et al., *Why Population Aging Matters*.

41. Daniel Callahan, "Medical Progress and Global Chronic Disease: The Need for a New Model," *Brown Journal of World Affairs* 20 (2013): 35.

42. James W. Vaupel, "Demographic Analysis of Aging and Longevity," *American Economic Review* (1998): 242–247.

43. Bongaarts, "How Long Will We Live?"

44. Neil K. Mehta and Virginia W. Chang, "Mortality Attributable to Obesity Among Middle-Aged Adults in the United States," *Demography* 46, no. 4 (2009): 851.

45. Paul Demeny, "Geopolitical Aspects of Population in the Twenty-First Century," *Population and Development Review* 38, no. 4 (2012): 693.

46. Donella H Meadows, Jørgen Randers, and Dennis L. Meadows, *The Limits to Growth: The 30-Year Update* (White River Junction, Vt.: Chelsea Green, 2004).

47. Andrew Clark and Claudia Senik, "Will GDP Growth Increase Happiness in Developing Countries?" IZA Discussion Paper No. 5595, March 2011, Bonn, Germany. http://ftp.iza .org/dp5595.pdf.

48. Ibid., cited in Clark and Senik, "Will GDP Growth Increase Happiness?"

49. Jagdish Bhagwati and Arvind Panagariya, *Why Growth Matters: How Economic Growth in India Reduced Poverty and the Lessons for Other Developing Countries* (New York: Public Affairs, 2013); Jean Drèze and Amartya Sen, *An Uncertain Glory: India and Its Contradictions* (Princeton: Princeton University Press, 2013).

50. Richard Heinberg, *The End of Growth: Adapting to Our New Economic Reality* (Gabriola, B.C.: New Society, 2011).

51. Ibid. (italics in original).

52. Ibid., 261.

53. Ibid., 6.

54. Ibid., 261 (italics in original).

55. Tim Jackson, *Prosperity Without Growth: Economics for a Finite Planet* (London: Earthscan, 2011), 235.

56. Ibid., 125–173.

57. Ibid., 200.

58. Robert J. Gordon, "Is US Economic Growth Over? Faltering Innovation Confronts the Six Headwinds" (NBER Working Paper No. 18315, Cambridge, Mass.: National Bureau of Economic Research, 2012).

59. Aron Gampel, "Global Growth—Better Luck Next Year," Scotiabank, August 2, 2013, www.gbm.scotiabank.com/English/bns_econ/globalviews130802.pdf.

60. Drèze and Sen, *An Uncertain Glory*, 20.

## 7. THE TECHNOLOGY FIX

1. Food and Agriculture Organization of the United Nations, *The State of Food Security in the World 2014* (Rome, 2014), 42–43.

2. Marilyn A. Brown and Benjamin K. Sovacool, *Climate Change and Global Energy Security: Technology and Policy Options* (Cambridge, Mass.: MIT Press, 2011), 36–37.

3. James Gustave Speth, *The Bridge at the Edge of the World: Capitalism, the Environment, and Crossing from Crisis to Sustainability* (New Haven, Conn.: Yale University Press, 2008), 34.

4. Eva Wollenberg et al., "Climate Change Mitigation and Agriculture: Designing Projects and Policies for Smalholder Farmers," in *Climate Change Mitigation and Agriculture*, ed. Eva Wollenberg, Marja-Liisa Tapio-Bistrom, Maryanne Grieg-Gran, and Alison Nihart (New York: Routledge, 2012), 3.

5. Ibid., 9 (italics in original).

6. Peter Rogers and Susan Leal, *Running Out of Water: The Looming Crisis and Solutions to Conserve Our Most Precious Resource* (New York: Palgrave Macmillan, 2010), 50–51.

7. Julia Pongratz et al., "Crop Yields in a Geoengineered Climate," *Nature Climate Change* 2, no. 2 (2012): 101–105.

8. Food and Agriculture Organization of the United Nations, *The State of Food and Agriculture, 2012* (Rome, 2012).

9. CGIAR (Consultative Group for International Agricultural Research Centers), Food and Agriculture Organization of the United Nations, Global Forum on Agricultural Research (GFAR), International Fund for Agricultural Development (IFAD), "Joint Statement for the ECOSOC Side Event on "Innovation Systems for Family Farming," July 3, 2013, https://library.cgiar.org/bitstream/ handle/10947/2845/Joint_statement_for_the _ECOSOC_side_event_on_Innovation_Systems_for_Family_Farming.pdf?sequence=1.

10. Monique Nuijten, "Food Security, Technology, and the Global Commons—'New' Political Dilemmas?" *Focaal* 2006, no. 48 (2006): vi.

11. World Health Organization, "Frequently Asked Questions On Genetically Modified Foods—May 2014" (Geneva: World Health Organization), http://www.who.int /foodsafety/areas_work/food-technology/faq-genetically-modified-food/en/

12. Artemis Dona and Ioannis S. Arvanitoyannis, "Health Risks of Genetically Modified Foods," *Critical Reviews in Food Science and Nutrition* 49, no. 2 (2009): 164–175.

13. Ibid., 172.

14. World Health Organization, "Food, Genetically Modified," 5.

15. European Union," Labeling of GMO Products: Freedom of Choice for Consumers," GMO Compass, October 28, 2011, www.gmo-compass.org/eng/regulation/labelling.

16. Quoted in UN Educational, Scientific, and Cultural Organization, *World Water Development Report 2012 (WWDR4): Managing Water Under Uncertainty and Risk* (Paris, 2012), 30.

17. U.S. Environmental Protection Agency (EPA), "Technology Fact Sheets," last updated August 8, 2014, http://water.epa.gov/scitech/wastetech.

18. Heather King, "The Water Industry: A Massive Market Bubbles to the Surface," *GreenBiz .com*, May 23, 2011, www.greenbiz.com/blog/2011/05/23/water-industry-massive-market -bubbles-surface.

19. Raz Godelnik, "5 Steps To Mitigate Water Risk And Manage Water Resources," February 15, 2013. http://www.triplepundit.com/2013/02/water-risk-manage-water-resources/

20. Associated Press, "FDA Approves Appetite-zapping Implant for Obese Patients," *New York Times*, January 14, 2015.

21. Associated Press, "Diabetes Drug Gains Approval for Treatment of Obesity," *New York Times*, December 24, 2014.

22. U.S. Food and Drug Administration, "FDA Approves Belviq to Treat some Overweight or Obese Adults," press announcement, June 27, 2012, www.fda.gov/newsevents/newsroom /pressannouncements/ucm309993.htm; "FDA Approves Weight-Management Drug Contrave," press announcement, September 10, 2014, http://www.fda.gov/NewsEvents /Newsroom/PressAnnouncements/ucm413896.htm; Andrew Pollack, "FDA Approves Qsymia, a Weight Loss Drug," *New York Times*, July 17, 2012.

23. E. Bartolini and E. McNeill. "Getting to Value: Eleven Chronic Disease Technologies to Watch," *New England Health Policy Institute*, June, 13, 2012, www.nehi.net /publications/30-getting-to-value-eleven-chronic-disease-technologies-to-watch/view.

24. National Institute for Health Care Management, "The Concentration of Health Care Spending" (NIHCM Foundation Data Brief, Washington, D.C., July 2012), www.nihcm .org/pdf/DataBrief3%20Final.pdf.

25. Daniel Callahan, *Taming the Beloved Beast: How Medical Technology Costs Are Destroying Our Health Care System* (Princeton: Princeton University Press, 2009).

26. Alan M. Garber and Dana P. Goldman, "The Changing Face of Health Care," in *Coping with Methuselah: The Impact of Molecular Biology on Medicine and Society*, ed. Henry J. Aaron and William B. Schwartz (Washington, D.C.: Brookings Institution, 2004), 105–123.

27. Thomas Bodenheimer, "High and Rising Health Care Costs. Part 2: Technologic Innovation," *Annals of Internal Medicine* 142, no. 11 (2005): 932.

28. Callahan, *Taming the Beloved Beast*.

29. Ibid., 90.

30. P. Lehoux, "The Duality of Health Technology in Chronic Illness: How Designers Envision Our Future," *Chronic Illness* 4, no. 2 (2008): 85–97.

31. Adam Sieminski, *International Energy Outlook 2013* (Washington, D.C.: U.S. Energy Information Agency, July 25, 2013).

32. Daniel Sarewitz and Richard Nelson. "Three Rules for Technological Fixes." *Nature* 456, no. 7224 (2008): 871–872.

33. David C. Mowery, Richard R. Nelson, and Ben R. Martin, "Technology Policy and Global Warming: Why New Policy Models Are Needed (or Why Putting New Wine in Old Bottles Won't Work)," *Research Policy* 39, no. 8 (2010): 5.

34. Ibid.

35. *Strategic Plan for the U.S. Climate Change Technology Program (CCTP)* (September 2006), 7. http://www.globalchange.gov/browse/reports/us-climate-change-technology -program-strategic-plan.

36. Eduardo Porter, "Unavoidable Answer for the Problem of Climate Change," *New York Times*, November 19, 2013.

37. "Nuclear Power in the 21st Century," *IAEA Bulletin* 54, no. 1 (2013): 16.

38. World Nuclear Association, "The Economics of Nuclear Power," www.world-nuclear.org /info/Economic-Aspects/Economics-of-Nuclear-Power. www.world-nuclear.org/info /Economics-of-Nuclear-Power, accessed October 2013.

39. Adam Davidson, "Welcome to Saudi Albany," *New York Times*, December 11, 2012.

40. Dieter Helm, *The Carbon Crunch: How We're Getting Climate Change Wrong—and How to Fix It* (New Haven, Conn.: Yale University Press, 2012), 199; Joe Nocera, "The Cuomo Cop-Out," *New York Times*, December 19, 2014.

41. International Energy Agency, www.iea.org.

42. William D. Nordhaus, *The Climate Casino: Risk, Uncertainty, and Economics for a Warming World* (New Haven, Conn.: Yale University Press, 2013).

43. Ibid.; William D. Nordhaus, "Energy: Friend or Enemy," *New York Review of Books* 58, no. 16 (2011): 30.

44. Helm, *Carbon Crunch*, 286.

45. Coral Davenport, "Large Companies Prepared to Pay Price on Carbon," *New York Times*, December 5, 2013.

46. Global Carbon Capture and Storage Institute, *The Global Status of CCS 2013: Summary Report* (Melbourne, Australia, 2013).

47. Cited in Joe Nocera, "A Real Carbon Solution," *New York Times*, March 16, 2013.

48. International Energy Agency. "World Energy Outlook 2013 Factsheet" (2013), http:// www.worldenergyoutlook.org/media/weowebsite/factsheets/weo2013_factsheets.pdf.

49. Patrick Falwell, "U.S. Department of Energy Investment in Carbon Capture and Storage (CCS): (Center for Climate and Energy Solutions, Arlington, Va., September 19, 2013).

50. Matthew L. Wald, "Despite Climate Concern, Global Study Finds Fewer Carbon Capture Projects," *New York Times*, October 11, 2013.

51. E. Kintisch, "Climate Policy. U.S. Carbon Plan Relies on Uncertain Capture Technology," *Science* 341, no. 6153 (2013): 1438–1439.

52. Richard K. Morse, "Cleaning Up Coal: From Culprit to Solution," *Foreign Affairs* Vol. 91, no. 4 (2012): 112.

53. Mizan A. Khan and S. Timmons Roberts, "Toward a Binding Adaptation Regime," In *Successful Adaptation to Climate Change: Linking Science and Practice in a Rapidly Changing World*, ed. Susanne C. Moser and Maxwell T. Boykoff (London: Routledge, 2013), 134.

54. Jon Barnett et al., "Reducing the Risk of Maladaptation in Response to Sea-Level Rise and Urban Water Scarcity," in *Successful Adaptation to Climate Change: Linking Science and Practice in a Rapidly Changing World*, ed. Susanne C. Moser and Maxwell T. Boykoff (London: Routledge, 2013), 37.

55. U.S. Energy Information Administration, "EIA Renewable Energy-Wind Data and Information" (Washington, D.C., January 2011), www.eia.gov/renewable/wind/wind .html.

56. John Theodore Houghton, *Global Warming: The Complete Briefing*, 4th ed. (Cambridge: Cambridge University, 2009), 253.

57. Brown and Sovacool, *Climate Change and Global Energy Security*, 129.

58. Dale Jamieson, *Reason in a Dark Time: Why the Struggle Against Climate Change Failed— And What It Means for our Future* (New York: Oxford University Press, 2014), 207.

59. Chris Green, "'Perspective' on Climate Change", in *Global Crises, Global Solutions*, 2d ed., ed. Bjørn Lomborg, (Cambridge: Cambridge University Press, 2009), 281.

60. Clive Hamilton, "Geoengineering: Our Last Hope, or a False Promise?" *New York Times*, May 27, 2013.

61. David W. Keith, "Geoengineering the Climate: History and Prospect 1," *Annual Review of Energy and the Environment* 25, no. 1 (2000): 245–284; Erin O'Donnell, "Buffering the Sun: David Keith and the Question of Climate Engineering," *Harvard Magazine*, July– August 2013, 36–40.

62. Keith, cited in O'Donnell, "Buffering the Sun," 37.

63. International Energy Agency, "World Energy Outlook 2013 Factsheet," (2013). http:// www.worldenergyoutlook.org/media/weowebsite/factsheets/weo2013_factsheets.pdf

64. National Oceanographic and Atmospheric Administration, "Record Greenhouse Gas Levels Impact Atmosphere and Oceans." Press Release no. 1002. September 9, 2014. www .wmo.int/pages/mediacentre/press_releases/ pr_1002_en.html.

65. Kassia Yanosek, "Policies for Financing the Energy Transition." *Daedalus* 141, no. 2 (2012): 94–104.

66. Peter H. Diamandis and Steven Kotler, *Abundance: The Future Is Better Than You Think* (New York: Free Press, 2012).

67. Walter Isaacson, *The Innovators: How a Group of Hackers, Geniuses, and Geeks Created the Digital Revolution* (New York: Simon & Schuster, 2014).

68. Michael Grubb, "Technology Innovation and Climate Change Policy: An Overview of Issues and Options." *Keio Economic Studies* 41, no. 2 (2004): 103–132.

69. Ibid., 117.

70. Brown and Sovacool, *Climate Change and Global Energy Security*, 146.

71. Ibid., 177.

72. Ernest J. Moniz, "Stimulating Energy Technology Innovation," *Daedalus* 141, no. 2 (2012): 82.

73. Hal Harvey, Franklin M. Orr Jr., and Clara Vondrich, "A Trillion Tons," *Daedalus* 142, no. 1 (2013): 8–25.

74. Quoted in David Ciplet, J. Timmons Roberts, and Mizan Khan, "The Politics of International Climate Adaptation Funding: Justice and Divisions in the Greenhouse," *Global Environmental Politics* 13, no. 1 (2013): 49.

75. Fiona Harvey, "Warsaw Climate Talks Set 2015 Target for Plans to Curb Emissions," *The Guardian*, November 24, 2013.

76. James Hansen et al., "Assessing 'Dangerous Climate Change': Required Reduction of Carbon Emissions to Protect Young People, Future Generations and Nature," *PloS One* 8, no. 12 (2013): e81648.

77. International Energy Agency, "World Energy Outlook2013 Factsheet," (2013). http://www.worldenergyoutlook.org/media/weowebsite/factsheets/weo2013_factsheets.pdf

78. David Leonhardt, "There's Still Hope for the Planet," *New York Times*, July 12, 2012.

79. Nordhaus, *Climate Casino*, 277.

## 8. A VOLATILE MIX

1. Wendy Koch, "Alaska Sinks as Climate Change Thaws Permafrost," *Journal News* (White Plains, N.Y.), October 10, 2013.

2. Tami C. Bond et al., "Bounding the Role of Black Carbon in the Climate System: A Scientific Assessment," *Journal of Geophysical Research: Atmospheres* 118, no. 11 (2013): 5380–5552.

3. Camilo Mora et al., "The Projected Timing of Climate Departure from Recent Variability," *Nature* 502, no. 7470 (2013): 183–187.

4. Justin Gillis, "Temperature Rising: Heat-trapping Gas Passes Milestone, Raising Fears," *New York Times*, May 11, 2013.

5. S. Pacala and R. Socolow, "Stabilization Wedges: Solving the Climate Problem for the Next 50 Years with Current Technologies," *Science* 305, no. 5686 (2004): 968.

6. Steven J. Davis et al., "Rethinking Wedges," *Environmental Research Letters* 8, no. 1 (2013): 011001.

7. Ibid., 3.

8. Andrew C. Revkin, "A Bigger Nod to Uncertainty in the Next Climate Panel Report on Global Warming Impacts; 3/25/14," *Dot Earth* (blog) NYTimes.com, March 25, 2014, http://dotearth.blogs.nytimes.com/2014/03/25/a-bigger-nod-to-uncertainty-in-the-next-ipcc-report-on-global-warming-impacts.

9. Andrew Freedman, "Obama's State of the Union Climate Mention Fits Pattern," climatecentral.org (January 29, 2014), http://www.climatecentral.org/news/obamas-latest-state-of-the-union-climate-mention-fits-pattern-17007.

10. R. A. Kerr, "Climate Change. In the Hot Seat," *Science* 342, no. 6159 (2013): 688–689.

11. Justin Gillis, "Science Linking Drought to Global Warming Remains Matter of Dispute," *New York Times*, February 17, 2014.

12. Daniel Sarewitz, "How Science Makes Environmental Controversies Worse." *Environmental Science & Policy* 7, no. 5 (2004): 385–403.

13. Laura Maxim and Jeroen P. van der Sluijs, "Quality in Environmental Science for Policy: Assessing Uncertainty as a Component of Policy Analysis," *Environmental Science & Policy* 14, no. 4 (2011): 482–492.

14. Ibid., 483–484.

15. Mike Hulme, *Why We Disagree About Climate Change: Understanding Controversy, Inaction and Opportunity* (Cambridge: Cambridge University Press, 2009).

16. Roger A. Pielke Jr., *The Climate Fix: What Scientists and Politicians Won't Tell You About Global Warming* (New York: Basic, 2010).

17. John M. Broder, "'Cap and Trade' Loses Its Standing as Energy Policy of Choice," *New York Times*, March 25, 2010.

18. Lisa Rosenbaum, "Invisible Risks, Emotional Choices—Mammography and Medical Decision Making," *New England Journal of Medicine* 371, no. 16 (2014): 1549–1552.

19. Riley E. Dunlap and Aaron M. McCright, "Organized Climate Change Denial," in *Oxford Handbook of Climate Change and Society*, ed. John S. Dryzek, Richard B. Norgaard, and David Schlosberg (Oxford: Oxford University Press, 2011), 144.

20. International Collective on Environment, Culture & Politics, "2000–2014 United States Newspaper Coverage of Climate Change or Global Warming (Media Monitoring of Climate Change or Global Warming)," http://sciencepolicy.colorado.edu/icecaps /research/media_coverage/usa/index.html. accessed Dec. 2014.

21. Quoted by Dunlap and McCright, "Organized Climate Change Denial:" 153.

22. Justin Gillis, "Verbal Warming: Labels in the Climate Debate," *New York Times*, February 17, 2015.

23. Ted Nordhaus and Michael Schellenberger, "Global Warming Scare Tactics," *New York Times*, April 8, 2014.

24. Andrew C. Revkin, "Other Voices: Earth Institute's Steven Cohen Seeks a Post-hysterical Approach to Climate Progress," *Dot Earth* (blog) NYTimes.com, April 7, 2014, http:// dotearth.blogs.nytimes.com/2014/04/07/earth-institutes-steven-cohen-seeks-a-post -hysterical-approach-to-climate-progress/?_r=0.

25. Andrew C. Revkin, "A Risk Analyst Explains Why Climate Change Risk Misperception Doesn't Necessarily Matter," *Dot Earth* (blog) NYTimes.com. April 16, 2014, http:// dotearth.blogs.nytimes.com/2014/04/16/a-risk-analyst-explains-why-climate-change -risk-misperception-doesnt-necessarily-matter/?_r=0.

26. Paul G. Bain et al., "Promoting Pro-environmental Action in Climate Change Deniers," *Nature Climate Change* 2 (2012): 600–603, www.climateaccess.org/sites/default/files /Bain_Promoting%20pro-environmental%20action.pdf.

27. Matthew C. Nisbet, *Nature's Prophet: Bill McKibben as Journalist, Public Intellectual and Activist* (Joan Shorenstein Center for Press, Politics, and Public Policy Discussion

Paper Series, D-78 March, Cambridge, Mass.: Kennedy School of Government, Harvard University, 2013), 3.

28. Andrew C. Revkin, "A Deeper Look at a Study Finding High Leak Rates from Gas Drilling," *Dot Earth* (blog) NYTimes.com, April 23, 2014, http://dotearth.blogs .nytimes.com/2014/04/23/a-deeper-look-at-a-study-finding-high-leak-rates-from-gas -drilling/#more-52091.

29. Paul Burstein, *American Public Opinion, Advocacy, and Policy in Congress: What the Public Wants and What It Gets* (New York: Cambridge University Press, 2014).

30. Jon A. Krosnick and Bo MacInnis, "Does the American Public Support Legislation to Reduce Greenhouse Gas Emissions?" *Daedalus* 142, no. 1 (2013): 26.

31. Ibid., 36.

32. Stephen Ansolabehere and David M. Konisky, "The American Public's Energy Choice," *Daedalus* 141, no. 2 (2012): 70.

33. Kelly Sims Gallagher, "Why and How Governments Support Renewable Energy," *Daedalus* 142, no. 1 (2013): 59–77.

34. Ralph Cavanagh, "How We Learned Not to Guzzle," *New York Times*, September 13, 2013.

35. Jon Birger Skjærseth, Guri Bang, and Miranda A. Schreurs, "Explaining Growing Climate Policy Differences Between the European Union and the United States," *Global Environmental Politics* 13, no. 4 (2013): 61–80.

36. Ibid., 76.

37. Ibid., 73.

38. Pew Research Center, "Deficit Reduction Declines As Policy Priority," January 27, 2014. http://www.people-press.org/2014/01/27/deficit-reduction-declines-as-policy -priority/.

39. http://www.people-press.org/2014/11/12/little-enthusiasm-familiar-divisions-after-the -gops-big-midterm-victory/#broad-support-for-stricter-emissions-limits.

40. So Young Kim and Yael Wolinsky-Nahmias, "Cross National Public Opinion on Climate Change: The Effects of Affluence and Vulnerability," *Global Environmental Politics* 14, no. 1 (2014): 79–106.

41. Ibid., 100–101.

42. Robert Wood Johnson Foundation. *F as in Fat: How Obesity Threatens America's Future* (Washington, D.C.: Trust for America's Health, 2010); NCD Alliance, "Healthy Planet, Healthy People: The NCD Alliance Vision for Health in the Post-2015 Development Agenda" (Geneva, 2013); World Health Organization, *Global Action Plan for the Prevention and Control of Noncommunicable Diseases: 2013–2020* (Geneva, 2013); Institute of Medicine, *Living Well with Chronic Disease: A Call for Public Health Action* (Washington, D.C.: National Academies Press, 2012).

43. Justin Gillis, "Climate Change Seen Posing Risk to Food Supplies," *New York Times*, November 2, 2013.

44. International Federation of Red Cross and Red Crescent Societies, *Annual Report 2013*, www.ifrc.org/annual-report-2013.

45. Pakistan Institute of Public Opinion, The Pakistani Affiliate of Gallup International. *Public Perception on Global Food Crisis - Findings from a Study in Pakistan and 25 Countries World-Wide.* Pakistan Institute of Public Opinion, The Pakistani Affiliate of Gallup International, 2008.

46. Cited in Johan F. M. Swinnen, Pasquamaria Squicciarini, and Thijs Vandemoortele, "The Food Crisis, Mass Media and the Political Economy of Policy Analysis and Communication," *European Review of Agricultural Economics* 38, no. 3 (2011): 409.

47. Ibid., 410.

48. Ibid., 428.

49. Elizabeth Mendes, "Gallup Poll: Americans' Concerns About Obesity Soar, Surpass Smoking," Gallup, July 18, 2012, www.gallup.com/poll/155762/americans-concerns-obesity-soar-surpass-smoking.aspx.

50. James A. Colbert and Jonathan N. Adler, "Sugar-sweetened Beverages—Polling Results," *New England Journal of Medicine* 368, no. 3 (2013): 1464–1466.

51. Associated Press–NORC Center for Public Affairs Research, "Obesity in the United States: Public Perceptions," 2013, www.apnorc.org/PDFs/Obesity/AP-NORC-Obesity-Release_2.pdf.

52. J. E. Oliver and T. Lee, "Public Opinion and the Politics of Obesity in America," *Journal of Health Politics, Policy and Law* 30, no. 5 (2005): 923–954.

53. Abigail Cope Saguy, *What's Wrong with Fat?* (New York: Oxford University Press, 2013).

54. Jenna Levy, "U.S. Obesity Rate Inches Up to 27.7% in 2014," Gallup, January 26, 2015, www.gallup.com/poll/181271/obesity-rate-inches-2014.aspx_2.

55. Globescan and Circle of Blue, *WaterViews: Water Issues Research* (Traverse City, Mich., August 17, 2009), www.circleofblue.org/waternews/wp-content/uploads/2009/08/circle_of_blue_globescan.pdf.

56. Stuart Leavenworth, "Why Journalists Need to Cover the Water Story," *NeimanReports*, March 15, 2005, http://niemanreports.org/articles/why-journalists-need-to-cover-the-water-story.

57. Rakesh Kalshian, "Mainstream News Reporting Ignores Critical Water Issues," *NeimanReports*. March 15, 2005, http://niemanreports.org/articles/mainstream-news-reporting-ignores-critical-water-issues.

58. Peter Brabeck-Letmathe, "More Attention in the Media to Global Water Risks," *Water Challenge* (blog), January 24, 2014, www.water-challenge.com/post/2014/01/22/More-attention-in-the-media-to-global-water-risks.aspx.

59. California Institute for Water Resources, http://ciwr.ucanr.edu.

60. Albert Gore, *Our Choice* (New York: Random House, 2009).

## 9. LAW AND GOVERNANCE

1. Mark Mazower, *Governing the World: The History of an Idea* (New York: Penguin, 2012), xii.

2. James Gustave Speth and Peter M. Haas, *Global Environmental Governance*, Foundations of Contemporary Environmental Studies (Washington, D.C.: Island Press, 2006), 134.

3. Kenneth W. Abbott and Duncan Snidal, "Why States Act Through International Formal Organizations," *The Journal of Conflict Resolution* 42, no. 1 (1998): 3. http://www.u.arizona.edu/~volgy/AbbottSnidal1998WhyStatesUseFormal.pdf.

4. Ibid., 5.

5. Peter Newell, "The Marketization of Global Environmental Governance: Manifestations and Implications," in *The Crisis of Global Environmental Governance: Towards a New Political Economy of Sustainability*, ed. Jacob Park, Ken Conca, and Matthias Finger (London: Routledge, 2008), 78.

6. William D. Nordhaus, "A New Solution: The Climate Club," *New York Review of Books* 62, no. 10 (2015): 36.

7. Mike Hulme, *Why We Disagree About Climate Change: Understanding Controversy, Inaction and Opportunity* (Cambridge: Cambridge University Press, 2009), 345.

8. Joseph A. Tainter, *The Collapse of Complex Societies*. New Studies in Archaeology. (Cambridge: Cambridge University Press, 1988), 214.

9. Roger A. Pielke Jr., *The Climate Fix: What Scientists and Politicians Won't Tell You About Global Warming* (New York: Basic, 2010), 104–109.

10. Stanley Reed, "After Failed Attempt in April, Europe Approves Emissions Trading System," *New York Times*, July 14, 2013.

11. John Timmer, "Wyoming Rejects Science Education Standards Over Climate Change," *Ars Technica*, March 16, 2014, http://arstechnica.com/science/2014/03/wyoming-rejects-science-education-standards-over-climate-change.

12. Jody Freeman, "Teaching an Old Law New Tricks," *New York Times*, May 29, 2014.

13. Timothy E. Wirth and Thomas A. Daschle, "A Blueprint to End Paralysis Over Global Action on Climate," *Environment 360*, May 19, 2014, http://e360.yale.edu/feature/a_blueprint_to_end_paralysis_over_global_action_on_climate/2766.

14. David Shorr, "Why the Glacial Pace of Climate Diplomacy Isn't Ruining the Planet," *Foreign Policy*, March 17, 2014.

15. William D. Nordhaus, *The Climate Casino: Risk, Uncertainty, and Economics for a Warming World* (New Haven, Conn.: Yale University Press, 2013), 239.

16. Anthony Giddens, *The Politics of Climate Change* (Cambridge: Polity, 2009), 151.

17. James Gustave Speth, *The Bridge at the Edge of the World: Capitalism, the Environment, and Crossing from Crisis to Sustainability* (New Haven, Conn.: Yale University Press, 2008), 147.

18. Maggie Black, Jannet King, and Robin Clarke, *The Atlas of Water: Mapping the World's Most Critical Resource* (London: Earthscan, 2009), 880–882.

19. United Nations Department of Economic and Social Affairs, "Transboundary Waters," www.un.org/waterforlifedecade/transboundary-waters.shtml, accessed October 10, 2014.

20. Christiana Z. Peppard, *Just Water: Theology, Ethics, and the Global Water Crisis* (Maryknoll, N.Y.: Orbis, 2014).

21. World Water Council, www.worldwatercouncil.org.

22. Robert O. Keohane and David G. Victor, "The Regime Complex for Climate Change," *Perspectives on Politics* 9, no. 1 (2011): 7.

23. Ken Conca, *Governing Water: Contentious Transnational Politics and Global Institution Building*. Global Environmental Accords (Cambridge, Mass.: MIT Press, 2006), 373.

24. Gwyn Prins et al., "The Hartwell Paper: A New Direction for Climate Policy After the Crash of 2009" (London School of Economics and Political Science, London, 2010). See chap. 1.

25. Frank A. Ward, "Forging Sustainable Transboundary Water-Sharing Agreements: Barriers and Opportunities," *Water Policy* 15, no. 3 (2013): 386.

26. United Nations Department of Economic and Social Affairs, "Transboundary Waters," www.un.org/waterforlifedecade/transboundary-waters.shtml, accessed October 10, 2014.

27. N General Assembly, "Sixty-Eighth Session—Agenda Item 87: The Law of Transboundary Aquifers," www.un.org/en/ga/sixth/68/TransAquifers.shtml, accessed 10/10/14.

28. Ward, "Forging Sustainable Transboundary Water-Sharing Agreements," 386–417.

29. Yoshihide Wada, "Non-sustainable Groundwater Sustaining Irrigation," *Global Water Forum*, February 13, 2012, www.globalwaterforum.org/2012/02/13/non-sustainable-groundwater-sustaining-irrigation.

30. Desirée C. Rabelo et al., "Citizenship Participation in Water Management Plans in the Doce River Basin, Brazil and Catalonia, Spain," *Water Policy* 16, no. 2 (2014): 205.

31. V. H. Honkalaskar, M. Sohoni, and U. V. Bhandarkar, "A Participatory Decision Making Process for Community-Level Water Supply," *Water Policy* 16, no. 1 (2014): 39.

32. Suzanne von der Porten and Rob C. de Loë, "Water Policy Reform and Indigenous Governance," *Water Policy* 16, no. 2 (2014): 222.

33. Michael Gilmont, "Decoupling Dependence on Natural Water: Reflexivity in the Regulation and Allocation of Water in Israel," *Water Policy* 16, no. 1 (2014): 79–101.

34. Rui Cunha Marques, Pedro Simões, and Sanford Berg, "Water Sector Regulation in Small Island Developing States: An Application to Cape Verde," *Water Policy* 15, no. 1 (2013): 153.

35. Joanne Chong, "Climate-Readiness, Competition and Sustainability: An Analysis of the Legal and Regulatory Frameworks for Providing Water Services in Sydney," *Water Policy* 16, no. 1 (2014): 1–18.

36. Rita Martins et al., "Assessing Social Concerns in Water Tariffs," *Water Policy* 15, no. 2 (2013): 193–211.

37. *The Economist* Intelligence Unit, "Global Food Security Index 2014: An Annual Measure of the State of Global Food Security, May 28, 2014," .http://foodsecurityindex.eiu.com /Resources.

38. Ibid.
39. Per Pinstrup-Andersen and Derrill D. Watson, *Food Policy for Developing Countries: The Role of Government in Global, National, and Local Food Systems* (Ithaca, N.Y.: Cornell University Press, 2011), 273.
40. Robert L. Paarlberg, *Food Politics: What Everyone Needs to Know* (New York: Oxford University Press, 2010), 177.
41. Ibid., 174–176.
42. ActionAid International USA, "Fueling the Food Crisis, the Cost to Developing Countries" (Washington, D.C., October 2012).
43. Ibid., 6.
44. Kimberly Ann Elliott, "U.S. Biofuels Policy and Global Food Price Crisis: A Survey of the Issues," in *Global Food Crisis, the: Governance Challenges and Opportunities*, ed. Jennifer Clapp and Marc J. Cohen, Studies in International Governance (Waterloo, Ont.: Wilfrid Laurier University Press, 2009):59–76.
45. Smarter Fuel Future, "Coming Together on Ethanol Mandates & Global Food Supply," September 12, 2013, http://smarterfuelfuture.org/blog/details/coming-together-on -ethanol-mandates-global-food-supply.
46. Ibid.
47. Lucie Edwards, "Food Fight: The International Assessment of Agricultural Knowledge, Science, and Technology for Development," *AgBioForum* 15, no. 1 (2012): art. 9.
48. UN Environment Programme, *International Assessment of Agricultural Knowledge, Science and Technology for Development (IAASTD): Synthesis Report with Executive Summary: A Synthesis of the Global and Sub-Global IAASTD Reports* (Nairobi, Kenya,, 2009).
49. Ibid.
50. Edwards, "Food Fight," 8.
51. Jonathan Latham, and Allison Wilson, "How the Science Media Failed the IAASTD," *Independent Science News*, April 7, 2008, www.independentsciencenews.org /environment/science-media-failed-the-iaastd.
52. Roger S. Magnusson, "Global Health Governance and the Challenge of Chronic, Non-communicable Disease," *Journal of Law, Medicine & Ethics* 38, no. 3 (2010): 499.
53. Bryan Thomas and Lawrence O. Gostin, "Tackling the Global NCD Crisis: Innovations in Law and Governance," *Journal of Law, Medicine & Ethics* 41, no. 1 (2013): 19.
54. Marc Suhrcke et al., *Chronic Disease: An Economic Perspective* (London: Oxford Health Alliance, 2006).
55. Lawrence O. Gostin, "Bloomberg's Health Legacy: Urban Innovator or Meddling Nanny?" *Hastings Center Report* 43, no. 5 (2013): 19.
56. Ibid., 23.
57. Thomas and Gostin, "Tackling the Global NCD Crisis."
58. Richard Sullivan et al., "Delivering Affordable Cancer Care in High-Income Countries," *Lancet Oncology* 12, no. 10 (2011): 933.
59. Daniel Callahan, "Medical Progress and Global Chronic Disease: The Need for a New Model," *Brown Journal of World Affairs* 20 (2013): 35.

60. George A. Bray, "History of Obesity," in *Obesity: Science to Practice*, ed. Gareth Williams and Gema Fruhbeck (Hoboken, N.J.: Wiley, 2009), 2.
61. "National Opinion Research Center," www.apnorc.org.
62. Boyd A. Swinburn et al., "The Global Obesity Pandemic: Shaped by Global Drivers and Local Environments," *Lancet* 378, no. 9793 (2011): 804.
63. Ng, Marie Ng, et al., "Global, Regional, and National Prevalence of Overweight and Obesity in Children and Adults During 1980–2013: A Systematic Analysis for the Global Burden of Disease Study 2013." *Lancet* 384, no. 9945 (2014): 766.
64. Thomas and Gostin, "Tackling the Global NCD Crisis."
65. Stephanie Strom, "Soda Makers Coca-Cola, PepsiCo and Dr Pepper Join in Effort to Cut Americans' Drink Calories," *New York Times*, September 24, 2014.
66. Richard Fausset, "Florida Food Stamp Bill Is Latest Attempt to Restrict Junk Food," *Los Angeles Times*, January 29, 2012.
67. Ruopeng An, "Effectiveness of Subsidies in Promoting Healthy Food Purchases and Consumption: A Review of Field Experiments," *Public Health Nutrition* 16, no. 7 (2013): 1215–1228.
68. Robert Wood Johnson Foundation, *F as in Fat: How Obesity Threatens America's Future* (Washington, D.C.: Trust for America's Health, 2010).
69. Boyd A. Swinburn, "Obesity Prevention: The Role of Policies, Laws and Regulations," *Australia and New Zealand Health Policy* 5 (2008): 14.
70. Sabrina Tavernise, "Children Are Eating Fewer Calories Study Finds," *New York Times*, February 21, 2013; Sabrina Tavernise, "Obesity Studies Tell Two Stories, Both Right," *New York Times*, April 14, 2014.
71. Rebecca M. Puhl and Chelsea A. Heuer, "Public Opinion About Laws to Prohibit Weight Discrimination in the United States," *Obesity* 19, no. 1 (2011): 74–82.
72. Janet D. Latner, Rebecca M. Puhl, and Albert J. Stunkard, "Cultural Attitudes and Biases Towards Obese Persons," in *Textbook of Obesity: Biological, Psychological and Cultural Influences*, ed. Sharon R. Akabas, Sally A. Lederman, and Barbara J. Moore, (Chichester, UK: Wiley-Blackwell, 2012), 48–49.
73. Ibid., see their citation on p. 49 and in note 1.
74. J. Stuber, S. Galea, and B. G. Link, "Smoking and the Emergence of a Stigmatized Social Status." *Social Science & Medicine* (1982) 67, no. 3 (2008): 420–430.
75. Michael R Lowe, "Self-Regulation of Energy Intake in the Prevention and Treatment of Obesity: Is it It Feasible?" *Obesity Research* 11, no. S10 (2003): 44S–59S.
76. Deuteronomy 32:10, 13b–14a, 15.
77. Sander L. Gilman, *Fat: A Cultural History of Obesity* (Cambridge: Polity, 2008), 78.
78. Ibid., 167.
79. Georges Vigarello, *The Metamorphoses of Fat: A History of Obesity*, European Perspectives: A Series in Social Thought and Cultural Criticism (New York: Columbia University Press, 2013).
80. John Stuart Mill, "On Liberty," in *Utilitarianism; On Liberty; Essay on Bentham*, ed. Mary Warnock (New York: New American Library, 1962), 135.

81. Ibid., 206.

82. Crystal L. Hoyt and Jeni L. Burnette, "Should Obesity Be a Disease?" *New York Times*, February 23, 2014.

83. Laura Dawes, *Childhood Obesity in America: Biography of an Epidemic* (Cambridge, Mass.: Harvard University Press, 2014), 234.

84. Bruce Jennings, "Beyond the Social Contract of Consumption: Democratic Governance in the Post-carbon Era," *Critical Policy Studies* 4, no. 3 (2010): 222–233.

85. Yi Zeng, "Effects of Demographic and Retirement-Age Policies on Future Pension Deficits, with an Application to China," *Population and Development Review* 37, no. 3 (2011): 553–569.

86. M. Prince et al., "The Global Prevalence of Dementia: A Systematic Review and Metaanalysis " *Alzheimer's & Dementia* 9, no. 1 (2013): 69.

87. Laurie Burkitt, "As Obesity Rises, Chinese Kids Are Almost as Fat as Americans; 5/29/14," *China Real Time*, May 29, 2014, http://blogs.wsj.com/chinarealtime/2014/05/29/as-obesity-rises-chinese-kids-are-almost-as-fat-as-americans.

88. Shannon Tiezzi, "China's Looming Water Shortage," *Diplomat*, November 30, 2014, http://thediplomat.com/2014/11/chinas-looming-water-shortage.

89. Food and Agriculture Organization of the United Nations, Regional Office for Asia and the Pacific, *Annex 3: Agricultural Policy and Food Security in China, in RAP Publication 1999/1 Poverty Alleviation and Food Security in Asia: Lessons and Challenges* (Bangkok, 2008).

90. Jennings, "Beyond the Social Contract of Consumption," 222–233.

91. Doris Kearns Goodwin, *No Ordinary Time: Franklin and Eleanor Roosevelt: The Home Front in World War II* (New York: Simon & Schuster, 1994), 45.

## 10. PROGRESS AND ITS ERRANT CHILDREN

1. George Packer, "The Uses of Division," *New Yorker*, August 11, 2014, 86.

2. Jeremy Brecher, Tim Costello, and Brendan Smith, "Globalization and Its Specter," in *Globalization: The Transformation of Social Worlds*, 3d ed., ed. D. Stanley Eitzen and Maxine Baca Zinn (Independence, Ky.: Cengage, 2012), 30.

3. G. A. Bray, S. J. Nielsen, and B. M. Popkin, "Consumption of High-Fructose Corn Syrup in Beverages May Play a Role in the Epidemic of Obesity," *American Journal of Clinical Nutrition* 79, no. 4 (2004): 537–543.

4. Ronald Inglehart and Wayne E. Baker, "Modernization, Cultural Change, and the Persistence of Traditional Values," *American Sociological Review* 65, no. 1 (2000): 19–51.

5. Ibid., 31.

6. Jared M. Diamond, *Collapse: How Societies Choose to Fail or Succeed* (New York: Viking, 2005); Joseph A. Tainter, *The Collapse of Complex Societies*, New Studies in Archaeology (Cambridge: Cambridge University Press, 1988).

7. Cited in Doris Kearns Goodwin, *No Ordinary Time: Franklin and Eleanor Roosevelt: The Home Front in World War II* (New York: Simon & Schuster, 1994).

8. Doris Kearns Goodwin, *Team of Rivals: The Political Genius of Abraham Lincoln* (New York: Simon & Schuster, 2005).

9. Floyd Norris, "Young Households Are Losing Ground in Income, Despite Education," *New York Times*, September 12, 2014.

10. Robert Nisbet, *History of the Idea of Progress* (New Brunswick, N.J.: Transaction, 1980, 1994).

11. Ibid., xii.

12. Ibid., xiii.

13. Margaret Meek Lange, "Progress," in *Stanford Encyclopedia of Philosophy*, Spring 2011 ed., article published February 17, 2011, http://plato.stanford.edu/archives/spr2011/entries/progress.

14. Nisbet, *History of the Idea of Progress*, 25.

15. Bob Goudzwaard, Josina Van Nuis Zylstra (translator). *Capitalism and Progress: A Diagnosis of Western Society*. (Milton Keynes, UK: Biblical & Theological Classics Library [Paternoster], 1997): xxiii. The Schumpeter quote is also on page xxiii.

16. Robert Bryce, *Smaller Faster Lighter Denser Cheaper: How Innovation Keeps Proving the Catastrophists Wrong* (New York: Public Affairs, 2014); Peter H. Diamandis and Steven Kotler, *Abundance: The Future Is Better Than You Think* (New York: Free Press, 2012); Diane Ackerman, *The Human Age: The World Shaped by Us* (New York: Norton, 2014).

17. Dani Rodrik, *The Globalization Paradox: Democracy and the Future of the World Economy* (New York: Norton, 2011), 233.

18. Jeffrey Sachs, *The Price of Civilization: Reawakening American Virtue and Prosperity* (New York: Random House, 2011), 87.

19. IPO Center, *Global Top 100 Companies by Market Capitalisation; 31 March 2015 Update*, PricewaterhouseCoopers, https://www.pwc.com/gx/en/audit-services/capital-market/publications/assets/document/pwc-global-top-100-march-update.pdf.

20. Sachs, *Price of Civilization*, 93ff.

21. Beth Macy, *Factory Man: How One Furniture Maker Battled Offshoring, Stayed Local—And Helped Save an American Town* (New York: Little, Brown, 2014).

22. Joseph E. Stiglitz, *Globalization and Its Discontents* (New York: Norton, 2009), 214.

23. Daniel Callahan and Angela A. Wasunna, *Medicine and the Market: Equity v. Choice* (Baltimore: Johns Hopkins University Press, 2006).

24. James Gustave Speth, *The Bridge at the Edge of the World: Capitalism, the Environment, and Crossing from Crisis to Sustainability* (New Haven, Conn.: Yale University Press, 2008), 170.

25. Stiglitz, *Globalization and Its Discontents*: 230–239.

26. George Soros, *George Soros on Globalization* (New York: Public Affairs, 2005), 33.

27. Per Pinstrup-Andersen and Derrill D. Watson, *Food Policy for Developing Countries: The Role of Government in Global, National, and Local Food Systems* (Ithaca, N.Y.: Cornell University Press, 2011), 279.

28. Robert O. Keohane and David G. Victor, "The Regime Complex for Climate Change" *Perspectives on Politics* 9, no. 01 (2011): 7.

29. Elinor Ostrom, *Governing the Commons: The Evolution of Institutions for Collective Action*, Political Economy of Institutions and Decisions (Cambridge: Cambridge University Press, 1990), 1.

30. Garrett Hardin, "The Tragedy of the Commons," *Science* 162, no. 3859 (1968): 1243.

31. Jeff Goodwin and James M. Jasper, *The Social Movements Reader: Cases and Concepts*, 2d ed., Blackwell Readers in Sociology, vol. 12 (Chichester, U.K.: Wiley-Blackwell, 2009), 3.

32. Hélène Ducros, "Localized Responses to Unsustainable Growth," *Global Environmental Politics* 14, no. 2 (2014): 122–128.

33. Stephen Mark Gardiner, *A Perfect Moral Storm: The Ethical Tragedy of Climate Change*, Environmental Ethics and Science Policy Series (New York: Oxford University Press, 2011); Melissa Lane, *Eco-Republic: What the Ancients Can Teach Us About Ethics, Virtue, and Sustainable Living* (Princeton: Princeton University Press, 2012).

34. Naomi Klein, *This Changes Everything: Capitalism vs. the Climate* (New York: Simon & Schuster, 2014).

## 11. THE NECESSARY COALITION

1. Henry Fountain and John Schwartz, "Climate Accord Relies on Environmental Policies Now in Place," *New York Times*, November 12, 2014.

2. Coral Davenport and Mark Landler, "U.S. to Give $3 Billion to Climate Fund to Help Poor Nations, and Spur Rich Ones," *New York Times*, November 14, 2014.

3. Gardiner Harris, "Coal Rush in India Could Tip Balance on Climate Change," *New York Times*, November 17, 2014.

4. Ellen Barry and Coral Davenport, "India Announces Plan to Lower Rate of Greenhouse Gas Emissions," *New York Times*, October 2, 2014.

5. U.S. Centers for Disease Control and Prevention, "Trends in Current Cigarette Smoking Among High School Students and Adults, United States, 1965–2011," last updated November 14, 2013, www.cdc.gov/tobacco/data_statistics/tables/trends/cig_smoking.

6. Brian D. Carter et al., "Smoking and Mortality—Beyond Established Causes," *New England Journal of Medicine* 372, no. 7 (2015): 631–640.

7. Nicholas Stern, *Stern Review on the Economics of Climate Change,*(London: HM Treasury, 2006).

8. Christopher J. Tassava, "The American Economy During World War II," *EH.net*, http://eh.net/encyclopedia/the-american-economy-during-world-war-ii, accessed March 2015.

9. Kelly D. Brownell, "Thinking Forward: The Quicksand of Appeasing the Food Industry," *PLoS Medicine* 9, no. 7 (2012): e1001254.

10. William H. Wiist, "The Corporate Play Book, Health and Democracy: The Snack Food and Beverage Industry's Tactics in Context," in *Sick Societies: Responding to the Global Challenge of Chronic Disease*, ed. David Stuckler and Karen Siegel (Oxford: Oxford University Press, 2011), 371.

11. Marion Nestle, *Food Politics: How the Food Industry Influences Nutrition and Health*, 3d ed. (Berkeley: University of California Press, 2007), 393.

12. *Pocket Oxford American Dictionary*, 2nd ed. (New York: Oxford University Press, 2008).

13. David Callahan, *Kindred Spirits: Harvard Business School's Extraordinary Class of 1949 and How They Transformed American Business* (Hoboken, N.J.: Wiley, 2002).

14. Lynn A. Stout, "The Problem of Corporate Purpose," *Issues in Governance Studies, Governance Studies at the Brookings Institution* 48 (June 2012):1–14.

15. Lester R. Brown, "The Rise and Fall of the Global Climate Coalition" (Earth Policy Institute, 2000), July 25, 2000. http://www.earth-policy.org/plan_b_updates/2000/alert6.

16. Christine Bader, *The Evolution of a Corporate Idealist: When Girl Meets Oil* (Brookline, Mass.: Bibliomotion, 2014).

17. George Kell, "Business to Play a Key Role on Carbon Pricing," *Huffington Post*, September 24 2014, www.huffingtonpost.com/georg-kell/business-to-play-a-key-ro_b_5876240.html.

18. Fred Pearce, "Monitoring Corporate Behavior: Greening or Merely Greenwashing?" *Environment 360*, January 27, 2014, http://e360.yale.edu/feature/monitoring_corporate_behavior_greening_or_merely_greenwash/2732.

19. Stephanie Strom, "Soda Makers Coca-Cola, PepsiCo and Dr Pepper Join in Effort to Cut Americans' Drink Calories," *New York Times*, September 24, 2014.

20. Joel Makower, "The Audacious Optimism of Climate Week; 9/29/14," *GreenBiz.com*, September 29, 2014, www.greenbiz.com/blog/2014/09/29/audacious-optimism-climate-week.

21. International Energy Agency, "Global Energy-Related Emissions of Carbon Dioxide Stalled in 2014," March 13, 2015, www.refworks.com/refworks2/default.aspx?r=references|MainLayout::init&lang=en.

22. NOAA. Global Summary Information—March 2015. National Centers for Environmental Information, National Oceanic and Atmospheric Administration www.ncdc.noaa.gov/sotc/summary-info/global/201503.

23. National Oceanic and Atmospheric Administration,"Greenhouse Gas Benchmark Reached: Global Carbon Dioxide Concentrations Surpass 400 Parts Per Million For The First Month Since Measurements Began," May 6, 2015, http://research.noaa.gov/ News/NewsArchive/LatestNews/ TabId/684/ArtMID /1768/ArticleID/11153/Greenhouse-gas-benchmark-reached-.aspx.

24. Jo Confino, "How Concerned Are CEOs About Climate Change: Not at All," *The Guardian*, January 20, 2015, www.theguardian.com/sustainable-business/2015/jan/20/global-warming-business-risks-government-regulation-taxes.

25. Joel Makower, "Two Steps Forward, The State of Green Business 2015," February 3, 2015, GreenBiz.com, http://www.greenbiz.com/article/state-green-business-2015; Joel Makower, Daniel Kelley, Richard Mattison, and James Salo, *State of Green Business Report, 2015* (Green Biz Group, 2015). www.greenbiz.com/report/state-green-business-report-2015.

26. SDG Compass, "The Guide for Business Action on the SDG: Executive Summary," http://sdgcompass.org/wp-content/uploads/2015/09/SDG_Compass_Guide_Executive _Summary.pdf.

27. "Walking the Walk: Firms Increasingly Believe That Saving the Planet Is Good for Business," *The Economist*, June 6, 2015; *New York Times* Editorial Board. "The Case For A Carbon Tax," *New York Times*, June 7, 2015.

# INDEX

Page numbers in italic refer to tables or figures.